C0-AOG-011

AGRARIAN LANDSCAPES IN TRANSITION

LONG-TERM ECOLOGICAL RESEARCH NETWORK SERIES
LTER Publications Committee

Grassland Dynamics: Long-Term Ecological Research in Tallgrass Prairie
Edited by
Alan K. Knapp, John M. Briggs, David C. Hartnett, and Scott L. Collins

Standard Soil Methods for Long-Term Ecological Research
Edited by
G. Philip Robertson, David C. Coleman, Caroline S. Bledsoe, and Phillip Sollins

Structure and Function of an Alpine Ecosystem: Niwot Ridge, Colorado
Edited by
William D. Bowman and Timothy R. Seastedt

Climate Variability and Ecosystem Response at Long-Term Ecological Sites
Edited by
David Greenland, Douglas G. Goodin, and Raymond C. Smith

Biodiversity in Drylands: Toward a Unified Framework
Edited by
Moshe Shachak, James R. Gosz, Steward T. A. Pickett, and Avi Perevolotsky

Long-Term Dynamics of Lakes in the Landscape: Long-Term Ecological Research on North Temperate Lakes
Edited by
John J. Magnuson, Timothy K. Kratz, and Barbara J. Benson

Alaska's Changing Boreal Forest
Edited by
F. Stuart Chapin III, Mark W. Oswood, Keith Van Cleve,
Leslie A. Viereck, and David L. Verbyla

Structure and Function of a Chihuahuan Desert Ecosystem: The Jornada Basin Long-Term Ecological Research Site
Edited by
Kris M. Havstad, Laura F. Huenneke, and William H. Schlesinger

Principles and Standards for Measuring Net Primary Production
Edited by
Timothy J. Fahey and Alan K. Knapp

Agrarian Landscapes in Transition: Comparisons of Long-Term Ecological and Cultural Change
Edited by
Charles L. Redman and David R. Foster

AGRARIAN LANDSCAPES IN TRANSITION

Comparisons of Long-Term Ecological and Cultural Change

Edited by

CHARLES L. REDMAN
DAVID R. FOSTER

2008

OXFORD
UNIVERSITY PRESS

Oxford University Press, Inc., publishes works that further
Oxford University's objective of excellence
in research, scholarship, and education.

Oxford New York
Auckland Cape Town Dar es Salaam Hong Kong Karachi
Kuala Lumpur Madrid Melbourne Mexico City Nairobi
New Delhi Shanghai Taipei Toronto

With offices in
Argentina Austria Brazil Chile Czech Republic France Greece
Guatemala Hungary Italy Japan Poland Portugal Singapore
South Korea Switzerland Thailand Turkey Ukraine Vietnam

Published by Oxford University Press, Inc.
198 Madison Avenue, New York, New York 10016

www.oup.com

Library of Congress Cataloging-in-Publication Data
Agrarian landscapes in transition : comparisons of long-term ecological
and cultural change / edited by Charles L. Redman and David R. Foster.
p. cm.—(Long-Term Ecological Research Network series)
Includes bibliographical references and index.
ISBN 978-0-19-536796-6
1. Agricultural ecology—United States. 2. Land use—United States.
I. Redman, Charles L. II. Foster, David R., 1954– III. Series.
S441.A327 2008
333.76—dc22 2007042097

9 8 7 6 5 4 3 2 1

Printed in the United States of America
on acid-free paper

Acknowledgments

The authors express their gratitude to the support of the National Science Foundation (NSF). This research has been supported by NSF Biocomplexity in the Environment grant no. DEB-0216560.

Harvard Forest would like to thank Audrey Barker Plotkin for coordinating activities associated with the Harvard Forest Long-Term Ecological Research (LTER) program. The research that forms the basis for our effort was supported by the A. W. Mellon Foundation and the NSF, especially the AgTrans Biocomplexity project, Harvard Forest LTER program, and Summer Ecology program for undergraduates (REU grant), which is coordinated by Aaron Ellison and Edythe Ellin. Our chapter is a contribution from the Harvard Forest Long-Term Ecological Research program.

Material presented in the Coweeta article is based on research supported by the NSF under cooperative agreements DEB-9632854 and DEB-0218001. We thank David Foster and Charles Redman for their comments on an early version of this manuscript. Andrew Hunt assisted with the collection of data on conservation organizations active in southern Appalachia. Any opinions, findings, conclusions, or recommendations expressed in the material are those of the authors and do not necessarily reflect the views of the NSF.

The research for the Shortgrass Steppe region has been supported by grant no. HD33554 from the National Institute of Child Health and Human Development. The Great Plains Population and Environment Project is a Global Land Use and Land Cover Change Project (see www.geo.ucl.ac.be/LUCC/lucc.html). We are grateful to William Myers and Lisa Isgett for their assistance, and to other members of the Great Plains team too numerous to mention.

The Kellogg Biological Station would like to thank the Michigan Agricultural Experiment Station and Kellogg Biological Station Long-Term Ecological Change Project. We are deeply indebted to the efforts of William Lovis and Jodie O'Gorman.

Research on the Konza region was supported by the Konza Prairie LTER V, also funded by the NSF. We are grateful to John Blair (principle investigator on Konza Prairie LTER V) for institutional support, and to John Harrington, Jr., for his valuable comments on the Konza chapter.

The research for Central Arizona–Phoenix (CAP) has been supported by NSF CAP LTER (DEB-9714833) and McDonnell Foundation grants. Many people have contributed to the thinking and data collecting built upon by this chapter. Of special note has been the work of Anne Gustafson, Sam Schmieding, David Bild, Glen Stewart, Nancy Grimm, and Grady Gammage, Jr.

More than any others, we are indebted to project coordinator Lauren Kuby, who shepherded the research teams from all corners of the United States to attain their objectives; and project editor Becky L. Eden, who brought the book to fruition by editing the text, designing the figures, and corralling the authors.

Contents

About the Contributors ix

Introduction 3

Charles L. Redman

1 Changing Agrarian Landscapes across America 16

A Comparative Perspective

Kenneth M. Sylvester and Myron P. Gutmann

2 New England's Forest Landscape 44

Ecological Legacies and Conservation Patterns Shaped by Agrarian History

David R. Foster, Brian Donahue, David Kittredge, Glenn Motzkin, Brian Hall, Billie Turner, and Elizabeth Chilton

3 Agricultural Transformation of Southern Appalachia 89

Ted L. Gragson, Paul V. Bolstad, and Meredith Welch-Devine

4 Dustbowl Legacies 122

Long-Term Change and Resilience in the Shortgrass Steppe

Kenneth M. Sylvester and Myron P. Gutmann

5 The Political Ecology of Southwest Michigan Agriculture, 1837–2000 152

Alan P. Rudy, Craig K. Harris, Brian J. Thomas, Michelle R. Worosz, Siena S. K. Kaplan, and Evann C. O'Donnell

6 Agrarian Landscape Transition in the Flint Hills of Kansas 206

Legacies and Resilience

Gerad Middendorf, Derrick Cline, and Leonard Bloomquist

7 Water Can Flow Uphill 238

A Narrative of Central Arizona

Charles L. Redman and Ann P. Kinzig

Conclusion 272

Ted L. Gragson

Index 279

About the Contributors

Harvard Forest Long-Term Ecological Research

David R. Foster is an ecologist and director of the Harvard Forest at Harvard University, where he is a faculty member in the Department of Organismic and Evolutionary Biology and principal investigator for the Harvard Forest LTER program. He is the author of *Thoreau's Country—Journey through a Transformed Landscape* (1999), *New England Forests through Time* (2000), and *Forests in Time—The Environmental Consequences of 1000 years of Change in New England* (2004).

Brian Donahue is associate professor of American Environmental Studies on the Jack Meyerhoff Fund at Brandeis University and environmental historian at Harvard Forest. He teaches courses on environmental issues, environmental history, and sustainable farming and forestry. He cofounded and directed Land's Sake, a nonprofit community farm in Weston, Massachusetts, and serves on that town's conservation and community preservation committees. Donahue's recent books, *Reclaiming the Commons: Community Farms and Forests in a New England Town* (2001) and *The Great Meadow: Farmers and the Land in Colonial Concord* (2004), examine the relationship between people and the land in New England. He aspires to be both a yeoman and a scholar.

David Kittredge is on the faculty in the Department of Natural Resources Conservation at the University of Massachusetts–Amherst, where he leads the undergraduate forestry program and serves as the state extension forester. He also participates in the Harvard Forest LTER site as a scientist and forest policy analyst. His research focuses on private woodland ownership trends and behaviors, and relevant public policy.

Glenn Motzkin is a plant ecologist at Harvard Forest whose work focuses on historical ecology and its application to conservation in New England. Motzkin is an avid field naturalist and serves as an ecological advisor to The Trustees of Reservations, and he is on an advisory committee of the Connecticut River Watershed Council.

Brian Hall is a graduate of the State University of New York (SUNY) at Plattsburgh and SUNY College of Environmental Science and Forestry in Syracuse, New York. He is a research assistant at Harvard Forest specializing in Geographic Information System–based analysis and data presentation. His work allows him to combine his background in plant ecology with his lifelong interest in the cultural history of New England.

B. L. Turner II is the Higgins professor of environment and society and director, Graduate School of Geography, Clark University. His research has ranged from ancient Mayan agriculture, to agricultural intensification among subsistence market farmers, to global environmental change focused on land change science.

Elizabeth Chilton received her PhD from the University of Massachusetts–Amherst in 1996. She was an assistant and then associate professor at Harvard University from 1996 to 2001 and is currently an associate professor and chair of anthropology at the University of Massachusetts–Amherst. Her research specialties include North American Indians, the archaeology of eastern North America, hunter-gatherers, the origins of agriculture, archaeological ceramics, geoarchaeology, and cultural resource management.

Coweeta Long-Term Ecological Research

Ted L. Gragson is a professor in the Department of Anthropology and adjunct professor in the Odum School of Ecology at the University of Georgia at Athens. He is also the lead principal investigator of the Coweeta LTER based in southern Appalachia. His primary research is in human decision making as it links individuals to their natural and social environments. He carries out this research by combining fieldwork with contemporary populations, using historical documentary archives, and modeling.

Paul V. Bolstad is a professor in the Department of Forest Resources at the University of Minnesota specializing in measuring and modeling ecosystem structure and function at a range of spatial scales. His research often focuses on ecosystem responses to human land use and management actions, and both legacy and contemporary effects of human disturbances and land use. He pursues this research with a combination of field, laboratory, and modeling work.

Meredith Welch-Devine is a PhD candidate in the Department of Anthropology at the University of Georgia. Her work focuses on the interface between large-scale conservation initiatives and small-scale land management institutions and their resulting governance issues. Her dissertation research centers on the creation of Natura 2000, a pan-European ecological network, and its implementation in the French Basque province of Soule.

Shortgrass Steppe Long-Term Ecological Research

Kenneth M. Sylvester is an assistant research scientist at the University of Michigan's Inter-university Consortium for Political and Social Research. A historian with interdisciplinary interests in agriculture and environment, social and economic change in Canada and the United States, he is the author of *The Limits of Rural Capitalism* (2001) and co-author of "Integrating the Biophysical and Social Sciences" in *Managing Agricultural Land for Environmental Quality: Strengthening the Science Base*, edited by Max Schnepf and Craig Cox (2007).

Myron P. Gutmann is professor of history and director of the Inter-university Consortium for Political and Social Research at the University of Michigan. He has broad interests in interdisciplinary history, especially population, health, immigration, and environment, in both Europe and the United States. He has published *War and Rural Life in the Early Modern Low Countries* (1980), *Towards the Modern Economy* (1988), and more than 60 articles and chapters.

Kellogg Biological Station Long-Term Ecological Research

Alan P. Rudy teaches in the Department of Sociology, Anthropology and Social Work at Central Michigan University. He received his PhD in sociology from the University of California, Santa Cruz, in 1995.

Craig K. Harris is an associate professor of sociology and rural sociology at Michigan State University, East Lansing. He received his PhD in mathematical sociology from the University of Michigan in 1978.

Brian J. Thomas is an assistant professor in the Department of Sociology, Saginaw Valley State University. He received his PhD in sociology from Michigan State University.

Michelle R. Worosz is an assistant professor at Michigan State University. She received her PhD in sociology from Michigan State University in 2006.

Siena S. K. Kaplan works at Environment America. She received her BA from Wellesley College in 2006.

Evann C. O'Donnell is a student at George Washington University Law School. She received her BA from Michigan State University in 2006.

Konza Long-Term Ecological Research

Gerad Middendorf is associate professor of sociology at Kansas State University. His research interests are in the areas of rural and environmental studies, the sociology of agriculture and food, and international development. His recent work has included a study of information needs of organic growers and retailers, and a study of agrarian landscape transition in eastern Kansas. He has published a number of articles and chapters on the implications of agricultural biotechnologies and on agricultural science and technology policy. He is co-author of *The Fight over Food* (2008), a book that examines the ways in which producers, consumers, and activists are challenging the global food system.

Derrick Cline received his BS and MA in sociology from Kansas State University. While a graduate student, he conducted research on land use and agrarian transition in eastern Kansas. His master's thesis was on the coevolution of society and invasive species (*Sericea lespedeza*) in the Kansas Flint Hills. Cline received his MA in 2006 and currently works in the Kansas City area as a research analyst.

Leonard Bloomquist was associate professor of sociology and head of the Department of Sociology, Anthropology, and Social Work at Kansas State University. Bloomquist's research included work on rural labor markets, with emphases on rural manufacturing and employment opportunities for different sociodemographic groups. He also conducted research on rural community development in Kansas, and conducted a multilevel analysis of prolonged population decline among rural localities in the Great Plains region. While at Kansas State, Bloomquist directed the university's Survey Research Laboratory and the Population Research Laboratory. He became head of the Department of Sociology, Anthropology and Social Work at Kansas State University in 2001. Dr. Bloomquist passed away in 2005.

Central Arizona–Phoenix Long-Term Ecological Research

Charles L. Redman received his BA from Harvard University, and his MA and PhD in anthropology from the University of Chicago. He taught at New York University and at SUNY–Binghamton before joining Arizona State University in 1983. Since then, he has served 9 years as chair of the Department of Anthropology and 7 years as director of the Center for Environmental Studies, and in 2004 he was chosen to be the Julie Ann Wrigley Director of the newly formed Global Institute of Sustainability. Redman's interests include human impacts on the environment, sustainable landscapes, rapidly urbanizing regions, urban ecology, environmental education, and public outreach. He is the author or co-author of 10 books, including *Explanation in Archaeology* (1971), *The Rise of Civilization* (1978), *People of the Tonto Rim* (1993), *Human Impact on Ancient Environments* (1999), and, most recently, *The Archaeology of Global Change* (2004). Redman is currently working on building upon the extensive urban environmental research portfolio of the Global Institute of Sustainability through the new School of Sustainability, for which he serves as director. The School is educating a new generation of leaders through collaborative learning, transdisciplinary approaches, and problem-oriented training to address the environmental, economic, and social challenges of the 21st century.

Ann P. Kinzig's research focuses broadly on ecosystem services and resilience. An associate professor in the School of Life Sciences at Arizona State University

(ASU), she carries out much of her work under the auspices of the ecoSERVICES Center in the College of Liberal Arts and Sciences. She is currently involved in three major research projects: (1) advancing conservation in a social context (www.tradeoffs.org), (2) the resilience of prehistoric landscapes in the American Southwest, and (3) assessments of ecosystem services, their valuation, and mechanisms for ensuring their continued delivery (www.diversitas-international.org). Kinzig's experience in science policy includes a year-long assignment (1998–1999) at the Office of Science and Technology Policy in the Executive Office of the President, where her portfolio included climate change policy, carbon science research, and energy issues. Kinzig received her PhD in energy and resources from the University of California at Berkeley in 1994.

AGRARIAN LANDSCAPES IN TRANSITION

Introduction

Charles L. Redman

The patterns humans impose on the earth, purposefully as well as inadvertently, through land-use change are fundamental determinants of local, regional, and global ecological processes that ultimately influence the sustainability of both biological and cultural landscapes, and thus human quality of life. These landscapes result from integrated socioeconomic and ecological dynamics playing out across potentially vast scales of space, time, and organizational complexity (Levin, 1999; Turner et al., 1990; Vitousek et al., 1997). Ecological systems have intrinsic temporal rhythms, driven by such things as generation time, age of reproduction, and disturbance frequencies. They also exhibit patterns on characteristic spatial scales, formed by such things as dispersal distance, topography, and interaction lengths. Moreover, these ecological systems also bear the signature of human institutions that act—either directly or indirectly—to alter the dominant spatial and temporal modes (e.g., suppressing fire frequencies or homogenizing landscapes) or to introduce new ones (e.g., 5-year planning cycles or rectangular state boundaries) (Carpenter and Gunderson, 2001; Pyne, 1997; Scheffer et al., 2001; Turner et al., 2002). At the same time, human institutions are shaped and influenced by the environmental rhythms and ecological arrangements of the biogeographical region in which they emerged (Berkes and Folke, 1998; Cronon, 1983; Diamond, 1997, 2005; Dove and Kammen, 1997; Ostrom et al., 1999). This reciprocal "imprinting" of scales means that scientists and managers cannot effectively parse landscapes into "natural" and "human" components, but instead must study them as an integrated whole (Kinzig et al., 2000; Michener et al., 2001; National Research Council, 1999; Liu et al., 2007). We look at how humans affect the landscape: How does the land deal with, and recuperate from, human-caused soil erosion and introduced species? How did the land deal with these factors before there was much human imprint on the land? How has the introduction of fossil-fuel energy, and thus mechanization, changed the landscape? What role do changing nutritional demands and food distribution systems play? Our theoretical objective in this volume and the research that it reflects is to understand what happens when humans impose their spatial and temporal signatures on ecological regimes and must then respond to the systems they have helped create, further altering the dynamics of the coupled system and the potential for ecological and social resilience.

We study this question within the context of agrarian transformations, both current and historical, because of their ubiquity and because of the tight coupling of human and environmental dynamics that are an inherent feature of agrarian landscapes (Geertz, 1963). The introduction, spread, and abandonment of agriculture represents the most pervasive alteration of the earth's environment during the past 10,000 years, affecting two thirds of the earth's terrestrial surface (Farina, 2000; Matson et al., 1997; Vitousek et al., 1986). The transitions of agrarian landscapes and lifeways continue to take many forms, ranging from abandonment to urban development to more intensified agriculture. In the United States alone, 105 acres of agricultural land go out of production every hour; about half of that is used for urban or suburban growth and the other half is used less intensely or actively conserved for its habitat values (USDA Policy Advisory Committee on Farm and Forest Land Protection and Land Use, 2001).

Many current conceptualizations of agrarian transformations, however, are based on simple linear assumptions—that is, people behave monolithically, and land-use decisions are governed by land rent, demographic pressures, and technological capabilities (Agarwal et al., 2001; Kinzig et al., 2000; Liu, 2001). The details of this linear dynamic are assumed to translate cleanly across space and time, applying equally well to regions with different ecological features, different development histories, or different natural resource institutions. In contrast, we conceptualize a more intricate and integrated cycle—of land-use change affecting landscapes, of altered landscapes affecting ecological processes, of both influencing the ways in which humans monitor and respond to their surroundings, and of human responses engendering further cycles of change (Fig. I.1). In this view, human institutions are ultimately products of their landscapes, and landscapes are products of the human institutions governing them (Duane, 1999; Peluso, 1992; Williams, 1980; Worster, 1984). The current state is intimately influenced by the iterative history of this cycle. We thus do not expect the details of agrarian transformations to hold constant across biogeographical regions or over time, although we do assert that variations can be understood and their patterns described.

To understand the richness, diversity, and complexity of agrarian landscapes and their transformations, then, we must monitor them at varying spatial and temporal scales, and place them in a context of former cycles of change, human perceptions of the lands and lifeways, and the emergence of institutions associated

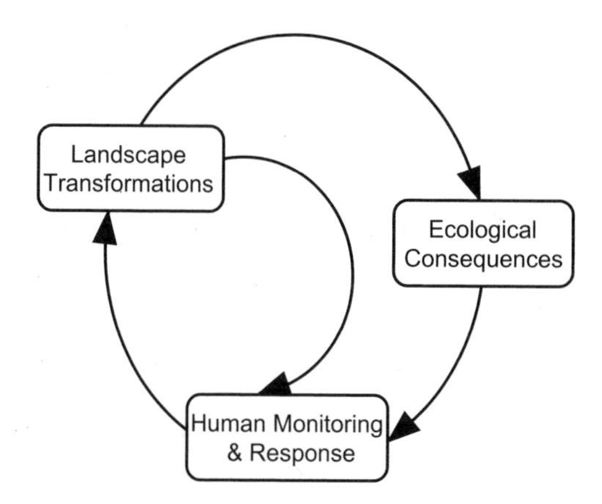

Figure I.1 Our general conceptualization of the cycle of agrarian transformations.

with natural resources. We thus investigated each of the six case studies in this volume to understand the three stages of that cycle:

1. How do human activities influence the spatial and temporal structures of agrarian landscapes? How does this vary over time and across biogeographical regions?
2. What are the ecological and environmental consequences of the resulting structural changes?
3. What are the human responses to both these structural and ecological changes, and how do these responses drive further changes in agrarian landscapes?

We are particularly interested in identifying and understanding the nonlinearities or "surprises" that emerge in this cycle. We want to know where they come from, how they affect feedback in the system, and how to avoid having them precipitate crises. We suspect, and the literature reinforces, that it is the lack of relating fast- and slow-moving processes, ignoring forces that seem too distant in time or unconnected from the system, and the mismatch of scales of monitoring the environment relative to making decisions about it that reduce resilience and precipitate crises (Carpenter and Gunderson, 2001; Diamond, 2005; Foster and Aber, 2002; Gunderson et al., 1995; Holling, 1973, 2001; Levin, 1999; Redman, 1999; Swetnam et al., 1999; Tainter, 1990). We propose to identify and describe the influence of several critical structures and dynamics on stability regimes within these coupled human–natural agrarian systems by promoting four significant innovations through our research:

1. Our approach is multiscalar—spanning temporal, spatial, and organizational scales—and emphasizes identification of potential critical "scales" or critical "cross-scale" interactions, elucidating the ways in which processes at one level of the hierarchy can constrain or influence processes at another.
2. We pay particular attention to long time spans, especially the influence of lags and legacies on current-day dynamics (e.g., in the ecological processes themselves, in the monitoring of change, or in human response to change).

3. We describe the strength and length of causal and closed loops in the human–environment interaction, correlating these feedback loops to shifts in stability regimes and changes in system resilience or vulnerability.
4. Our framework is comparative to elicit the more general processes and relationships that drive the patterns observed in particular cases. These comparisons are cross-site, cross-cultural, and cross-biogeographical.

As a practical test of our findings, we examine our approach and insights in the context of conservation activities in each case study, including an emphasis on creation of preserves and ecoregional planning (Groves et al., 2002). Throughout this volume, we define conservation broadly to include not only the preservation of "natural patches" and environmental quality across a changing landscape, but conservation of the desirable cultural qualities of landscapes, including those described as working landscapes.

Case Studies

To implement this ambitious approach to gain a new understanding of the intricately coupled human and natural systems embodied in agrarian landscapes, we take advantage of the enormous background research and continuing inquiries of the National Science Foundation's (NSF's) Long-Term Ecological Research (LTER) network. The value of using LTER sites stems from their spatial distribution and the duration and richness of data collection. Chapters 2 through 7 report on six LTER sites that represent different major biogeographical regions of the United States, varying local cultures and institutions, and contrasting agrarian landscape transformations (Fig. I.2). Long-Term Ecological Research studies encompass both localized and landscape-level phenomena, and rapid and longer term processes. We complement these scales by presenting data on a regional scale and over a longer duration, from such sources as the U.S. Census and agricultural census, archival records, and remote sensing. Our intention is to develop and implement a theoretical approach and practical framework that would be adopted by the entire LTER network, other relevant regional studies both here and abroad, and natural resource managers. Although we have centered our research at LTER sites, the work represents a significant extension of LTER studies. First, the site-specific LTER monitoring and analyses are supplemented with extensive regional data, allowing a broader understanding of ecological patterns and dynamics. Second, we extend LTER data sets temporally, at least to the beginning of the 20th century and often much earlier. Third, there has not generally been a focus on agricultural research within the LTER network, although many of the 24 sites have substantial areas that are, or once were, agrarian landscapes (Foster et al., 2003; Grimm and Redman, 2004). Lastly, and most important, this volume and associated research define a fundamentally new role for integrative and interdisciplinary science within the LTER network and beyond, by forging new collaborations among natural and social scientists.

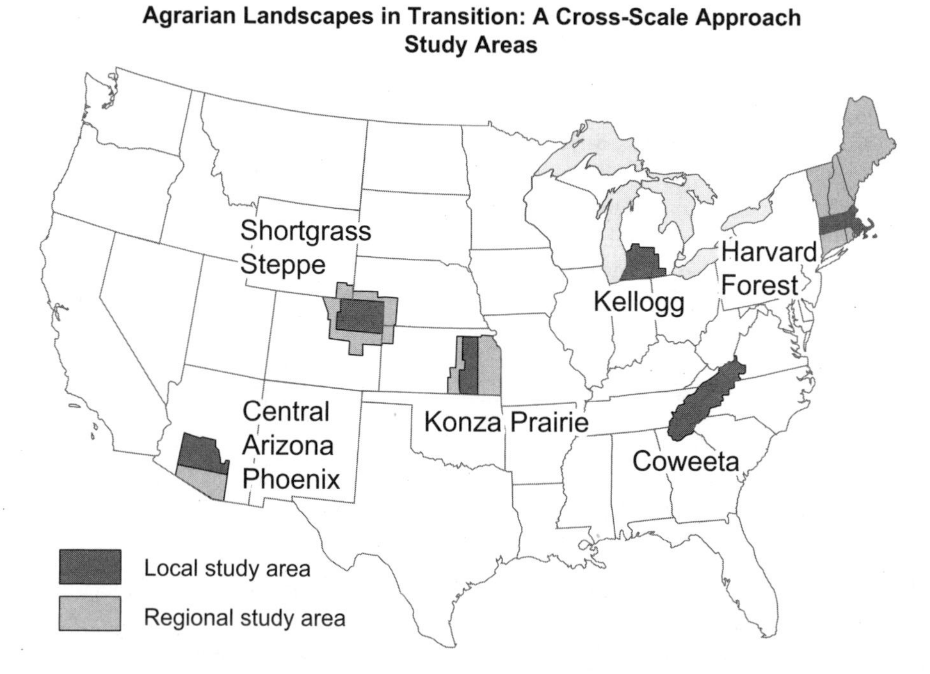

Figure I.2 The six LTER study sites.

Chapter 2—Harvard Forest: New England's Forest Landscape: Ecological Legacies and Conservation Patterns Shaped by Agrarian History

The dramatic reduction in agriculture in New England during the past 150 years generated a wave of land-cover change as forest cover increased from less than 30% to 70–95% in many regions (Foster and O'Keefe, 2000). The reestablishment of forest ecosystem characteristics progressed unevenly, with compositional, structural, and functional attributes exhibiting different lags in development. In all cases, however, the modern distribution of vascular plant species, levels of forest biomass, and soil structure, chemistry, and fertility are strongly conditioned by legacies of varied land-use history (Compton et al., 1998; Foster et al. 1998a,b; Motzkin et al., 1996). The scale and grain of this landscape conditioning are controlled by the physical environmental template, geographical location relative to population centers, and the specific cultural traditions of the regional population, which vary in subtle fashion. In general, however, the broad pattern has been for a homogenization of ecological characteristics at the site scale (resulting from the uniformity in land use) and at the regional scale (resulting from the broad-scale similarity of changes in land use and cover), and for a more patchy and heterogeneous structure with abrupt ecological discontinuities at a landscape scale (resulting from the small-grained and patchy landownership pattern) (Foster and Aber, 2002).

These changing landscape conditions and patterns have generated distinctly different approaches to conservation and management, largely driven by individual value systems and the extent to which the legacies and lags are interpreted correctly (Foster and O'Keefe, 2000; Kittredge et al., 2003). Each emerging tradition in conservation has different ecological consequences (Boose et al., 2001; Foster et al., 1997, 1999). This narrative explores when and under what circumstances different conservation strategies emerge, and what they might do to the stability regimes of the landscape. Harvard Forest scientists have assembled regional archaeological, historical, ecological, and environmental data; developed a statewide overlay of forest harvesting for the past 20 years and analyzed the resulting harvested areas for carbon stocks, invasive species, forest dynamics, and browsing by large ungulates (e.g., deer and moose); and analyzed the regional patterns of land protection—timing, ownership, spatial pattern, and conservation effectiveness.

Chapter 3—Coweeta: Agricultural Transformation of Southern Appalachia

The Coweeta chapter examines land use and agrarian transformation cycles for southern Appalachia from prehistory through the present. By virtue of its geographical position, southern Appalachia serves as a natural laboratory for evaluating across diverse gradients the proportional contribution of socioeconomic and biophysical processes to the structure and function of ecosystems (Bailey, 1996; Markusen, 1987; Whittaker, 1966). In consequence of this advantage, Coweeta LTER studies have strived to demonstrate how current ecological conditions, including aquatic and terrestrial community structure, nutrient pools, and water quality, are adequately explained only by also considering former land use.

The pervasiveness of land-use-related legacies in southern Appalachia is taken for granted across a variety of disciplines, but many questions still remain. For example, there is at least a 50-year lag between the recovery of invertebrate and fish assemblages in forested Appalachian streams relative to the recovery of riparian forests in the same areas (Jones et al., 1999). Stating that the lag is a consequence of the complex interaction between human and natural forces fails to identify concrete steps that could be taken toward facilitating or accelerating the recovery of streams and forests from a conservation standpoint. To contribute toward such an objective, Coweeta scientists have assembled archaeological, historical, and biophysical information for the 60,000-km^2 southern Appalachian region and have used this information to begin to develop a comprehensive understanding of the human–natural dynamic in southern Appalachia.

The Coweeta LTER looks at fundamental issues in the historical ecology of southern Appalachia by reconstructing when and where particular natural and human events occurred. It also addresses fundamental issues about the contemporary gentrification of southern Appalachia that, combined with the legacy of past land-use practices on contemporary terrestrial and aquatic ecosystems, has important implications for the future of the region (Wear and Bolstad, 1998). Mere stories about the past or the relation of humans to the natural world provide little guidance in the whirlpool of prophecy that decision makers at all levels are asking social and ecological scientists for help with in southern Appalachia.

Bold decisions must be made during the next few years. Will they be reactive or anticipatory? Will they focus on rescuing biophysical systems to the exclusion of socioeconomic systems? Will they attempt to reconcile protection and restoration with development and livelihood? By quantifying the spatial heterogeneity in disturbance legacies and the temporal heterogeneity of disturbance trajectories, researchers on the Coweeta LTER are calculating the duration and magnitude of consequences at different organizational levels. These will then be used to develop forecast scenarios of future social and ecological responses with the objective of building scalable estimates for processes of importance to decision makers responding to local and regional conditions. A common theme to the Coweeta LTER investigations is moving understanding about coupled human–natural systems from stereotypes toward process and explanation.

Chapter 4—Shortgrass Steppe: Dustbowl Legacies: Long-Term Change and Resilience in the Shortgrass Steppe

This chapter explores the history of drought, agriculture, landscape transformation, and human perception in the U.S. Great Plains, where more than a century of land-use change has coincided with climate variability at timescales that vary from the seasonal to decadal. This process is generally well-known to the ecological and environmental history communities, but the specific mechanisms of change and feedback are still contested (Antle et al., 2001).

The history of 20th-century droughts and their broad cycles of response and counterresponse are often told: Land-use change in the 1920s and 1930s led to drought and dust storms in the 1930s (Worster, 1979). Land abandonment and improved agricultural practices resulted in the drought of the 1950s having a less severe effect on the land, but, at the same time, more migration occurred because of a dynamic national economy (Gutmann and Cunfer, 1999; Hurt, 1981). In addition, the long-term environmental consequences brought about by land-use change in the 1950s are likely to be more meaningful than those of the 1930s because the technology of the 1950s (irrigation, new seed varieties) has already altered weather patterns in the semiarid grasslands (Epstein et al., 1999). Human perceptions of these changes, from local farmers to national policymakers, have varied and evolved in response to environmental knowledge, the speed of change, and political and social considerations. This chapter examines the critical causal relationships and feedback loops governing these responses and their influence on stability regimes.

To document these changes, researchers assembled a range of historical and biophysical information for the High Plains of northeastern Colorado and vicinity. Their report focuses on change during the prehistory of the shortgrass and the relative continuity observed in agricultural land use after the first 30 years of settlement. Agricultural land use remains commercially important in the region, and the report makes use of the agricultural and population census data to analyze the drivers of change in the 26-county, 52,800-km^2 study area.

Chapter 5—Kellogg Biological Station: The Political Ecology of Southwest Michigan Agriculture, 1837–2000

Kellogg Biological Station (KBS) has developed an agroecological history of southwest Michigan. Among the qualitative and quantitative resources used to craft their narrative are agricultural and population census data, Michigan Agricultural Experiment Station and Extension bulletins and circulars, mappings of historical ecological phenomena (soils, forests), Centennial Farm lists and records (with the intent of future interviews and archival work), and recovery of transportation and industrial histories of the region. Integrating the relation of these data to the contested politics of the region's diverse forms of conservation efforts—whether driven by environmental movements, agricultural organizations, or governmental agencies—is an important aspect of the study. The conceptual foundation of the KBS approach is one that sees agriculture as drawing on historical, immediate, and future social and ecological developments. Southwest Michigan is a fairly industrialized region of progressive farmers and progressive conservationists and, although the focus is on the consequences for industrial society and regional ecologies of farm practices, understanding the relationship among these forces is vitally important and difficult to conceive.

Chapter 6—Konza: Agrarian Landscape Transition in the Flint Hills of Kansas: Legacies and Resilience

The Flint Hills of east-central Kansas, the setting of the Konza LTER, contain the largest remaining area of unplowed tallgrass prairie in North America. This chapter explores the changes in the agricultural and settlement patterns of the region from early European American settlement to the present, linking these changes to transitions in the agrarian landscape. Major land-use regimes and transitions include (1) the seminomadic, mixed cropping, and hunting patterns of the Kansa Indians; (2) European American settlement, including the establishment and subsequent decline of general farming, extensive enclosure of grasslands, and the emergence of large-scale grazing and beef production by the late 19th century; (3) the expansion of the agricultural economy in the late 19th and early 20th centuries; (4) drought and depression during the interwar years; (5) agricultural intensification of the post–World War II period; and (6) a set of current issues discussed under the rubric of conservation, including urban edge development, fire suppression, invasive species, and the institutional context within which land-use decisions are made. To document these changing land-use regimes, researchers reviewed historical and current land-use data from the agricultural census and Kansas State Board of Agriculture from the late 19th century onward, including cropping patterns and livestock production. Other data include a review of literature on Native American land-use patterns, population census data, remotely sensed land-cover data, and face-to-face interviews with farmers/ranchers on more recent changes in land-use practices.

Chapter 7—Central Arizona–Phoenix (CAP): Water Can Flow Uphill: A Narrative of Central Arizona

Among the most compelling coupled natural–human systems in the arid West is that of water and its human use. Availability of water for irrigated agriculture and municipal growth is subject to the vagaries of climate. Humans attempt to manage this variability, as well as that produced from flooding, by controlling impoundments and water releases, and by otherwise heavily modifying the hydrosystem (Reisner, 1986; Worster, 1985). The absolute necessity of supplemental water for farming and human settlement has made its allocation a priority for governments at all levels. It is a powerful driver of the economy and is pivotal for the continuing growth in the regional population (Gammage, 1999). What makes this case study fascinating is that each level of government monitors and allocates water according to different spatial units, and there are differing temporal lags in monitoring the availability of water (Carter et al., 2000; Merideth, 2001). These differences are compounded by the differential willingness of each sector to pay for available water, with residential users willing to pay far more than farmers. Despite this market imbalance, agriculture still consumes 80% of the water in Arizona.

At a regional scale, land-use change and newly engineered water sources have allowed the study area to more than double in population while reversing groundwater depletion. At the more "human" scale of the landscape, outcomes are far more varied, with some local water tables dropping and riparian areas going dry (Grimm and Redman, 2004). The driving forces and cascading influences associated with patterns of water availability and use operate at varying scales of geography, with lags in response determined by nature and the legal system, and are embedded at the center of an economic system that not only operates on a rapid frequency, but often prices water well into the future.

This chapter highlights the critical interactions among climate change, jurisdictional scales, human monitoring and response, and consequences for agrarian transformations. The authors examine institutional responses at different levels (political as well as social) by examining the correlation between significant events (e.g., destructive floods, world wars) and subsequent adjustments in laws, water-planning strategies, and land-use decisions.

The CAP LTER has gathered data, maps, photographs, and remotely sensed images on current and historic trends (from the 1870s onward) in the Arizona study area, emphasizing agricultural activities, population, water, land use, land cover, and the region's conservation history. The agricultural information gathered includes the extent and types of agricultural land, economic returns, and available and emerging technologies. The data gathered on population describe the total population, rural versus urban divisions, and the extent of the workforce engaged in agricultural activities. Central Arizona–Phoenix has examined water availability and use, as well as water management. Lastly, researchers have characterized the extent of urban settlements, public versus private lands, desert versus "converted" lands, and irrigated lands, as well as detailed the region's conservation history.

Objectives

The framework used in this research is comparative to elicit the more general processes and relationships that drive the patterns observed in particular cases. These comparisons will be cross-site, cross-cultural, and cross-biogeographical. The ultimate goal is to identify and quantify the ways in which agrarian transformations differ across biogeographical regions and across time, and how these variations can be used to generate explanations of cross-scale socioecological patterns. We hope the insights derived from this volume can inform the establishment of future agricultural regimes, improve the sustainability of cities as they grow on former farmlands, and help develop tools, strategies, and new ideas on land and habitat conservation and restoration on former farmlands.

References

Agarwal, C., C. M. Green, J. M. Grove, T. P. Evans, and C. M. Schweik. 2001. *A review and assessment of land-use change models: Dynamics of space, time, and human choice.* CIPEC collaborative report series no. 1. Bloomington, Ind.: Center for the Study of Institutions Population, and Environmental Change, Indiana University.

Antle, J. M., S. M. Capalbo, E. T. Elliot, H. W. Hunt, S. Mooney, and K. H. Paustian. 2001. "Research needs for understanding and predicting the behavior of managed ecosystems: Lessons from the study of agroecosystems." *Ecosystems* 4(8): 723–735.

Bailey, R. G. 1996. *Ecosystem geography.* New York: Springer-Verlag.

Berkes, F., and C. Folke (eds.). 1998. *Linking social and ecological systems: Management practices and social mechanisms for building resilience.* New York: Cambridge University Press.

Boose, E. R., K. E. Chamberlin, and D. R. Foster. 2001. "Landscape and regional impacts of hurricanes in New England." *Ecological Monographs* 71: 27–48.

Carpenter, S. R., and L. H. Gunderson. 2001. "Coping with collapse: Ecological and social dynamics in ecosystem management." *BioScience* 51(6): 451–457.

Carter, R. H., P. Tschakert, and B. J. Morehouse. 2000. *Assessing the sensitivity of the Southwest's urban water sector to climatic variability.* The Climate Assessment Project for the Southwest (CLIMAS) Report Series CL1–00. Tucson, Ariz.: Institute for the Study of Planet Earth, University of Arizona.

Compton, J. E., R. D. Boone, G. Motzkin, and D. R. Foster. 1998. "Soil carbon and nitrogen in a pine-oak sand plain in central Massachusetts: Role of vegetation and land-use history." *Oecologia* 116: 536–542.

Cronon, W. 1983. *Changes in the land: Indians, colonists, and the ecology of New England.* New York: Hill and Wang.

Diamond, J. 1997. *Guns, germs, and steel: The fates of human societies.* New York: W. W. Norton.

Diamond, J. 2005. *Collapse: How societies choose to fail or succeed.* New York: Viking.

Dove, M., and D. Kammen. 1997. "The epistemology of sustainable resource use: Managing forest products, swidden, and high-yielding variety crops." *Human Organization* 1: 91–101.

Duane, T. 1999. *Shaping the Sierra: Nature, culture, and conflict in the changing West.* Berkeley, Calif.: University of California Press.

Epstein, H. E., I. C. Burke, and W. K. Lauenroth. 1999. "Response of the Shortgrass Steppe to changes in rainfall seasonality." *Ecosystems* 2(2): 139–150.

Farina, A. 2000. "The cultural landscape as a model for the integration of ecology and economics." *BioScience* 50(4): 313–320.

Foster, D. R., and J. Aber (eds.). 2002. *Forests in time. Ecosystem structure and function as a consequence of history.* New Haven, Conn.: Yale University Press.

Foster, D. R., J. Aber, R. Bowden, J. Melillo, and F. Bazzaz. 1997. "Forest response to disturbance and anthropogenic stress." *BioScience* 47: 437–445.

Foster, D. R., M. Fluet, and E. R. Boose. 1999. "Human or natural disturbance: Landscape dynamics of the tropical forests of Puerto Rico." *Ecological Applications* 9: 555–572.

Foster, D. R., D. Knight, and J. Franklin. 1998a. "Landscape patterns and legacies resulting from large infrequent forest disturbance." *Ecosystems* 1: 497–510.

Foster, D. R., G. Motzkin, and B. Slater. 1998b. "Land-use history as long-term broad-scale disturbance: Regional forest dynamics in central New England." *Ecosystems* 1: 96–119.

Foster, D. R., and J. O'Keefe. 2000. *New England forests through time: Insights from the Harvard Forest dioramas.* Cambridge, Mass.: Harvard University Press, Harvard Forest, Petersham.

Foster, D. R., F. Swanson, J. Aber, D. Tilman, N. Bropakw, I. Burke, and A. Knapp. 2003. "The importance of land-use and its legacies to ecology and environmental management." *BioScience* 53(1): 77–88.

Gammage, G. 1999. *Phoenix in perspective: Reflections on developing the desert.* The Herberger Center for Design Excellence. Tempe, Ariz.: Arizona State University, College of Architecture and Environmental Design.

Geertz, C. 1963. *Agricultural involution: The process of ecological change in Indonesia.* Berkeley, Calif.: University of California Press.

Grimm, N. B., and C. L. Redman. 2004. "Approaches to the study of urban ecosystems: The case of central Arizona–Phoenix." *Urban Ecosystems* 7: 199–213.

Groves, C. R., D. B. Jensen, L. L. Valutis, K. H. Redford, M. L. Shaffer, J. M. Scott, J. V. Baumgartner, J. V. Higgins, M. W. Beck, and M. G. Anderson. 2002. "Planning for biodiversity conservation: Putting conservation science into practice." *BioScience* 52(6): 499–512.

Gunderson, L. H., C. Holling, and S. S. Light (eds.). 1995. *Barriers and bridges to the renewal of ecosystems and institutions.* New York: Columbia University Press.

Gutmann, M. P., and G. Cunfer. 1999. *A new look at the causes of the Dust Bowl.* Publication no. 99-1. Lubbock, Texas: International Center for Arid and Semiarid Land Studies.

Holling, C. S. 1973. "Resilience and stability of ecological systems." *Annual Review of Ecology and Systematics* 4: 1–23.

Holling, C. S. 2001. "Understanding the complexity of economic, ecological, and social systems." *Ecosystems* 4(5): 390–405.

Hurt, D. 1981. *The Dust Bowl: An agricultural and social history.* Chicago, Ill.: Nelson-Hall.

Jones, E. B. D., III, G. S. Helfman, J. O. Harper, and P. V. Bolstad. 1999. "The effects of riparian deforestation on fish assemblages in southern Appalachian streams." *Conservation Biology* 13: 1454–1465.

Kinzig, A. P., J. Antle, W. Ascher, W. Brock, S. Carpenter, F. S. Chapin III, R. Costanza, K. Cottingham, M. Dove, H. Dowlatabadi, E. Elliot, K. Ewel, A. Fisher, P. Gober, N. Grimm, T. Groves, S. Hanna, G. Heal, K. Lee, S. Levin, J. Lubchenco, D. Ludwig, J. Martinez-Alier, W. Murdoch, R. Naylor, R. Norgaard, M. Oppenheimer, A. Pfaff, S. Pickett, S. Polasky, H. R. Pulliam, C. Redman, J. P. Rodriguez, T. Root, S. Schneider,

R. Schuler, T. Scudder, K. Segersen, R. Shaw, D. Simpson, A. Small, D. Starrett, P. Taylor, S. Van Der Leeuw, D. Wall, and M. Wilson. 2000. *Nature and society: An imperative for integrated environmental research.* Report of a workshop to the National Science Foundation. Tempe, Ariz.

Kittredge, D. B., A. O. Finley, and D. R. Foster. 2003. "Timber harvesting as ongoing disturbance in a landscape of diverse ownership." *Forest Ecology and Management* 180: 425–442.

Levin, S. A. 1999. *Fragile dominion: Complexity and the commons.* Reading, Mass.: Perseus Books.

Liu, J., T. Dietz, S. R. Carpenter, M. Alberti, C. Folke, E. Moran, A. Pell, P. Deadman, T. Kratz, J. Lubchenco, E. Ostrom, Z. Ouyang, W. Provencher, C. Redman, S. Schneider, and W. Taylor. 2007. "Complexity of coupled human and natural systems." *Science* 317: 1513–1516.

Liu, J. 2001. "Integrating ecology with human demography, behavior, and socioeconomics: Needs and approaches." *Ecological Modeling* 140: 1–8.

Markusen, A. R. 1987. *Regions: The economics and politics of territory.* Totowa, N.J.: Rowman and Littlefield.

Matson, P. A., W. J. Parton, A. G. Power, and M. J. Swift. 1997. "Agricultural intensification and ecosystem properties." *Science* 277: 504–509.

Merideth, R. 2001. *A primer on climatic variability and change in the Southwest.* Tucson, Ariz.: University of Arizona, Udall Center for Studies in Public Policy and the Institute for the Study of Planet Earth.

Michener, W. K., T. J. Baerwald, P. Firth, M. A. Palmer, J. L. Rosenberger, E. A. Sandlin, and H. Zimmerman. 2001. "Defining and unraveling biocomplexity." *BioScience* 51(12): 1018–1023.

Motzkin, G., D. R. Foster, A. Allen, and J. Harrod. 1996. "Controlling site to evaluate history: Vegetation patterns of a New England sand plain." *Ecological Monographs* 66: 345–365.

National Research Council. 1999. *Our common journey: A transition toward sustainability.* Board on Sustainable Development, Policy Division. Washington, D.C.: National Academy Press.

Ostrom, E., J. Burger, C. Field, R. B. Norgaard, and D. Policansky. 1999. "Revisiting the commons: Local lessons, global changes." *Science* 284: 278–282.

Peluso, N. 1992. *Rich forests, poor people: Resource control and resistance in Java.* Berkeley, Calif.: University of California Press.

Pyne, S. J. 1997. *Fire in America: A cultural history of wildland and rural fire.* Seattle, Wash.: University of Washington Press.

Redman, C. L. 1999. *Human impact on ancient environments.* Tucson, Ariz.: University of Arizona Press.

Reisner, M. P. 1986. *Cadillac Desert: The American West and its disappearing water.* New York: Viking.

Scheffer, M., S. Carpenter, J. Foley, C. Folke, and B. Walker. 2001. "Catastrophic regime shifts in ecosystems." *Nature* 413: 591–596.

Swetnam, T. W., C. D. Allen, and J. L. Betancourt. 1999. "Applied historical ecology: Using the past to manage for the future." *Ecological Applications* 9(4): 1189–1206.

Tainter, J. 1990. *The collapse of complex societies.* Cambridge, U.K.: Cambridge University Press.

Turner, B. T., W. C. Clark, R. W. Kates, J. F. Richards, J. T. Matthews, and W. B. Meyer. 1990. *The earth as transformed by human action.* Cambridge, U.K.: Cambridge University Press.

Turner, B. L., D. R. Foster, and J. Geoghegan (eds.). 2002. *Land change science and tropical deforestation: The final frontier in southern Yucatan.* New York: Oxford University Press.

USDA Policy Advisory Committee on Farm and Forest Land Protection and Land Use. 2001. *Maintaining farm and forest lands in rapidly growing areas.* Report to the Secretary of Agriculture. Washington, D.C.: U.S. Department of Agriculture.

Vitousek, P. M., P. R. Ehrlich, A. H. Ehrlich, and P. A. Matson. 1986. "Human appropriation of the products of photosynthesis." *BioScience* 36(6): 368–373.

Vitousek, P. M., H. A. Mooney, J. Lubchenco, and J. M. Melillo. 1997. "Human domination of earth's ecosystems." *Science* 277: 494–499.

Wear, D. N., and P. V. Bolstad. 1998. "Land use changes in southern Appalachian landscapes: Spatial analyses and forecast evaluation." *Ecosystems* 1: 575–594.

Whittaker, R. H. 1966. "Forest dimensions and production in the Great Smoky Mountains." *Ecology* 47: 103–121.

Williams, R. 1980. "Ideas of nature," pp. 67–85. In: *Problems in materialism and culture: Selected essays.* London, U.K.: Verso.

Worster, D. 1979. *Dust Bowl: The Southern Plains in the 1930s.* New York: Oxford University Press.

Worster, D. 1984. "History as natural-history: An essay on theory and method." *Pacific Historical Review* 53(1): 1–19.

Worster, D. 1985. *Rivers of empire: Water, aridity, and the growth of the American West.* New York: Oxford University Press.

1

Changing Agrarian Landscapes across America

A Comparative Perspective

Kenneth M. Sylvester
Myron P. Gutmann

Over a mere five or six human generations, agriculture has all but disappeared from rural landscapes in the eastern half of the United States. During an equally brief period, agriculture has transformed forests, valleys, prairies, and plains in the interior of the continent. Economic models have explained the shift in terms of the lower costs of land, the larger scale of farming, and better connections to export markets, via river transport at first and then by rail (Fitzgerald, 2003; Gardner, 2002; Hart, 2003). Enormous gains in agricultural productivity since the green revolution have accelerated these trends, ensuring that the physical extent of agricultural land peaked in the United States in 1950, and the pace of abandonment has quickened (Theobald, 2001; Theobald et al., 2005). In the information age, distance to nearby population centers is an increasingly unimportant factor in predicting the prevalence of agricultural land use. With a nationally integrated market and export-driven demand shaping land-use patterns, the production locations have shifted to landscapes with fertile soils; flat, open terrain; and favorable climates. Still, we know that landscapes in the six areas examined have not escaped the legacies of prior patterns of development. Each was managed in different and path-dependent ways during the past 130 years. Choices framed by the timing of settlement, cultural inheritances, and institutional arrangements continue to shape the overall sustainability of ecosystems long after the initial transformation of landscapes.

The legacies of these distinct agricultural systems are explored in this chapter primarily through the lens of the agricultural census. Scientists have expressed growing interest in understanding the effect of historical land use on ecosystem dynamics. Land-use data have been used to drive ecosystem models capable of simulating changes in carbon budgets, nutrient cycling, and the overall sustainability of ecosystems (Matson et al., 1997; Parton et al., 2005; Ramankutty and Foley, 1999; Ramankutty et al. 2002). Historical land-use data are an important tool not only for understanding the scale of anthropogenic impacts, but also for developing a better understanding of the ecosystem services provided by working agrarian landscapes and successional landscapes. The primary goal of this chapter is to compare historical patterns that have dominated land use among the study areas and then evaluate the broad effects of past practice.

The data needed to inform this exercise are available mainly from published U.S. agricultural censuses. Between 1870 and 1920, the data are summarized every 10 years in county-level tabulations, and roughly every 5 years thereafter. Information is reported on the amount of land harvested by crop, and the overall magnitudes of grazing and total farmland. The temporal and spatial scale of the information requires a number of simplifying assumptions. We are comparing areas of the United States that were transformed during very different historical eras and that had very different rates of development, stabilization, and decline. Information about tillage practices, crop varieties, planting and harvest dates, crop harvest practices, and fertilizer application for dominant crop rotations can be surmised from a variety of historical sources. In a recent paper, Parton et al. use the prescriptive literature of the U.S. Department of Agriculture (USDA) and various federal and state extension services to calibrate a simulation exercise using the CENTURY model (Parton et al., 1993, 2005). Our intention here is to develop regional study area comparisons that can point to further detailed investigations. The chief simplifying assumption is that regional-level data faithfully describe dominant cropping systems. Far more research is needed to understand whether the regional findings discussed here actually scale to local or household levels.

Demographic, Social, and Economic Change in Six Regions of the United States

Population change in the six study areas has followed several broad national trends, including the rapid urbanization of the postwar years. The pace of urban growth between 1940 and 1960 decisively altered the social context in which agricultural landscapes are embedded, turning farm and rural folk into minority populations in all the study areas except southern Appalachia. Rates of natural increase, traditionally higher in the countryside, also stalled in the 1950s, and the continued population growth has come increasingly from a reverse migration of urban residents to nonmetropolitan counties late in the 20th century (Johnson and Beale, 1992, 1998, 2002; Johnson and Fuguitt, 2000; Johnson et al., 2005). The timing of change in each region has not simply mirrored industrial growth. The story is more complex. Industrial growth was far more uniform than population change. Long-running statistical series from the census indicate that, in terms of a simple measure of the volume of manufacturing activity, like the number of manufacturing establishments, each region shared periods of expansion

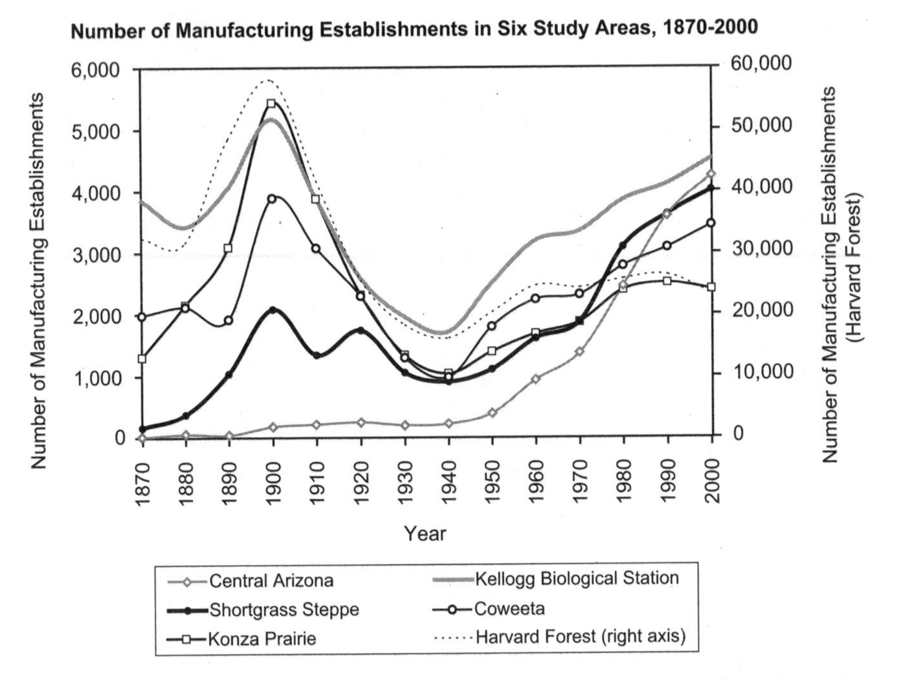

Figure 1.1 Number of manufacturing establishments in six study areas, 1870–2000. Based on summary county data reported in U.S. Department of Commerce, *Census of manufacturing* (1880, 1890, 1900, 1910, 1920, 1930, 1940) and *County and city data book* (1947, 1949, 1950, 1952, 1956, 1962, 1967, 1970, 1972, 1977, 1983, 1988, 1994, 2000).

and contraction simultaneously. The trends visible in Figure 1.1 reflect a well-known story about the first mass production economy based on clothing, food processing, steel making, and rail transportation. A second expansion beginning in the 1940s—based on automobiles, petroleum, chemicals, plastics, a postwar baby boom, increasing consumer demand, and government guarantees in home, farm, and export finance—drove patterns of rural emigration (Rosenberg, 2003; Shuman and Rosenau, 1972; Wells, 2003).

However, early industrialization in New England meant that rural depopulation had deeper historical roots. The loss of female labor in particular to textile mills in southern New England limited the land-use alternatives available to farm families that practiced mixed husbandry prior to industrialization—raising corn, small grains, and livestock (Dublin, 1981; Hareven and Langenbach, 1978). In southern New England, Donahue (2004) suggests that as industrialization began, farmers close to Boston turned to raising beef, and in Vermont (where manufacturing was less prevalent) dairying became more common (Barron, 1984, 1997; Jager, 2004). By the beginning of the 20th century, rural population loss was quite advanced in southern New England. The geography of this population change is illustrated in Figure 1.2 for four dates in the 20th century. The maps display the percent change

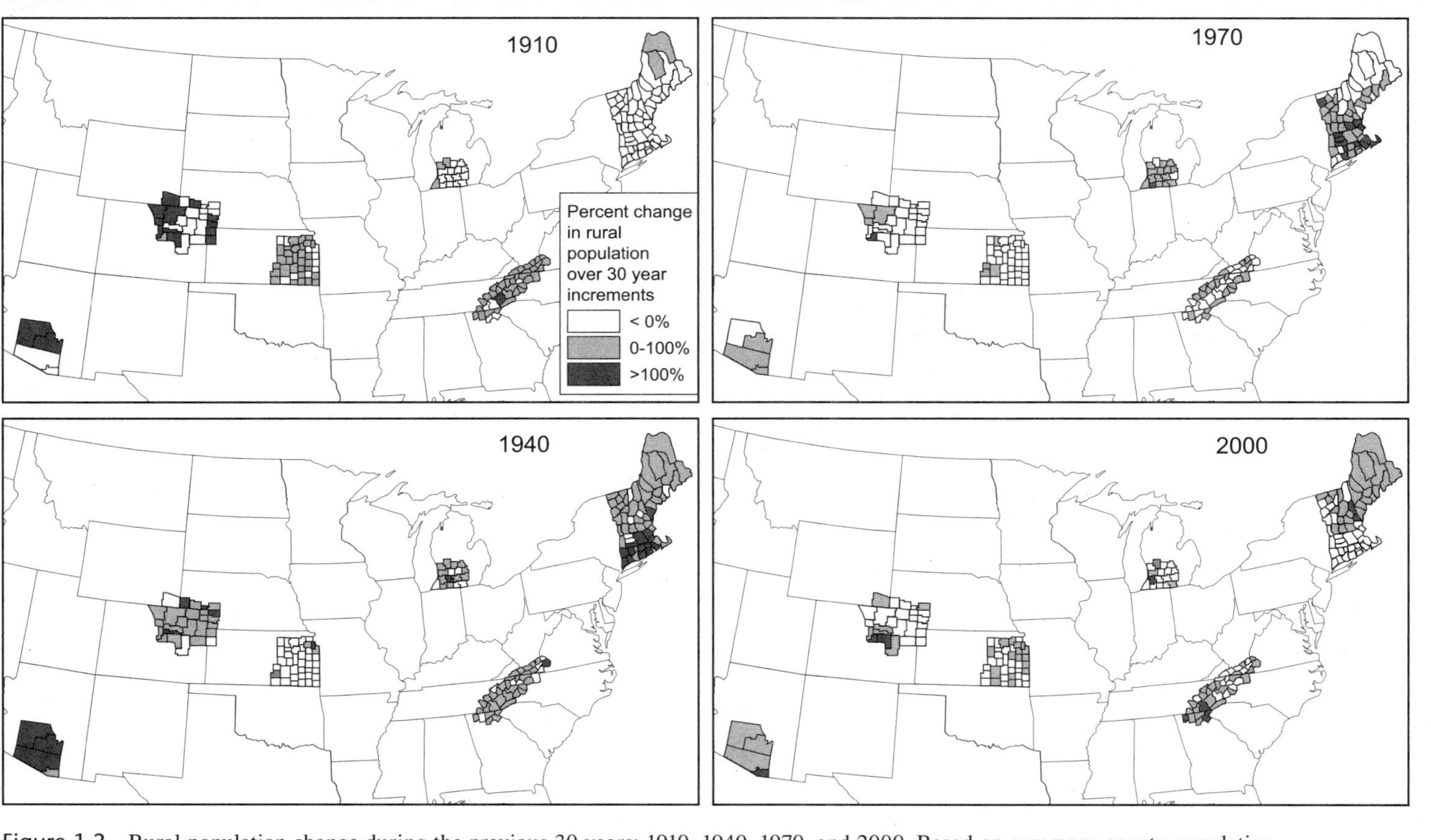

Figure 1.2 Rural population change during the previous 30 years: 1910, 1940, 1970, and 2000. Based on summary county population data from U.S. Department of Commerce (1880b, 1910b, 1940b, 1970b, 2000b).

in the rural population 30 years prior to each census. In 1940, counties in southern New England were already experiencing a historical forerunner of exurban growth, as former urban dwellers took advantage of commuting trains and automobiles to relocate to the increasingly postagrarian landscapes of Connecticut and Massachusetts. By 1970, the more familiar postwar exodus from rural life is visible across the country, as population losses were common across all study areas. In eastern Kansas, the losses were already visible in 1940 and became generalized across all study areas in the three decades preceding the 1970 census. By 2000, rural population "rebounds" were concentrated close to urban centers in the study areas: Grand Rapids in southwest Michigan, Atlanta in southern Appalachia, Denver in eastern Colorado, Kansas City and Wichita in eastern Kansas, Tucson and Phoenix in Arizona, and in New England, in an urban fringe that extended as far as southern New Hampshire and southern Maine.

Before the mid century, the timing of local population change differed less because of proximity to urban centers than the timing of original settlement. Rural population numbers began to climb in absolute terms (Fig. 1.3) from the early part of the 20th century in the longest settled regions: in New England, southern Appalachia, and southwestern Michigan. In western study areas, rural populations peaked when agricultural land use came close to its maximum extent: in eastern Kansas at the beginning of the 20th century, and in eastern Colorado and central Arizona in the 1950s and 1960s. Nevertheless, after 1945, rural

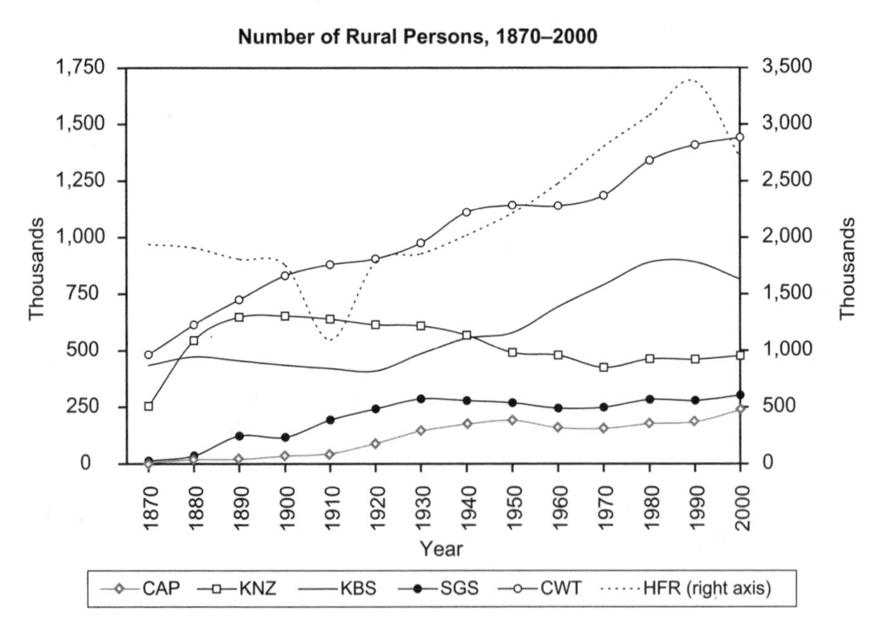

Figure 1.3 Number of rural persons, 1870–2000. CAP, Central Arizona–Phoenix; CWT, Coweeta; HFR, Harvard Forest; KBS, Kellogg Biological Station; KNZ, Konza Prairie; SGS, Shortgrass Steppe.

populations were embedded in rapidly urbanizing societies. The postwar trends affected New England and southern Appalachia least. In New England because the early pattern of exurban growth stabilized the proportion of the population that was nonurban, and in southern Appalachia because urbanization did not explode after the war, but continued a slow but steady increase. The postwar urbanization boom was more typical of the dramatic decrease in the percent of persons living in rural settings in central Arizona, eastern Colorado, eastern Kansas, and southwest Michigan.

Evolution of Land in Farms, Numbers of Farms, and Changes in Farm Size

The six study areas are also representative of several broad land-use trajectories in the past century—particularly, the abandonment of farmland, the growth of residential development at the urban fringe, and declines in the diversity of land use. All occurred much earlier in the northeastern United States, where abandonment expanded the scope for natural succession and for industrial and residential development on former croplands and pastures. The six study areas have also faced an expansion of suburbs, road networks, and industrial development beyond the urban fringe. Hobby farms, vacation homes, and resorts are restructuring rural landscapes around the United States, pressing on planning agendas, begging for answers to questions of which farmland to preserve. In the information age, exurban growth is increasingly free to seek access to natural amenities, shifting from metropolitan counties to nonmetropolitan counties, and the volume of agricultural production is often unrelated to population density, particularly where mechanization has displaced family labor (Brown et al., 2005; Finnegan et al., 2000; Huston, 2005; Maizel et al., 1998; Waisanen and Bliss, 2002).

The six study areas reached peak levels of agricultural land use on distinctly different timescales. The length of these stages varied according to the history of indigenous agriculture and European colonization. Agriculture was more central to central Arizona peoples than to any other ancient North Americans, but horticulture was a part of the traditions of indigenous peoples for several centuries in New England; southern Appalachia; and, to a lesser extent, the woodland–prairie peoples of southwestern Michigan and the grassland dwellers of eastern Kansas and eastern Colorado. Agriculture was more evident at the time of European–native contact in New England and southern Appalachia, and coexisted with European colonization through the early 19th century (Fig. 1.4).

After European colonization began, however, the transformation of landscapes became more extensive and stages of development briefer. In New England, more than 150 years separated the maximum extent of farmland from the time of initial settlement. In southern Appalachia, farmland peaked 80 years after European Americans moved into former Cherokee homelands. In southwestern Michigan, it took half a century for settlers to identify the maximum extent of agricultural land. In eastern Kansas, the timescale was shortened to two generations. Further west, semiarid and arid environments slowed the pace of change. In northeastern

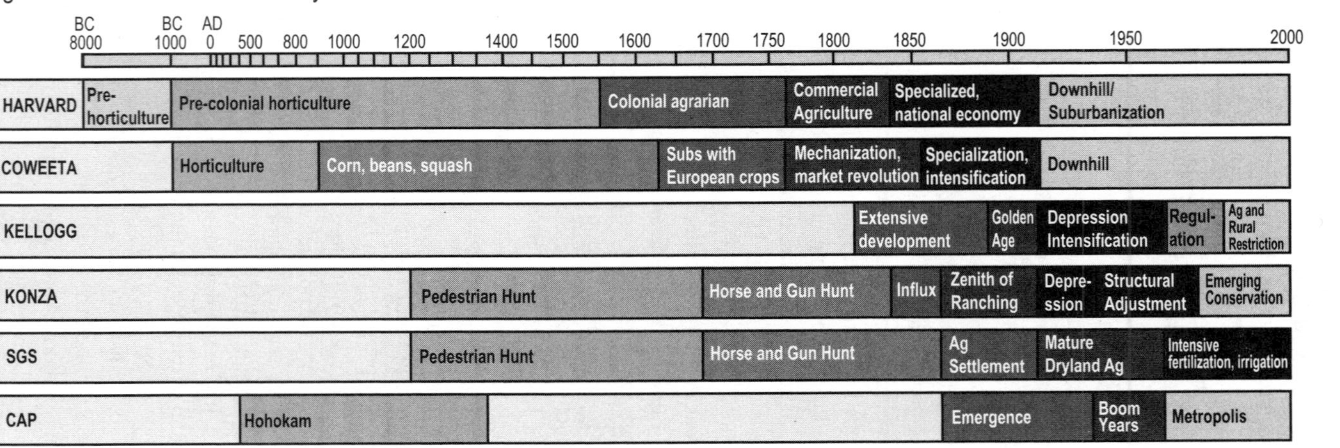

Figure 1.4 Agricultural transition stages and agricultural transition time line by site. Ag, agricultural; CAP, Central Arizona–Phoenix; SGS, Shortgrass Steppe; Subs, subsistence.

Colorado, agricultural land reached its full extent 70 years after settlement began, and in central Arizona, it took 80 years for agriculture to reach its peak.

In Massachusetts, detailed historical investigation using state census and tax assessment records has shown that agricultural settlement, begun during the 17th century, did not peak before 1830 (Hall et al., 2003). Although the earliest settlements were concentrated along the eastern seaboard in the Plymouth and Massachusetts Bay Colonies, the Connecticut River Valley afforded access to the western interior, from early settlements in New Haven, Windsor, Springfield, Longmeadow, and Agawam (Cronon, 1983). But if settlement identified arable lands early, development proceeded deliberately, in part because of the resistance of native peoples to European settlement, the pace of immigration to colonial America, and, to a lesser extent, the proprietary land grant system, the village-centered organization of settlement, and the use of the metes and bounds system to distribute new lands (Cronon, 1983; Hubbard, 1803; Vaughan, 1999). The culture of the time also prefigured a measured pace of development, because (as many local histories have demonstrated) farm families did not experience full integration into the marketplace until the 19th century (Bushman, 1998; Clark, 1990; Donahue, 2004; Kulikoff, 2000; Vickers, 1990).

These same patterns are evident in southern Appalachia, where the search for arable lands negotiated a series of deep steep-sided valleys, dissected by numerous streams. Growth came as a result of expanding numbers of small-holding farm families that dominated the upland south and were generally not part of the plantation system (Hahn, 1983; Hofstra, 2004; Salstrom, 1994). It was these yeoman farmers who spilled over into the Blue Ridge when lands were no longer available in the upland south: to the east in the Piedmont in North Carolina and Virginia, and to the east and to the west in the Great Valley of eastern Tennessee. The northwestern part of the Blue Ridge began to be settled by approximately 1780, but the southwestern portion was still home to the Cherokee, who had adjusted quite successfully to the presence of Europeans, adopting several non-native foods after making sustained contact with Europeans and Africans beginning in 1670. Watermelon, peaches, apples, horses, pigs, and chickens were especially prized by the Cherokee. By the mid 18th century, the Cherokee participated in growing trade in cattle and hogs, working as drovers tending to herds of cattle and hogs that ranged free in unfenced forests, and supplied meat to major Atlantic sea ports. By 1819, pressure to expel the Cherokee from their homelands was partially realized when a large tract of land was purchased. This was the first major step along a path that led to wholesale removal of the Cherokee to Indian territory west of the Mississippi by Andrew Jackson in 1838 [Garrison, 2002; Remini, 2002; U.S. Congress (21st 1st session) and Evarts, 1830].

The measured pace of growth is evident in the land use visible from the federal census of agriculture. Land in farms across southern Appalachia peaked around 1890, and then began a steady decline (Fig. 1.5). Resources that were prized for so long and came at such a heavy price were, in the end, more of a refuge from the wider market economy than a point of entry. We can see these dynamics in the steady downward drift in farm size during the historical period. From 1880 until 1940, average farm size continued to decline and the numbers of farms increase

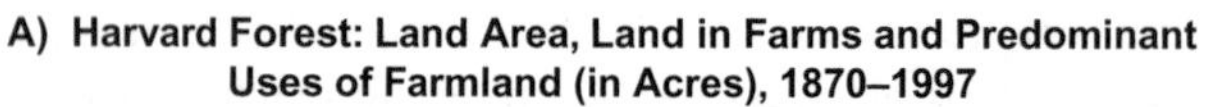

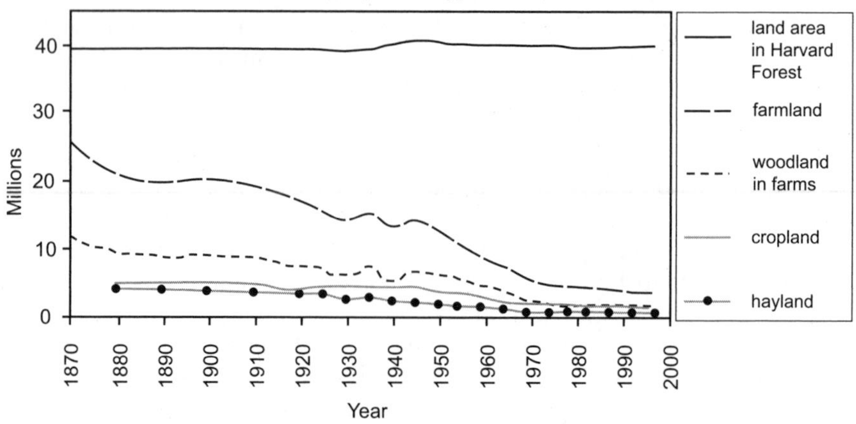

B) Coweeta: Land Area, Land in Farms and Predominant Uses of Farmland (in Acres), 1870–1997

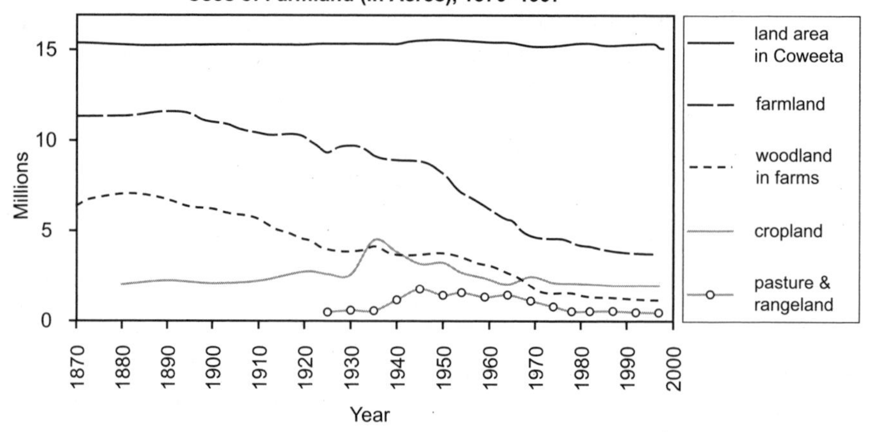

Figure 1.5 (A, B) Land area, land in farms, and predominant uses of farmland (in acres), 1870–1997, for the Harvard Forest (A) and Coweeta (B) regions. U.S. Department of Commerce (1870a, 1880a, 1890a, 1900a, 1910a, 1920a, 1925, 1930a, 1935, 1940a, 1950, 1954, 1959, 1964, 1969, 1974, 1978, 1982, 1987, 1992); U.S. Department of Agriculture (1997).

when Depression-era resettlement programs and urbanization began to reshape life chances in southern Appalachia. Southern Appalachia was the only study area where this pattern existed. In every other region of the country, farm size has increased steadily through the modern era.

Further west in the prairie–forest savannahs of southwestern Michigan (the Kellogg Biological Station region), population settlement left few landscapes untouched. First settled by European Americans in the 1820s, nearly all of southwestern Michigan was recorded as land in farms when the question was first posed in the federal agricultural census in 1870. Although the proportion in

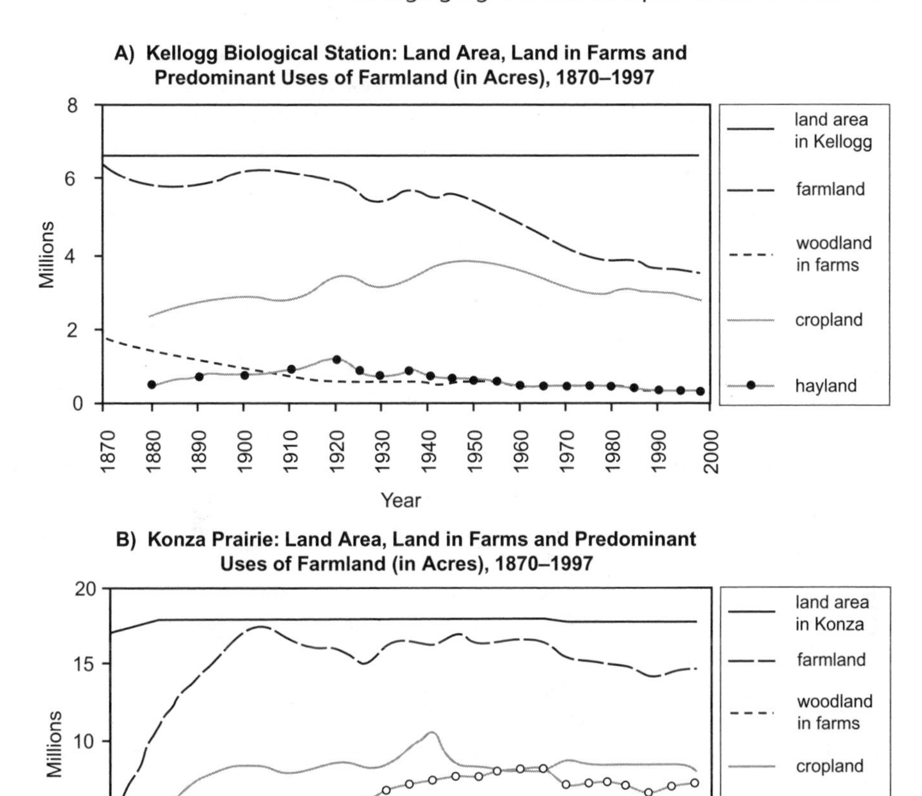

Figure 1.6 (A, B) Land area, land in farms, and predominant uses of farmland (in acres), 1870–1997, for the Kellogg Biological Station (A) and Konza Prairie (B) regions. U.S. Department of Commerce (1870a, 1880a, 1890a, 1900a, 1910a, 1920a, 1925, 1930a, 1935, 1940a, 1950, 1954, 1959, 1964, 1969, 1974, 1978, 1982, 1987, 1992); U.S. Department of Agriculture (1997).

southern Appalachia never exceeded 75% at its peak, nearly all land in southwestern Michigan (some 96%) represented land in farms in the 1890 agricultural census (Fig. 1.6) (U.S. Department of Commerce, 1890a). Not well documented prior to 1870, the transformation of southwestern Michigan reflected the growth of Chicago's hinterland, as the commerce in grain focused on the West's new metropolis after the mid century. With access to rail, European American farmers raised grains on a scale that eclipsed the small plots of Potawatomies and other native peoples who had raised corn around Lake Michigan for generations

(Cronon, 1991). Southwestern Michigan is both an extension of broad northern patterns of agricultural development and the first region discussed here that bears the modernizing stamp of the public land survey system. Average farm size remained bounded by the dimensions of the 68-ha (160-acre) quarter-section parcel well into the 20th century. As late as the region's so-called Golden Age (1900–1920), reported farm scale did not increase much beyond the quarter-section parcel. At the same time, land use in southwestern Michigan remained diverse and intensive. Although the land in farms has declined since 1945 (reaching a mere 52% of land area in 1997), farms are larger and cropped more intensively (U.S. Department of Commerce, 1945a, 1997). In 1925, roughly 60% of the land in farms in the region was cropped, and the proportion has steadily increased. In the 1987, 1992, and 1997 agricultural censuses, an average of 80% of the region's farmland was reported as cropland (U.S. Department of Commerce, 1925, 1987, 1992, 1997).

The pace of land transformation was better documented in the census when European American settlement reached the grasslands of eastern Kansas. Not settled in earnest until after the Civil War, only a quarter of the Konza Prairie study area was in private hands as farmland in 1870. But by 1900, virtually all (some 96%) of the eastern third of the state was reported as land in farms (U.S. Department of Commerce, 1870a, 1900a). In eastern Kansas, the transformation to cropland was never as complete as it had been in the Midwest. The Flint Hills prevented a similar plow-out of the tallgrass prairie that extended from eastern Kansas, northern Missouri, the Dakotas, southern Minnesota, Iowa, and Illinois. The uplands in east-central Kansas are punctuated by limestone outcroppings, making tillage difficult in many steeply sloped landscapes (Knapp et al., 1993). Nevertheless, cropland expanded steadily in the forest–grassland mosaic east of the Flint Hills and in the treeless plains to the west, where terrain is flatter. The ratio of cropland to pasture in the Konza Prairie study area has remained unchanged for generations (Fig. 1.6). After spiking to a high of 64% of land in farms in 1940, cropland has rarely exceeded 55% of the land in farms during the second half of the 20th century, and pastureland occupies a relatively fixed proportion of land use on farms—an average of 47% of land in farms since 1940 (U.S. Department of Commerce, 1940a).

By comparison, the pace of change in semiarid and arid study areas was slower, reflecting the inexperience of settlers with climate conditions of the High Plains of eastern Colorado and desert conditions in the Phoenix basin. Water regimes have controlled the scale of development in both areas more than the humid areas in the eastern half of the United States. In eastern Colorado, population was concentrated in the South Platte River watershed where several gravity flow systems were built in the late 19th and early 20th centuries (Tyler, 1992; Wohl, 2001, 2004). This expanded the area of cropland with access to irrigation water. Land in farms did not peak until after the invention of the horizontal centrifugal pump permitted wells in the High Plains to be sunk deeper than 50 ft. Even so, the proportion of cropland has remained relatively stable during the groundwater era. Land use in the Phoenix basin is tied very closely to water availability. Modern agriculture concentrates along the Salt River, which delivers (on average) more than 1 million acre-ft. of water per year (Graybill et al., 2006). Cropland has

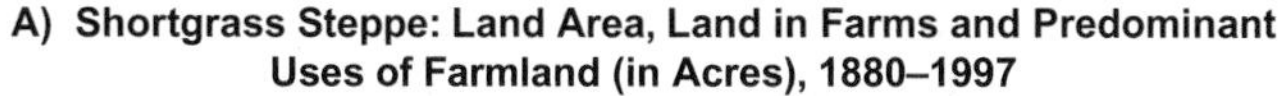

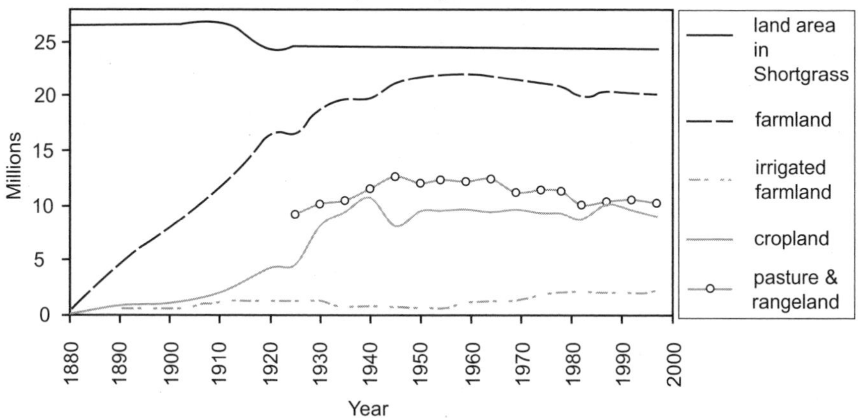

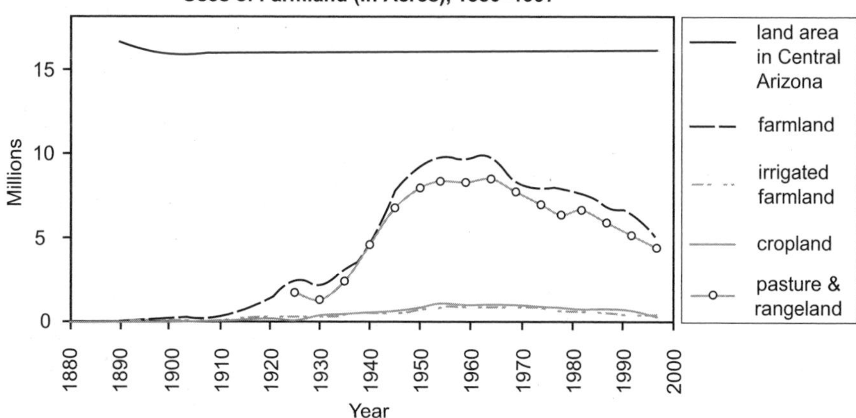

Figure 1.7 (A, B) Land area, land in farms, and predominant uses of farmland (in acres), 1870–1997, for the Shortgrass Steppe (A) and central Arizona (B) regions. U.S. Department of Commerce (1870a, 1880a, 1890a, 1900a, 1910a, 1920a, 1925, 1930a, 1935, 1940a, 1950, 1954, 1959, 1964, 1969, 1974, 1978, 1982, 1987, 1992); U.S. Department of Agriculture (1997).

never reached far beyond farms with access to irrigation (Fig. 1.7). The greatest proportion of land in farms in central Arizona is used as pasture and range, which has declined in step with farmland since peaking in 1964 (U.S. Department of Commerce, 1964).

Today, the distribution of agriculture over the United States is highly predictable in relation to environmental conditions. But historically, arable lands in close proximity to navigable waterways helped to define where early colonial settlements concentrated in the eastern half of the country (Curtin et al., 2001; Hofstra,

2004; Mires, 1993). These constraints were loosened as transportation improved during the 19th century. Long-distance trade began to shift agriculture to the interior of the continent. Farm populations did not immediately collapse in the east. Land no longer in farms has remained in private hands, with many more owners than in the agrarian past. During the late 20th century, the Harvard Forest, Coweeta, and increasingly Kellogg Biological Station regions share this trajectory. But signs of greater intensity are distinctly modern. Across each study area, despite considerable differences in timing of settlement and the types of agriculture practiced, a decisive decline in the number of farms and growth in average farm size occurred during the mid 20th century. Driven by postwar urbanization and green revolution technologies, the countryside lost population everywhere at the same time. Does this mean that historical practices were abandoned or path dependencies made irrelevant?

The Evolution of Land Used for Crops

Much of the literature examining the effects of past land use has tried to identify the physical extent of agricultural land use. Seminal work in the northeastern United States linking historical land use and modern forest composition has demonstrated the importance of understanding sequence and extent of past agricultural activity (Abrams, 1995; Burgi et al., 2000; Foster at al., 1998; Hall et al., 2003; Whitney, 1994). The relative abundance of long-lived tree taxa (e.g., beech, sugar maple, hemlock, yellow birch) has declined in ecological regions with widespread alteration of the landscape, and faster growing species (red maple, black, gray or white birch, poplar, and cherry) have increased. These studies indicate that the relative effect of environmental versus historical factors is strongly dependent on the scale of analysis. At broad geographical scales, despite differences in specific crops or land-use practices, the amount of land cleared for tillage, hay, pasture, or woodland remained relatively constant at the town level in Massachusetts from 1800 to 2000. The environmental variation within New England did not permit differentiation in agricultural practice to affect the extent of disturbance.

Nevertheless, the kind of crop mixtures that prevailed in these landscapes did change enormously and must be considered in context with other landscapes across the country to understand the legacies of past land use. In New England, what was in colonial times a form of mixed husbandry dominated by corn and small grains (wheat, oats, rye, and barley) eventually gave way to a crop system, as we see in summaries of the regional data, dominated by hay, corn, potatoes, and oats (Fig. 1.8). We can detect some of the change in the spatial distribution of these crops over time. At the end of the 19th century, for example, potatoes and corn were still evenly distributed across New England counties, but by 1920 they were increasingly concentrated as a proportion of farmland in counties nearest metropolitan areas (like Fairfield and New Haven, Connecticut; and Newport and Bristol, Rhode Island) and in far-flung Aroostook, Maine, which developed a specialization in potatoes. As a cropping system, the rotations that prevailed were simplified to a corn–oats system in southern New England. Corn served as silage

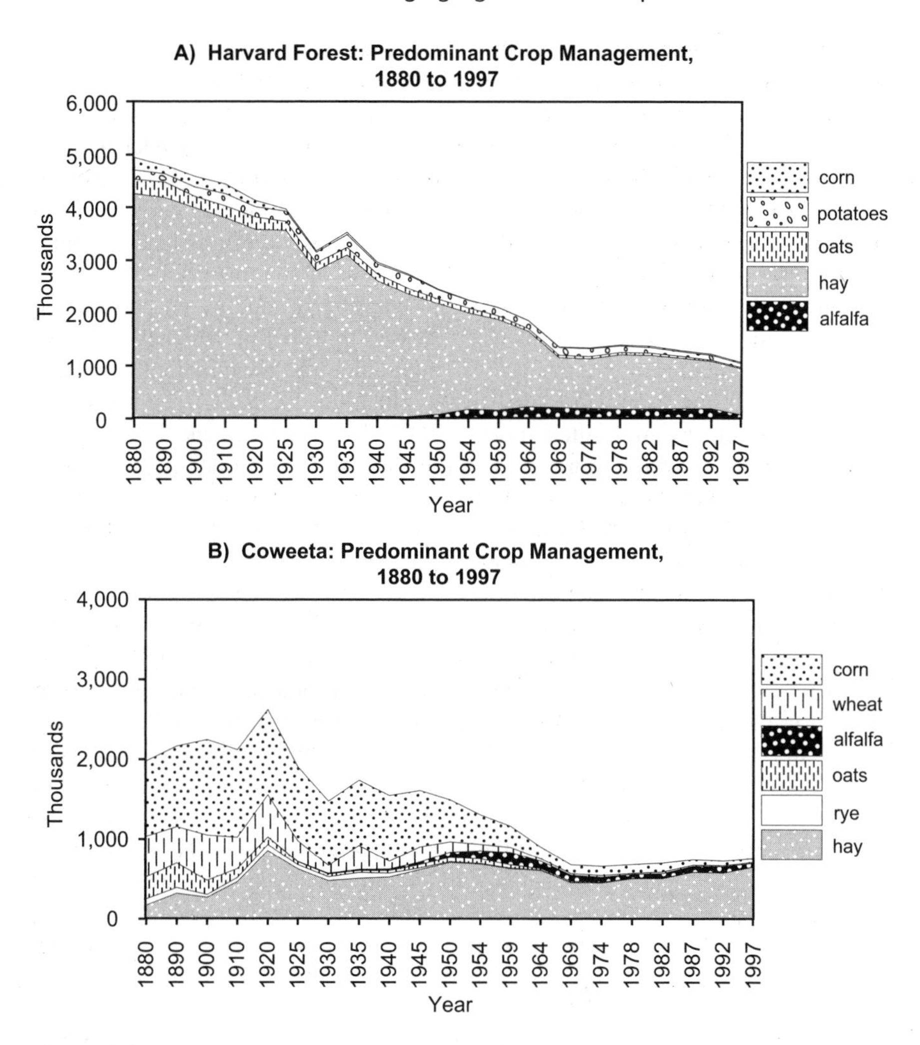

Figure 1.8 (A, B) Predominant crop management, 1880–1997, for the Harvard Forest (A) and Coweeta (B) regions. U.S. Department of Commerce (1870a, 1880a, 1890a, 1900a, 1910a, 1920a, 1925, 1930a, 1935, 1940a, 1950, 1954, 1959, 1964, 1969, 1974, 1978, 1982, 1987, 1992); U.S. Department of Agriculture (1997).

for dairy cattle and oats were important as horse feed before automobile ownership was more widespread in the 1920s. Eventually, dairying was increasingly concentrated in northwestern Vermont, along the eastern shore of Lake Champlain. Declining corn acreages reflected this loss and the reality that it was cheaper to buy corn silage from the Midwest. Competition from other regions in the country made agricultural land use in New England far less diverse in the modern era.

By comparison, the tillage system in the southern Appalachians retained its focus on cereals until the late 19th century before experiencing a similar spatial concentration of production in the 1920s. Wheat and rye, corn and oats, all played reinforcing roles in a rotation scheme that remained more diverse than in the northeast. More removed from metropolitan centers, less industrialized than the north, the incentives to specialize were not as immediate in the rural south, and southern Appalachia retained a culture based on local exchange longer into the 20th century (Egnal, 1998; Jones, 2002; Kulikoff, 2000; Morgan, 2001; Walker, 2000). Just as the proportion of cropland devoted to wheat and corn began to decline in the 1920s, corn reached its peak as a proportion of cropland, accounting for 35% of the tillage system (U.S. Department of Commerce, 1930a). From the mid 20th century until the present, the intensity of grazing on pasture and hayland, and in woodland in farms has increased in response to steady increases in beef herds and in response to a postwar boom-and-bust cycle in poultry raising. The collapse of the farming of small grains since the 1960s has meant that disturbance phenomena are limited almost exclusively to the maintenance of pasture and hayland in a management system that is far less diverse than more commercially oriented agrarian landscapes in the interior of the continent.

Further west, hard-earned folk wisdoms (Brookfield, 2001; Medin and Atran, 1999)—adapted from imported European practices—were applied to more forgiving landscapes. In southwestern Michigan, eastern Kansas, and eastern Colorado, settlers found ways to reinvent the diversity necessary to make family-scale agriculture work in the east. The same template of European American agriculture—now with corn firmly integrated into the repertoire of food and fodder crops—was applied to the forest–prairie savannah of southwestern Michigan in the 19th century. Landscapes in the Kellogg Biological Station region were a mix of wheat and corn, moving in temporal magnitude with rye and oats, and a proportion of land in hay that increased steadily until 1920 (Fig. 1.9). A transitional period between 1920 and 1960 ushered in the now-dominant midwestern corn–soy rotation, but other nitrogen-fixing legumes like alfalfa made an early appearance in the 1920s, probably in response to the scientific advocacy of agricultural extension programs. Alfalfa has persisted in this system, probably as a complement to wheat, and soy has grown in magnitude to keep pace with the scale of corn production, which reached a peak of 1.2 million acres in 1982.

The region surrounding Konza Prairie did not experience a similar beginning. Corn dominated early tillage, as settlers responded to the unleashing of nitrogen during the early plow-out of the plains. In many places the fertility of prairie soils soon convinced farmers in eastern Kansas to expand dramatically cropland devoted to staple cultivation. Oats were never harvested in magnitudes sufficient to serve as a restorative rotation or a winter cover crop, but eventually, by the early 20th century, successive droughts tempered the widespread devotion to corn. After 1920, corn retreated to the northern part of the state, where cooler temperatures could take advantage of an average rainfall of between 1,000 and 1,100 mm (40–44 in.) per year. After the settlement era, farm practice evolved in a more sustainable direction. With wheat dominating tillage after 1920, alfalfa and oats were sown in greater acreages. A particularly surprising finding was the marked increase the acreage devoted to soy in the region, suggesting that parts of eastern Kansas have adopted the double-cropping wheat–soybean system prevalent in Arkansas, Mississippi, and Alabama (Kyei-Boahen and Zhang, 2006).

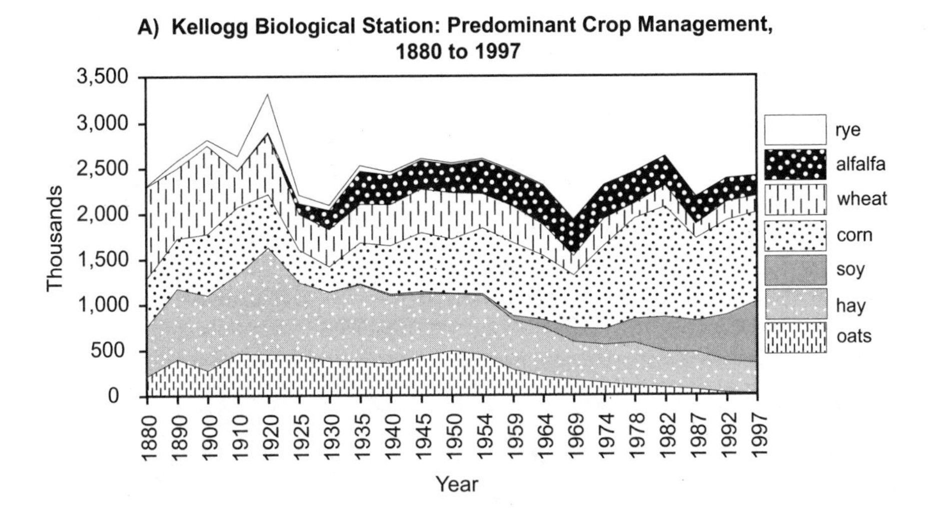

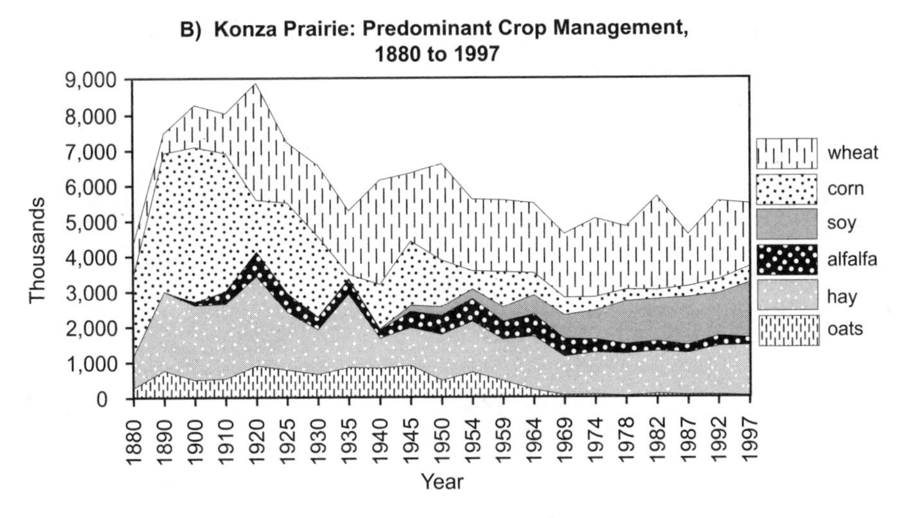

Figure 1.9 (A, B) Predominant crop management, 1880–1997, for the Kellogg Biological Station (A) and Konza Prairie (B) regions. U.S. Department of Commerce (1870a, 1880a, 1890a, 1900a, 1910a, 1920a, 1925, 1930a, 1935, 1940a, 1950, 1954, 1959, 1964, 1969, 1974, 1978, 1982, 1987, 1992); U.S. Department of Agriculture (1997).

In the Shortgrass Steppe, cropland expanded slowly as European Americans experimented with dryland cropping methods. The small-grains template was very much in evidence as the proportion of land devoted to crops slowly but steadily increased. Corn was stubbornly cultivated as a staple and a seasonal fodder for cattle, but was balanced by wheat production (Fig. 1.10). The physical extent of corn relied on gravity flow irrigation during the early settlement period, and it was the widespread adoption of winter wheat varieties, bred in the semiarid climate of the Russian steppes, that permitted acreage to expand to the High Plains

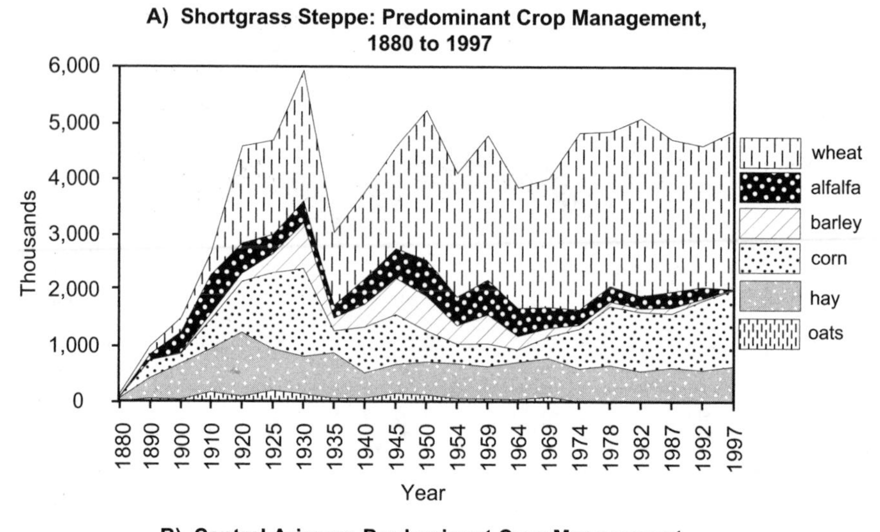

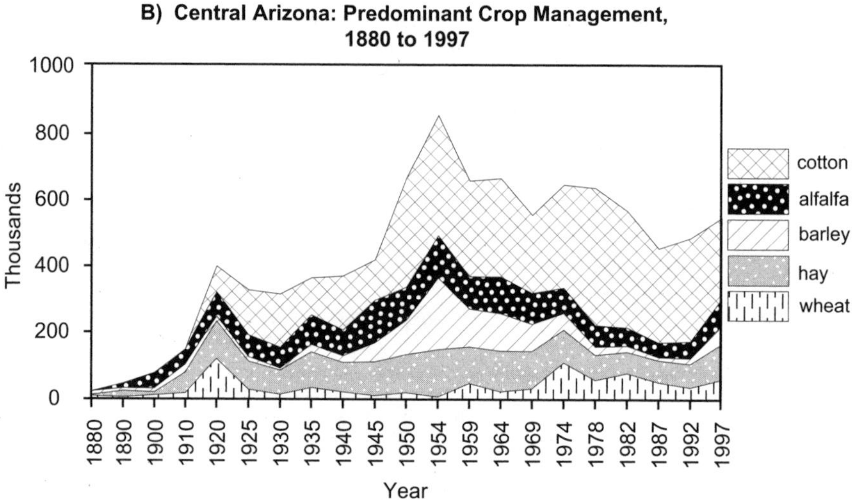

Figure 1.10 (A, B) Predominant crop management, 1880–1997, for the Shortgrass Steppe (A) and central Arizona (B) regions. U.S. Department of Commerce (1870a, 1880a, 1890a, 1900a, 1910a, 1920a, 1925, 1930a, 1935, 1940a, 1950, 1954, 1959, 1964, 1969, 1974, 1978, 1982, 1987, 1992); U.S. Department of Agriculture (1997).

(Kirshenmann, 2002). Again, the diversity of the crop system is striking viewed at this scale. Oats, barley, and alfalfa were all introduced early during the 20th century to fill out corn and wheat rotations. It wasn't until the postwar era that continuous wheat cultivation dominated the cropping profile of the region. With the expansion of irrigated acreage since the 1950s, continuous corn rotations are also increasingly the norm. The ecological implications of this intensification for sustainability are a source of concern. Irrigated cropping has been shown to reach near-equilibrium levels of soil carbon and nitrogen mineralization faster than dryland cropping, and continuous cropping is better for maintaining soil carbon if tillage practice minimizes soil disturbance. Nevertheless, even if the atmospheric contributions of agriculture to global change are in better balance, intensification can have negative local consequences, including increased soil erosion, reduced biodiversity, pollution of groundwater, and the eutrophication of rivers and lakes (Matson et al., 1997; Parton et al., 2005; Tilman et al., 2002). More research is needed in the evolution of how dominant management practices and the growing scale of farm proprietorship have transformed the dimensions of working landscapes (Belfrage et al., 2005; Kirshenmann, 2002; Langley-Turnbaugh and Keirstead, 2005). The overall diversity of production has stabilized since the 1960s, but the cereals template brought by small farmers to the High Plains has been radically transformed by industrial farm methods.

In central Arizona, the cereals template fit well, even in the Sonoran Desert, taking advantage of winter rains and ancient irrigation canals along the Salt River. Demand came first from the U.S. Cavalry post at Fort McDowell, established to contain the Apache, but the Desert Land Act of 1877 cemented the economic incentives necessary for setters to reexcavate canals abandoned by the Hohokam during the 14th and 15th centuries, and build new ones. For the first 50 years of agricultural development, European Americans did little to revive the Mesoamerican food complex—corn, beans, and squash—that had characterized the region's agriculture in ancient times. Instead, they imported wheat, barley, and alfalfa, a workable subset of European American cultivation that served the purposes of a growing frontier population. Dramatic expansion of agriculture, however, awaited a more reliable flow of water and the reintroduction of another Mesoamerican crop: cotton (Bayman, 2001; Redman, 1999). With the completion of the Roosevelt Dam in 1911, 60 miles upstream from Phoenix, and the disruption of supplies during World War I, it was the Egyptian variety—the long-staple pima cotton so essential in airplane fabric, balloons, and cord tires—that underwrote cotton's return to the Salt River Valley.

Remarkably, however, alfalfa retained its function as a restorative cover crop in rotation with cotton. Barley also made a return to farm fields of central Arizona in the mid 1950s, no doubt in response to a surge in cattle and poultry holdings. Both developments took advantage of the explosive urbanization of the Phoenix basin, as the city's population tripled between the 1950 and 1960 censuses (Gammage, 1999, p. 46). But the expansion of cropland and pasture that accompanied urban expansion did not last. Urban and exurban development encroached on grazing lands, as pastureland began a steady decline during the early 1960s. Cotton production reached a peak in 1978 and has declined steadily since, as well. The boom period was predicated squarely on green revolution technologies, and the regional cropping system has only begun to show signs of more diversity since the late 1980s, as barley, alfalfa, and wheat acreages have increased.

Livestock

Many of the changes in livestock production in the United States during the past half century have come in response to urbanization and increasing consumption

(Princen, 2005; Princen et al., 2002). Higher labor force participation and higher discretionary incomes during the postwar years allowed Americans to consume more animal protein in their diets. Beef led the way, growing from a per-capita consumption of 50 lb. in 1950 to 95 lb. in 1970 (Hart, 2003). Chicken overtook beef and pork in the 1990s, reaching a per-capita consumption of 96 lb. in 2002. The mixed farms that earlier in the century produced a little bit of everything and sold their marketable surpluses in local or regional markets were displaced in terms of sales by more specialized and commercial operations that grew to meet the demands for animals with tender and thicker flesh and consistent presentation on supermarket shelves. The concentration of livestock in increasingly larger farm enterprises that accompanied the drive to mass production has also shifted cereals production to the feeding of animals. Even though the volume of its trade makes the United States the largest agricultural exporter in history, with roughly 20% to 25% of its grain corn harvest, one third of its soybeans, and 40% to 50% of its wheat regularly sold abroad, it is still estimated that roughly 70% of the United States' cereal and legume harvest was fed to animals in the 1990s (Smil, 2001).

Each of these new animal-raising enterprises are far more specialized and spatially concentrated than the animal husbandry that preceded them. Livestock are not raised and fed on mixed farms that integrate their grazing activity and manure output into cereal production, but are bred on contract by smaller calving, farrowing, or hatching operations before being shipped to larger feeding and finishing operations. Hart (2003) describes the resulting concentration of animals as a new macrogeography of American agriculture. By century's end, most farms in the United States had sold their chickens, milk cows, and hogs. Hart (2003) argues that the core areas of cereals production in the United States—found in the Corn Belt states of Ohio, Michigan, Illinois, Wisconsin, Iowa, Missouri, Nebraska, Minnesota, and South Dakota—now produce the feed that nourish cattle, poultry, and hogs in Maryland, North Carolina, northern Georgia, northern Alabama, Arkansas, Oklahoma, the Texas panhandle, northeastern Colorado, and central Arizona. A third region in this new macrogeography is found in California, Florida, and the Northeast, and is increasingly focused on producing vegetables, fruits, nursery and greenhouse products, and other highly specialized crops. Beef cattle are the only exception to the larger trend toward spatial concentration, because they are easier to raise on small part-time farms (Hart, 2003; Hoppe and Kork, 2005). Between 32% and 43% of what the USDA refers to as *limited-resource*, *retirement*, *residential/lifestyle*, and *low-sales farms* specialize in beef cattle—particularly cow–calf operations—which require less attention and are more compatible with off-farm employment (Cash, 2002).

Most of the beef raised on American farms no longer comes from large ranches in the West, but from farms east of the Mississippi. These changes were perceptible in the persistence of cattle in New England farming during the 19th century. Sheep were also an important feature of New England agricultural tradition, but flocks began a steady decline in the 1870s. By the 1920s, local demand for wool was undercut by a relocation of textile manufacturing to the South, and supplies that came from sheep raisers in the Plains and the arid West (Brisbin, 1959; Delfino and Gillespie, 2005; Gemming, 1979; Jager, 2004). Low agricultural prices in the

1920s also forced farmers to reduce corn acreages and scale back hog inventories. Farms in Middlesex and Worcester counties in Massachusetts, near Boston, were the only areas in the New England region with inventories of more than 20,000 hogs in the 1920s. By 1950, Middlesex farmers had increased hog inventories to 34,000, but the growth was sustained mostly with feed from outside the region. Corn acreage during the same period declined from 1,682 acres to 273 acres in Middlesex (U.S. Department of Commerce, 1920a, 1930a, 1950a). Small farmers responded to the loss of farm income (declining beef, sheep, and dairy production) by turning to poultry. Most farm families had maintained small flocks of laying hens before World War II. Farm women usually managed the poultry and gathered the eggs to barter with country merchants for store-bought goods (Jager, 2004; McMurry, 1995; Walker, 2000). When these hens were past their laying days, families baked them for Sunday dinner or sold them live into urban markets. With the change in consumer tastes after the war, a greater demand arose for dressed chickens, already slaughtered and prepared for grocery displays. This led to the breeding of chickens that grew faster with less feed, known in the trade as *broilers* (Hart, 2003).

Northeastern Georgia, Arkansas, and Maryland were broiler-producing areas that received a boost from the War Food Administration in 1942, when it ordered dressed chickens for the armed forces (Sawyer, 1971). These contracts spurred the early development of vertical integration in the poultry business. Operators tired of the insecure supply of chicks and decided to start their own large-scale hatcheries, and then worked on developing feed mills. Eventually, most of the firms also moved into marketing and distributing their broilers. The South was receptive to the broiler trade for many reasons, not the least of which was an infestation of the boll weevil that killed cotton crops across the South during the Depression (Hart, 2003). The South also provided lower startup costs. Building materials and labor were cheaper, on the farm and in the feed mills and processing plants. Over the long term, these advantages help to explain why the South captured most of the broiler industry.

However, during the immediate postwar period, farmers in every study area examined here experimented with broiler production. The growth in poultry inventories was actually more cautious in southern Appalachia than in New England. Inventories doubled in New England between 1940 and 1974, but the growth of poultry inventories did eventually reach the upland farms in the Coweeta study area after 1950, when inventories doubled in just 20 years. Even the central Arizona and Shortgrass Steppe regions joined in the trend, experiencing their own tripling of inventories respectively between 1950 and 1969 and between 1969 and 1987 (U.S. Department of Commerce, 1950a, 1969, 1987). Each of these areas eventually declined in the face of competition from the integrated operations of big producers, centered in Arkansas and Maryland. The exception to the larger pattern of boom and bust occurred in southwestern Michigan, where a higher proportion of poultry inventories have been involved in egg production. Much of the most recent production has in fact been concentrated in Allegan County, Michigan, which reported 2,143,903 laying hens and 2,420,666 broilers in the 2002 census. These totals represent 30% and 60% percent, respectively, of the state inventories of both kinds of poultry (U.S. Department of Agriculture, 2002).

The most lasting impact of the trend toward livestock agribusiness and the jump in scale of production is the divorce of animal husbandry from crop agriculture, with all the negative consequences that the separation implies. Hog farming seems to be the only form of livestock concentration that is still integrated meaningfully into the crop systems of family farms. As larger hog operations grew in the 1960s, they contributed to the transformation of traditional crop rotations. The midwestern Corn Belt tended to follow a 3-year rotation of corn, small grains, and hay. Corn was usually followed by winter wheat or oats during the second year of the rotation, and clover during the third. Increasingly, alfalfa took the place of clover in the midwestern system. From very early in the development of the region's agriculture, Corn Belt farmers fed most of their crops to their livestock, but could shift more into markets if prices were good. Wheat usually ended up in the market, but could also be fed to livestock. Oats were an important feedstock for horses, and hay was an important fodder (Hart, 2003). Crop farmers were the key to expanding the business, and hog producing therefore tended to be concentrated where pasture was limited. Crop farmers who agreed to feed hogs on contract helped to expand the business in the same way that small farmers participated in the expansion of poultry processing. Most farmers explained the benefits of the change as a way to keep their children engaged in the farm enterprise, adding an activity that required more labor and added revenue, and made use of the manure to lower fertilizer costs and keep corn and soybean yields high. The outcomes, as we see in the Kellogg Biological Station region cropping data, have been an increasing focus on corn and soybeans, and a reduction of pasture in the land that remains in farms in southwestern Michigan.

But the recoupling of animal husbandry and cropping remains unlikely in most of the study areas. There is a growing mismatch in the overall scale of cropland and livestock farming in these regions. Recent declines in livestock inventories are a reflection of this (Fig. 1.11). Without the local feed crops to sustain smaller breeder and feeding operations, large-scale operators have come to dominate the confinement regime. The concentration of animals in larger operations means that they are grazing for far less of their much-shortened lives. The loss of herbivory from grassland ecosystems is one important consequence of the change (Gibon, 2005). The concentration of animal wastes also represents a growing problem. Manure management is improving in large-scale livestock facilities, but the spatial concentration means that nutrients are not distributed as widely as they once were, and they pose environmental risks to soil formation and water quality. The effects are very uneven (Acosta-Martinez et al., 2004; Corkal et al., 2004; Osterberg and Wallinga, 2004). Some crop farms may have much-improved access to manure, and others are too far away from the centers of livestock production to benefit from the industrial scale and availability of biologically friendly fertilizer. Ultimately, regions situated at the heart of diverse and commercially oriented cereal agriculture areas—like Konza, Kellogg Biological Station, and Shortgrass Steppe—are in the best position to make the most sustainable use of these restructured resources. With little or no decline in farmland or cropland, livestock numbers are actually closer to historical carrying capacities in the West than in the postagrarian landscapes of the eastern United States (Fig. 1.11).

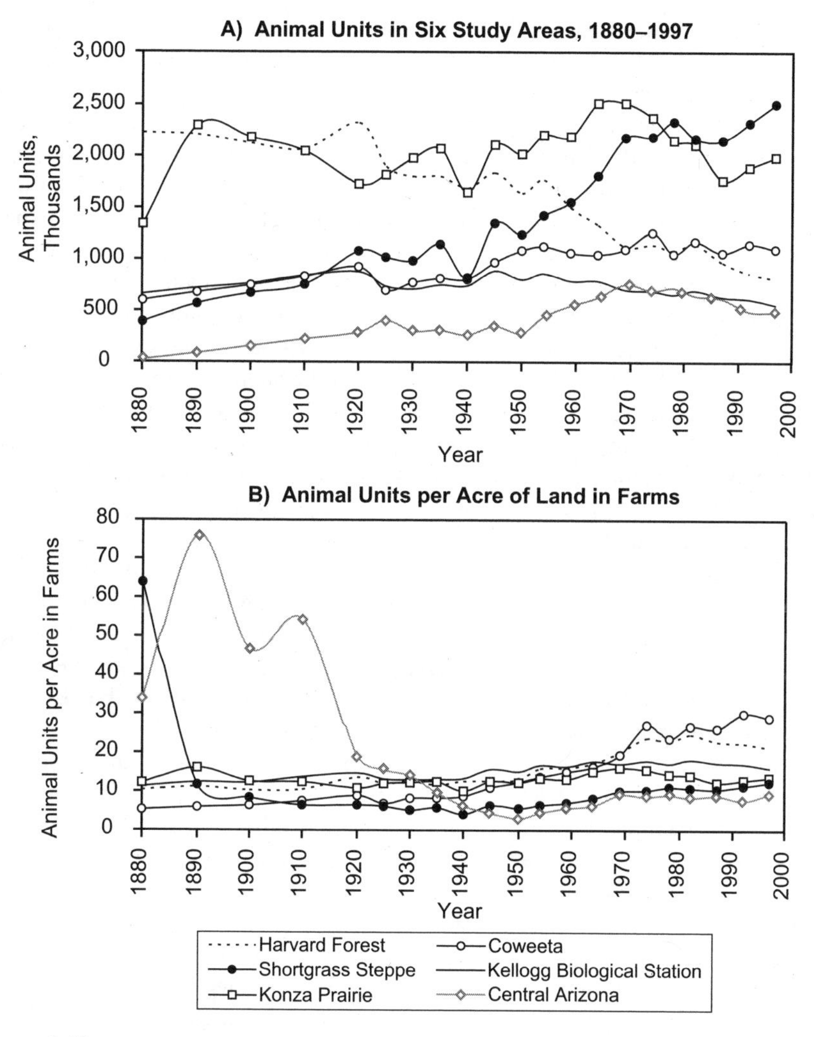

Figure 1.11 (A, B) Animal units in six study areas (A) and animal units per acre of land in farms (B), 1880–1997. U.S. Department of Commerce (1870a, 1880a, 1890a, 1900a, 1910a, 1920a, 1925, 1930a, 1935, 1940a, 1950, 1954, 1959, 1964, 1969, 1974, 1978, 1982, 1987, 1992); U.S. Department of Agriculture (1997).

Implications

In historical terms, then, the ecological impacts of livestock concentration are relatively minor when compared with the movement of cereal production into the center of the continent. This shifting geography was and is the single largest dynamic reshaping landscapes in the regions studied. Few agrarian experiences in the world rival the rapid movement of cereal agriculture across the territories of the United States between 1870 and 2000. In Europe, the struggle to sort out which land was ultimately marginal for agricultural purposes lasted centuries, and was complicated by social structures that concentrated the ownership of land in very few hands and restricted the movement of ordinary people (Moriceau, 2002; Pollard, 1997). In the United States, freehold tenures and family proprietorship established a common settlement vernacular across the interior of the continent. Less mediated by the state or wealthy landowning classes, the agrarian experience was freer to explore ecological niches and to monitor the response of upland, lowland, prairie, and plains to a small-grains complex brought by Europeans to the Americas. A reciprocal imprinting of a European repertoire of small grains—wheat, rye, barley, oats, and cultivated hays (clover)—and domesticated livestock with the North American landscape began with the adoption of maize as a staple fodder and cereal crop. Corn yielded more plant mass than small grains and was easier to work in small fields in hilly terrain or on steep slopes. Yet, stubbornly, once maize was adopted as part of the system, European Americans applied the same template to every biogeography across the continent, adapting to each only in subtle and gradual fashion.

Part of the advantage of the more discrete regional spaces analyzed here, we argue, is that the scale of observation allows for a meaningful discussion of temporal trends. A national scale analysis is too large to capture phenomena that can be related to local agents of change. The bioregional scale we use is sufficiently aggregated to recognize connections to wider processes, but is scaled close enough to local patterns of change to frame paths of development. The agricultural census provides us with the specific dimensions of those transformations within bioregions and reinforces the importance of understanding what came before the present in every setting. Each has followed paths that have been responsive to broader national trends, but the agriculture of each region has been specific to the biogeography that facilitated agrarian change.

In New England, patterns of stewardship and care that were necessary to make the land productive during the colonial era demonstrated that the land could be used sustainably within the framework of an 18th-century economy and a state confined to the eastern seaboard. But in the context of a continental nation, the hilly terrain and stony soils of New England demanded too great an effort to remain viable as grain surpluses from midwestern prairies grew during the early 19th century. The turn to a pastoral economy was decisive in New England because population densities and rural industry pulled labor away from the internal economies of mixed farms. In southern Appalachia, by contrast, distance from major population centers ensured that the transition to a pastoral economy and industrialization waited until well into the 20th century. In each case, upland geographies did not prefigure the duration of particular agrarian regimes; however, once integrated into the larger economy, neither region could escape the pastoral turn that biogeography set out for agrarian change.

By contrast, in the Midwest and Plains, the fit between the cereals and biogeography was more seamless. Lowland, prairie, and plains geographies, with rich soils and temperate climates, ensured that agriculture would have a stable existence. But, one of the major surprises that emerged from the analysis is how diverse production remains in these commercially important agricultural regions. Mixed farming has been a consistent feature of farming in the Midwest and the eastern Plains since settlement began. In eastern Kansas, corn was grown to excess between the 1880s and 1920s, but decades of intermittent rains and falling yields eventually broke farmers of corn culture and led to a switch to wheat, as the biogeography itself encouraged a more sustainable path of development. Even in more challenging environments, like the semiarid and arid West, where the fit between cereals and biogeography was problematic, mixed farming framed early development. Only with the introduction of cotton in Arizona in the 1950s has a genuine monoculture emerged (Carriere et al., 2003).

This review of historical patterns indicates that we know too little about the internal dynamics of farm systems to make definitive judgments. Nevertheless, the regional summaries are suggestive of the ecological impacts over the long term. They indicate that biogeography modified human agricultural systems slowly and that agrarian landscapes were far from permanent. They also indicate that diverse production was a common component of agrarian change as it moved across the continent. Despite the varying intensity of commercial change, which affected the pace of change at different stages of national development, each agrarian transformation renewed—rather than abandoned—traditions of mixed farming. Until the mid 20th century, most farms combined livestock raising and grain growing. The collapse of this basic signature of European American agriculture since then implies that the most negative and intense ecological impacts rest in our own time.

References

Abrams, M. D., and C. M. Ruffner. 1995. "Physiographic analysis of witness tree distribution (1765–1798) and present forest cover through north central Pennsylvania." *Canadian Journal of Forest Research* 25: 659–668.

Acosta-Martinez, V., T. M. Zobeck, and V. Allen. 2004. "Soil microbial, chemical and physical properties in continuous cotton and integrated crop-livestock systems." *Soil Science Society of America Journal* 68: 1875–1884.

Barron, H. S. 1984. *Those who stayed behind: Rural society in nineteenth-century New England.* Cambridge, U.K.: Cambridge University Press.

Barron, H. S. 1997. *Mixed harvest: The second great transformation in the rural North, 1870–1930.* Chapel Hill, N.C.: University of North Carolina Press.

Bayman, J. M. 2001. "The Hohokam of southwest North America." *Journal of World Prehistory* 13: 257–311.

Belfrage, K., J. Björklund, and L. Salomonsson. 2005. "The effects of farm size and organic farming on diversity of birds, pollinators, and plants in a Swedish landscape." *Ambio* 34: 582–588.

Brisbin, J. S. 1959. *The beef bonanza; or, How to get rich on the plains, being a description of cattle-growing, sheep-farming, horse-raising, and dairying in the West.* Norman, Okla.: University of Oklahoma Press.

Brookfield, H. C. 2001. *Exploring agrodiversity.* New York: Columbia University Press.

Brown, D. G., K. M. Johnson, T. R. Loveland, and D. M. Theobald. 2005. "Rural land use trends in the coterminous United States, 1950–2000." *Ecological Applications* 15: 1851–1863.

Burgi, M., E. W. B. Russell, and G. Motzkin. 2000. "Effects of post-settlement human activities on forest composition in north-eastern United States: A comparative approach." *Journal of Biogeography* 27: 1123–1138.

Bushman, R. L. 1998. "Markets and composite farms in early America." *William and Mary Quarterly* 55: 351–374.

Carriere, Y., C. Ellers-Kirk, M. Sisterson, L. Antilla, M. Whitlow, T. J. Dennehy, and B. E. Tabashnik. 2003. "Long-term regional suppression of pink bollworm by *Bacillus thuringiensis* cotton." *PNAS* 100: 1519–1523.

Cash, J. A. 2002. "Where's the beef? Small farms produce majority of cattle." *Agricultural Outlook,* USDA ERS AGO-288: 21–24.

Clark, C. 1990. *The roots of rural capitalism: Western Massachusetts, 1780–1860.* Ithaca, N.Y.: Cornell University Press.

Corkal, D., W. C. Schutzman, and C. R. Hilliard. 2004. "Rural water safety from the source to the on-farm tap." *Journal of Toxicology and Environmental Health, Part A: Current Issues* 67: 1619–1642.

Cronon, W. 1983. *Changes in the land: Indians, colonists, and the ecology of New England.* 1st ed. New York: Hill and Wang.

Cronon, W. 1991. *Nature's metropolis: Chicago and the Great West.* New York: W. W. Norton.

Curtin, P. D., G. S. Brush, and G. W. Fisher. 2001. *Discovering the Chesapeake: The history of an ecosystem.* Baltimore, Md.: Johns Hopkins University Press.

Delfino, S., and M. Gillespie. 2005. *Global perspectives on industrial transformation in the American South.* Columbia, Mo.: University of Missouri Press.

Donahue, B. 2004. *The Great Meadow: Farmers and the land in colonial Concord.* New Haven, Conn.: Yale University Press.

Dublin, T. 1981. *Farm to factory: Women's letters, 1830–1860.* New York: Columbia University Press.

Egnal, M. 1998. *New world economies: The growth of the thirteen colonies and early Canada.* New York: Oxford University Press.

Finnegan, N. J., E. T. Sundquist, P. J. Waisanen, N. B. Bliss, and M. E. Budde. 2000. *Using modern land cover maps and historical data to estimate historical land cover for the conterminous U.S.* Presented at the American Geophysical Union 2000 fall meeting, San Francisco, Calif., December 15–19.

Fitzgerald, D. K. 2003. *Every farm a factory: The industrial ideal in American agriculture.* New Haven, Conn.: Yale University Press.

Foster, D. R., G. Motzkin, and B. Slater. 1998. "Land-use history as long-term broad-scale disturbance: Regional forest dynamics in central New England." *Ecosystems* 1: 96–119.

Gammage, G. J. 1999. *Phoenix in perspective: Reflections on developing the desert.* Tempe, Ariz.: The Herberger Center for Design Excellence, College of Architecture and Environmental Design, Arizona State University.

Gardner, B. L. 2002. *American agriculture in the twentieth century: How it flourished and what it cost.* Cambridge, Mass.: Harvard University Press.

Garrison, T. A. 2002. *The legal ideology of removal: The southern judiciary and the sovereignty of Native American nations.* Athens, Ga.: University of Georgia Press.

Gemming, E. 1979. *Wool gathering: Sheep raising in old New England.* New York: Coward, McCann and Geoghegan.

Gibon, A. 2005. "Managing grassland for production, the environment and the landscape: Challenges at the farm and the landscape level." *Livestock Production Science* 96: 11–31.

Graybill, D. A., D. A. Gregory, G. S. Funkhouser, and F. L. Nials. 2006. "Long-term streamflow reconstructions, river channel morphology, and aboriginal irrigations systems along the Salt and Gila rivers," pp. 69–123. In: J. S. Dean and D. E. Doyle (eds.), *Environmental change and human adaptation in the ancient Southwest.* Salt Lake City, Utah: University of Utah Press.

Hahn, S. 1983. *The roots of southern populism: Yeoman farmers and the transformation of the Georgia Upcountry, 1850–1890.* New York: Oxford University Press.

Hall, B., G. Motzkin, D. R. Foster, M. Syfert, and J. Burk. 2003. "Three hundred years of forest and land-use change in Massachusetts." *Journal of Biogeography* 29: 1319–1335.

Hareven, T. K., and R. Langenbach. 1978. *Amoskeag: Life and work in an American factory–city.* 1st ed. New York: Pantheon Books.

Hart, J. F. 2003. *The changing scale of American agriculture.* Charlottesville, Va.: University of Virginia Press.

Hofstra, W. R. 2004. *The planting of New Virginia: Settlement and landscape in the Shenandoah Valley.* Baltimore, Md.: Johns Hopkins University Press.

Hoppe, R. A., and P. Kork. 2005. "Large and small farms: Trends and characteristics," pp. 5–21. In: D. E. Banker and J. M. MacDonald (eds.), *Structural and financial characteristics of U.S. farms: 2004 family farm report.* Washington, D.C.: Economic Research Service, USDA.

Hubbard, W. 1803. *A narrative of the Indian wars in New-England, from the first planting thereof in the year 1607, to the year 1677: Containing a relation of the occasion, rise and progress of the war with the Indians, in the southern, western, eastern and northern parts of said country.* Stockbridge, Mass.: Heman Willard.

Huston, M. A. 2005. "The three phases of land-use change: Implications for biodiversity." *Ecological Applications* 15: 1864–1878.

Jager, R. 2004. *The fate of family farming: Variations on an American idea.* Hanover, N.H.: University Press of New England.

Johnson, K. M., and C. L. Beale. 1992. "Natural population decrease in the United States." *Rural Development Perspectives* 8: 8–15.

Johnson, K. M., and C. L. Beale. 1998. "The rural rebound." *Wilson Quarterly* 22: 16–27.

Johnson, K. M., and C. L. Beale. 2002. "Nonmetro recreation counties: Their identification and rapid growth." *Rural America* 17: 12–19.

Johnson, K. M., and G. V. Fuguitt. 2000. "Continuity and change in rural migration patterns, 1950–1995." *Rural Sociology* 65: 27–49.

Johnson, K. M., P. R. Voss, R. B. Hammer, G. V. Fuguitt, and S. McNiven. 2005. "Temporal and spatial variation in age-specific net migration in the United States." *Demography* 42: 791–812.

Jones, L. A. 2002. *Mama learned us to work: Farm women in the New South.* Chapel Hill, N.C.: University of North Carolina Press.

Kirshenmann, F. 2002. "Scale: Does it matter?" pp. 91–97. In: A. Kimbrell (ed.), *Fatal harvest: The tragedy of industrial agriculture.* Washington, D.C.: Island Press.

Knapp, A. K., J. T. Fahnestock, S. P. Hamburg, L. B. Statland, T. R. Seastedt, and D. S. Schimel. 1993. "Landscape patterns in soil–plant water relations and primary production in tallgrass prairie." *Ecology: A publication of the Ecological Society of America* 74: 549–560.

Kulikoff, A. 2000. *From British peasants to colonial American farmers.* Chapel Hill, N.C.: University of North Carolina Press.

Kyei-Boahen, S., and L. Zhang. 2006. "Early-maturing soybean in a wheat–soybean double-crop system: Yield and net returns." *Agronomy Journal* 98: 295–301.

Langley-Turnbaugh, S. J., and D. R. Keirstead. 2005. "Soil properties and land use history: A case study in New Hampshire." *Northeastern Naturalist* 12: 391–402.

Maizel, M., R. D. White, S. Gage, L. Osborne, R. Root, S. Stitt, and G. Muehlbach. 1998. "Historical interrelationships between population settlement and farmland in the conterminous United States, 1790 to 1992," pp. 5–12. In: T. D. Sisk (ed.), *Perspectives on the land use history of North America: A context for understanding our changing environment.* Washington, D.C.: U.S. Geological Survey.

Matson, P. A., W. J. Parton, A. G. Power, and M. J. Swift. 1997. "Agricultural intensification and ecosystem properties." *Science* 277: 504–509.

McMurry, S. A. 1995. *Transforming rural life: Dairying families and agricultural change, 1820–1885.* Baltimore, Md.: Johns Hopkins University Press.

Medin, D. L., and S. Atran. 1999. *Folkbiology.* Cambridge, Mass.: MIT Press.

Mires, P. B. 1993. "Relationships of Louisiana colonial land claims with potential natural vegetation and historic standing structures: A GIS approach." *Professional Geographer* 45: 342–350.

Morgan, K. 2001. *Slavery and servitude in colonial North America: A short history.* Washington Square, N.Y.: New York University Press.

Moriceau, J.- M. 2002. *Terres mouvantes: Les campagnes françaises du féodalisme á la mondialisation, 1150–1850.* Paris: Fayard.

Osterberg, D., and D. Wallinga. 2004. "Addressing externalities from swine production to reduce public health and environmental impacts." *American Journal of Public Health* 94: 1703–1708.

Parton, W. J., M. P. Gutmann, S. Williams, M. Easter, and D. Ojima. 2005. "Ecological impact of historical land-use patterns in the Great Plains: A methodological assessment." *Ecological Applications* 15: 1915–1928.

Parton, W. J., J. M. O. Scurlock, D. S. Ojima, T. G. Gilmanov, R. J. Scholes, D. S. Schimel, T. Kirchner, H.-C. Menaut, T. Seastedt, E. Garcia Moya, A. Kamnalrut, and J. L. Kinyamario. 1993. "Observations and modeling of biomass and soil organic matter dynamics for the grassland biome worldwide." *Global Biogeochemical Cycles* 7(4): 785–809.

Pollard, S. 1997. *Marginal Europe: The contribution of marginal lands since the Middle Ages.* Oxford: Oxford University Press.

Princen, T. 2005. *The logic of sufficiency.* Cambridge, Mass.: MIT Press.

Princen, T., M. Maniates, and K. Conca. 2002. *Confronting consumption.* Cambridge, Mass.: MIT Press.

Ramankutty, N., and J. A. Foley. 1999. "Estimating historical changes in global land cover: Croplands from 1700 to 1992." *Global Biogeochemical Cycles* 13(4): 997–1027.

Ramankutty, N., J. A. Foley, and N. J. Olejniczak. 2002. "People on the land: Changes in global population and croplands during the 20th century." *Ambio* 31: 251–257.

Redman, C. L. 1999. *Human impact on ancient environments.* Tucson, Ariz.: University of Arizona Press.

Remini, R. V. 2002. *Jackson versus the Cherokee Nation.* Chicago, Ill.: Chicago Historical Society.

Rosenberg, S. 2003. *American economic development since 1945: Growth, decline, and rejuvenation.* Basingstoke, U.K.: Palgrave Macmillan, Houndmills.

Salstrom, P. 1994. *Appalachia's path to dependency: Rethinking a region's economic history, 1730–1940.* Lexington, Ky.: University Press of Kentucky.

Sawyer, G. 1971. *The agribusiness poultry industry: A history of its development.* New York: Exposition Press.

Shuman, J. B., and D. Rosenau. 1972. *The Kondratieff wave.* New York: World Pub.

Smil, V. 2001. *Enriching the earth: Fritz Haber, Carl Bosch, and the transformation of world food production.* Cambridge, Mass.: MIT Press.

Theobald, D. M. 2001. "Land use dynamics beyond the American urban fringe." *Geographical Review* 91: 544–564.

Theobald, D. M., T. Spies, J. Kline, B. Maxwell, N. T. Hobbs, and V. Dale. 2005. "Ecological support for rural land-use planning." *Ecological Applications* 15: 1906–1914.

Tilman, D., K. G. Cassman, P. A. Matson, R. Naylor, and S. Polasky. 2002. "Agricultural sustainability and intensive production practices." *Nature* 418: 671–677.

Tyler, D. 1992. *The last water hole in the West: The Colorado–Big Thompson Project and the Northern Colorado Water Conservancy District.* Niwot, Colo.: University Press of Colorado.

U.S. Congress (21st 1st session) and J. Evarts. 1830. Speeches on the passage of the bill for the removal of the Indians delivered in the Congress of the United States. Boston: Perkins and Marvin.

U.S. Department of Agriculture. 1997, 2002. *Census of agriculture.* Washington, D.C.: National Agricultural Statistics Service.

U.S. Department of Commerce. 1870a, 1880a, 1890a, 1900a, 1910a, 1920a, 1925, 1930a, 1935, 1940a, 1945a, 1950a, 1954, 1959, 1964, 1969, 1974, 1978, 1982, 1987, 1992, 1997. *Census of agriculture.* Washington, D.C.: Bureau of the Census.

U.S. Department of Commerce. 1870, 1880, 1890, 1900, 1910, 1920, 1930, 1940. *Census of manufactures.* Washington, D.C.: Bureau of the Census.

U.S. Department of Commerce. 1880b, 1910b, 1940b, 1970b, 2000b. *Census of population.* Washington, D.C.: Bureau of the Census.

U.S. Department of Commerce. 1947, 1949, 1950, 1952, 1956, 1962, 1967, 1970, 1972, 1977, 1983, 1988, 1994, 2000. *County and city data book.* Washington, D.C.: Bureau of the Census.

Vaughan, A. T. 1999. *New England encounters: Indians and Euro-Americans ca. 1600–1850: Essays drawn from* The New England Quarterly. Boston, Mass.: Northeastern University Press.

Vickers, D. 1990. "Competency and competition: Economic culture in early America." *William and Mary Quarterly* 47: 3–29.

Waisanen, P. J., and N. B. Bliss. 2002. "Changes in population and agricultural land in conterminous United States counties, 1790 to 1997." *Global Biogeochemical Cycles* 16: 84-1–84-19.

Walker, M. 2000. *All we knew was to farm: Rural women in the upcountry South, 1919–1941.* Baltimore, Md.: Johns Hopkins University Press.

Wells, W. C. 2003. *American capitalism, 1945–2000: Continuity and change from mass production to the information society.* Chicago, Ill.: Ivan R. Dee.

Whitney, G. G. 1994. *From coastal wilderness to fruited plain: A history of environmental change in temperate North America, 1500 to the present.* New York: Cambridge University Press.

Wohl, E. E. 2001. *Virtual rivers: Lessons from the mountain rivers of the Colorado Front Range.* New Haven, Conn.: Yale University Press.

Wohl, E. E. 2004. *Disconnected rivers: Linking rivers to landscapes.* New Haven, Conn.: Yale University Press.

2

New England's Forest Landscape

Ecological Legacies and Conservation Patterns Shaped by Agrarian History

David R. Foster
Brian Donahue
David Kittredge
Glenn Motzkin
Brian Hall
Billie Turner
Elizabeth Chilton

After a regionwide two-century period of deforestation and agrarian expansion, the dramatic reduction in agriculture in New England during the past 150 years generated a wave of land-cover change. Forest cover increased from less than 30% to more than 75% in many regions. Despite supporting one of the densest human populations in the nation, New England is among the most heavily forested regions in the United States. The story of this remarkable landscape transformation is one of recovery of nature, the legacy of past events in the details of modern ecosystems, and opportunity matched by challenge for conservation.

The reestablishment of forest ecosystem characteristics progressed unevenly, with compositional, structural, and functional attributes exhibiting different lags in development. In all cases, however, the modern distribution of vascular plant species, levels of forest biomass, and soil structure, chemistry, and fertility are strongly conditioned by legacies of a varied land-use history. The scale and grain of this landscape conditioning is controlled by the physical environmental template (e.g., topography, glacial geology, soils), geographical location relative to population centers, and the specific cultural traditions of the regional population, which varies in subtle fashion. In general, however, the broad pattern has been for a homogenization of ecological characteristics at the site scale (resulting from uniformity in land use) and at the regional scale (resulting from broad-scale similar changes in land use and land cover), and for the development of a more patchy and heterogeneous structure characterized by abrupt ecological discontinuities at a landscape scale (resulting from the small-grained and patchy landownership and land-use pattern).

This changing landscape condition and pattern has generated distinctly different approaches to conservation and management, largely driven by individual value systems and the extent to which the legacies and lags are not only interpreted but also interpreted correctly. Each of these emerging traditions in conservation is based on different attitudes toward the history of agrarian transition and yields contrasting management strategies and ecological consequences. Four major traditions may be identified: (1) a preservationist approach in which either the forest landscape is read as near natural and therefore warranting complete protection, or a "rewilding" approach is taken toward the secondary forest and its developing natural attributes—represented by the wildlands approach and the approach seen in core areas in The Nature Conservancy (TNC) matrix forest; (2) an ancient natural landscape approach in which specific habitats and species assemblages are interpreted as relicts or descendants of pre-European landscapes, maintained prehistorically by fire and Native American activity and warranting active fire restoration and management today (this is a major approach to uncommon and high-priority grassland, shrubland, heathland, and savanna adopted by TNC, the National Park Service, and Heritage Programs); (3) a cultural landscape approach in which the ubiquitous legacies of European land use are recognized, and the maintenance and restoration of specific cultural landscapes and species assemblages [e.g., such as those mentioned in (2)] are sought by mimicking or reintroducing intensive traditional agricultural practices such as sheep grazing and mowing (which is a new perspective adapted by regional organizations and paralleling European conservation practice); and (4) a resource-based approach that seizes on the great annual wood increment in this region to argue for increased extraction, which is the working landscape approach that may well provide benefits to (1) and (3).

The landscape of New England has undergone one of the most remarkable histories of transformation worldwide. Once extensively covered with mature and old-growth forest, the land was cleared for agriculture, remaining forest areas were cut extensively for diverse wood products, and then, nearly as rapidly as it was cleared, the forest rebounded in extent and maturation after the regional decline in agriculture. This story is not unique in world history, as one can find roughly contemporaneous histories elsewhere in the eastern United States and portions of northwestern Europe, and an even greater landscape transformation in the Yucatan peninsula after the collapse of the Mayan empire. In New England there is a richness of scientific and historical details that offer great insights into major ecological, social, and conservational questions. In this chapter we explore this history and its human and ecological consequences, beginning approximately

500 years before European arrival. We combine social, biological, and physical science perspectives to address a series of broad fundamental and applied questions:

- What are the major physical, social, and biological drivers of change and how have these interacted through time?
- What are the ecological responses and consequences of the environmental changes that have occurred?
- How have these historical dynamics conditioned or constrained subsequent human activity and ecological dynamics?
- How can conservation and natural resource management integrate this history of ecological and social change into effective management strategies?
- What social, ecological, and conservation issues emerge from this historical–ecological consideration of the region's history?

Study Region

The Harvard Forest study focuses on three nested study regions: the New England region, the state of Massachusetts, and, in a few cases, subregions in central or coastal Massachusetts (Fig. 2.1A). The state of Massachusetts is the central focus because it captures much of the physical, biological, and cultural variation of New England. It provides a convenient scale for examining important cultural and environmental processes, and yet represents a feasible area for data collection and analysis. The Massachusetts study area is also relevant to a number of our major research programs (e.g., the Harvard Forest LTER; NSF funded), Harvard University National Institutes of Global Environmental Change (Department of Energy), Clark University Human Environment Research Observatory (NSF funded), and the Regional Forest Conservation Study (funded by TNC, the A. W. Mellon Foundation, and the USDA). In this work, broader regional contexts (e.g., New England, the entire United States, our global setting) and more local scales (e.g., town and family scales) are considered when they are pertinent to particular issues and drivers of change.

New England Region

The New England region (Maine, New Hampshire, Vermont, Massachusetts, Connecticut, and Rhode Island) displays considerable variation in vegetation and flora, natural disturbance regimes, and cultural history. The region has been strongly modified by many episodes of glaciation, and landscape patterns of soils, stream drainage, and topography have developed through interactions between the bedrock geology and the erosional and depositional history of the most recent glacial period, which ended approximately 15,000 years ago. New England is a predominantly hilly region of broad highlands ranging from 200 to 500 m above sea level (a.s.l.), with narrow valleys and a few broad lowlands and river valleys that extend below 200 m in elevation. On a broad scale and across local

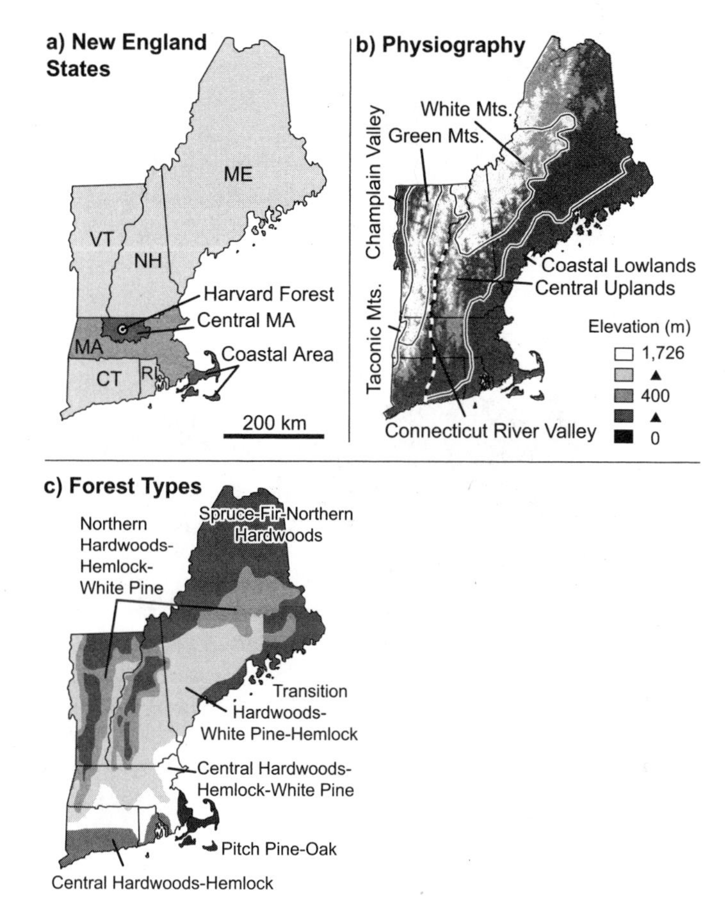

Figure 2.1 (A–C) A map of New England showing state boundaries and scales of research conducted by the Harvard Forest (A), elevation and physiographic areas (B), and forest vegetation zones (C) as described by Westveld and the Committee on Silviculture, New England Section, Society of American Foresters (1956).

landscapes, much of this variation occurs through alternating valleys and uplands that trend north to south because of the structure of the underlying bedrock.

Physiographic Divisions

Major physiographic areas that we have investigated in detail include the Green Mountain Uplands, Connecticut Valley, Central Uplands, and Coastal Lowlands (Fig. 2.1B).

The *Green Mountain Uplands* extend the length of New England. Metasedimentary and metavolcanic rocks constitute the bulk of these uplands. In general, this region has more productive soils and therefore a more diverse flora than the Central Uplands, including the area surrounding the Harvard Forest and the White Mountains.

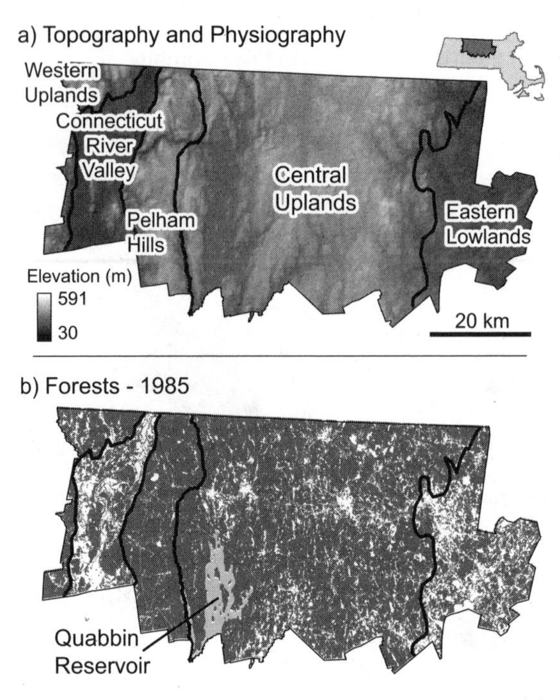

Figure 2.2 (A, B) The central Massachusetts region showing topography and physiographic areas (A) and 1985 forest cover (dark gray) and the Quabbin Reservoir (B). Based on data from Foster et al. (1998b).

The *Connecticut Valley* separates the two large upland regions in New England: the Green Mountains and the Central Uplands. It is underlain primarily by sandstone and shale to the south, and metasedimentary and metavolcanic rocks to the north.

To the east and north, the *Central Uplands* constitute the largest physiographic region and include the White Mountains, with the tallest peaks in New England, and the intensive study areas of the Harvard Forest LTER program (Fig. 2.2). The rocks of this region are variable but tend to be of metasedimentary and metavolcanic origin and produce acidic soils of low nutrient status.

The *Coastal Lowlands* form a 60- to 100-km-wide belt that extends from the shores of New Jersey to the central coast of Maine. Relief is generally low, the bedrock is highly variable, and the contact with the adjoining highland areas to the north and west is typically abrupt. The extensive coastal plain of New England is largely submerged off the Atlantic coast, where it forms the Continental Shelf.

Morainal and outwash deposits that have been modified by coastal processes since the last glaciation compose the areas from Cape Cod southward through the islands of Nantucket and Martha's Vineyard in Massachusetts, Block Island in Rhode Island, and Long Island in New York state. These coastal areas are dominated by sandy soils, low elevation, and varied relief.

Strong gradients in precipitation, temperature, and length of growing season across New England are driven largely by elevation and latitude. Mean annual temperature ranges from 11 °C in southern Connecticut to 4 °C in the northern highlands of Vermont, New Hampshire, and Maine, whereas precipitation ranges from 880 to 1,250 mm and is distributed fairly evenly throughout the year.

Forest Vegetation

The forests of this environmentally complex region can be described fairly well by a classification system and map that includes five vegetation zones: spruce–fir–northern hardwoods, northern hardwoods–hemlock–white pine, transition hardwoods, central hardwoods, and pitch pine–oak [Fig. 2.1C (Westveld and the Committee on Silviculture, New England Section, Society of American Foresters, 1956)]. This map agrees well with recent reconstructions based on tree data recorded by surveyors at the time of regional land settlement.

Central Massachusetts Subregion

The central Massachusetts area comprises portions of four counties in the north-central part of the state and encompasses physical, biological, and cultural gradients that vary across three major physiographic regions that differ in relief, geology, soils, land-use history, and climate. The broadest physiographic area is the Central Uplands (and its Pelham Hills subarea) (Fig. 2.2A), which in north-central Massachusetts is characterized by north–south trending hills and narrow valleys. The acidic bedrock is overlain by thin glacial till on the uplands and deeper and more level outwash, alluvial deposits, and peats in the narrow valleys. Soils are acid sandy loams of low nutrient status.

With the exception of developed areas, lakes, and marshes, the Central Uplands are forested, and there are few remaining farms. Upland villages with low population density are scattered across the forested hills, whereas larger, postindustrial towns and small cities border some of the major streams. The Quabbin Reservoir (10,000 ha; Fig. 2.2B), which was created in the 1930s and now provides drinking water for approximately 40% of the Massachusetts population living in the Boston metropolitan area, is surrounded by approximately 25,000 ha of land owned and managed by state agencies. The Quabbin Reservation forms the largest piece of an extensive, although loosely affiliated, conservation partnership—the North Quabbin Regional Land Partnership—which is composed of state agencies, nonprofit conservation organizations, and educational institutions, including the Harvard Forest. Most of the land in the area and, indeed, in New England is in small private ownership. Much of the forested land is actively managed for wood products and is extensively used for diverse recreation.

To the west is the Connecticut River Valley. The level to rolling plains at 30 to 75 m a.s.l. are underlain by sedimentary bedrock and support level deposits of outwash, alluvium, and glacial lake sediments. Soils range from excessively well-drained, sandy outwash to poorly drained, silty, floodplain sediments. A series of bedrock ridges composed mainly of volcanic basalt (*traprock*) emerges through the valley bottom and reaches a maximum height of 400 m a.s.l. Its rich and fertile soils, level terrain, ease of river navigation, and long settlement history by Native Americans and Europeans have led to a modern cover of extensive farmland; concentrated urban, industrial, and residential areas; and discontinuous forests. In contrast, the traprock ridges remain largely wooded. The diverse environments from valley to ridgetop support a remarkable array of unusual plant assemblages. Along with its great cultural heritage, these make the Connecticut River Valley a priority area for state and national conservation.

To the east, the Central Uplands grade gradually into the Eastern Lowland, which is part of the extensive Coastal Lowland. This area of hills, gentle relief, and meandering rivers set in broad valleys ranges from 40 to 200 m a.s.l. Acidic bedrock is overlain by till, broad glacial lake sediments, alluvium, and marine deposits. The region grades eastward from rural, agricultural, and forested areas into the densely populated suburban and high-technology region adjoining Boston.

The physiographic variation across central Massachusetts yields subtle gradients in environment, history, and vegetation. The Connecticut Valley and Eastern Lowland have low elevations, gentle relief, and mild climates, with summer and winter temperatures that average 2 to 3 °C warmer than the intervening Central Uplands. In response to this climatic variation, southern plant species decline on the uplands, and the northern hardwoods–hemlock forest extends southward from New Hampshire onto this area of higher elevation. Broadly, the Connecticut Valley is lower, has more nutrient-rich soils, and is more agricultural than the Eastern Lowlands.

Precontact Native Land Use in New England

Any consideration of ecology or conservation in the New England landscape needs to consider the lengthy history of human activity and environmental change. Both have shaped the nature of the land and the vegetation, and both figure strongly in real-world discussions and decisions concerning conservation management and restoration.

Despite a long history of archaeological, historical, and paleoecological research, many fundamental questions remain regarding the activities of pre-European peoples. Addressing these questions provides a basis for extensive interdisciplinary collaboration within the Harvard Forest group. Major issues include the following:

- What were the environmental and social constraints to the size and subsistence patterns of native populations, and how did these patterns vary geographically and temporally?
- To what extent did human activities, including horticulture, hunting, and fire, modify the natural environment, vegetation patterns, and faunal abundance?
- What is the basis for the apparent discrepancy between archaeological data and ethnographic sources regarding population size, distribution,

subsistence patterns, and ecological impacts? Can these be reconciled through comparative multidisciplinary studies?
- Does our understanding of pre-European ecological and cultural patterns assist modern ecological interpretations and conservation planning?

Initial Colonization

The first people to arrive in the New England region encountered a boreal landscape that was dramatically different than today. The region's oldest archaeological sites date to 11,000 to 12,000 years before present (BP), a few thousand years after glacial ice had melted and large proglacial lakes had drained (Dincauze, 1990). The retreating glacier left large amounts of mineral sediments that filled river valleys and streams, and as a consequence of this and a higher water table overall, rivers flowed up to 60 m higher than they do today. Thus, many of the earliest archaeological sites have been buried or destroyed by postglacial river down-cutting.

Archaeological evidence for these first paleo-Indians indicates that they were hunter-gatherers living in fairly small family groups (20–50) who moved their encampments to exploit seasonally available resources.

Plant and animal life during this period were dynamic in response to deglaciation and relatively rapid climate change; the forests were dominated by northern conifer species, such as spruce and fir, and were interspersed with areas of patchy tundra (Dincauze, 1990; Foster and Zebryk, 1993; Gaudreau, 1988; Gaudreau and Webb, 1985; Lindbladh et al., 2007). Climatic warming and ameliorating conditions allowed white pine, spruce, fir, and temperate deciduous trees to migrate into southern New England at the time humans were arriving in the region (McWeeney, 2003).

Paleo-Indians apparently subsisted on a wide variety of plants and animals, including caribou, giant beaver, hawthorn seeds, wild grapes, and migratory birds (Dincauze, 1990; Chilton, 1999, 2002; Dincauze and Jacobsen, 2001). They relied on stone tools for a variety of tasks such as wood working, food processing, hunting, fishing, canoe building, and wigwam construction.

The Archaic Period

Through the Archaic period (11,000 to 3000 BP) native peoples continued to rely on hunting and gathering, but under greatly changing conditions. The environment ameliorated to temperate conditions, and the boreal forests were replaced first by white pine and oak forests, and then by a diverse forest of broad-leaf and conifer species (Dincauze, 1990; Foster and Zebryk, 1993).

During this period, people clearly adapted to diverse subregions: forested uplands, fertile valley bottoms, rich coastal environments, and interior wetlands. Evidence from many thousands of archaeological sites across New England indicate that population sizes grew, stone tools continued to predominate, and social groups occupied seasonal settlements within well-defined homelands. Fluted spear points were replaced by a diverse array of point types—side notched, stemmed, corner notched—many of which exhibit subregional stylistic differences. The earliest evidence of houses dates to this period, presumably because evidence of earlier dwellings simply has not survived. It includes small, round structures that may have sheltered small, extended family groups. From archaeological evidence and ethnographic analogy, activity is interpreted as changing seasonally, with large groups coming together for social gatherings and to exploit concentrated resources, but splitting up into smaller groups to use widely dispersed resources (e.g., large spring fish runs vs. small winter hunting camps). These diverse site types are well represented in the archaeological record of the region.

Most archaeological sites of the Archaic period date to the Late Archaic period (7000 to 3000 BP), presumably as a result of an increase in population density as well as better preservation resulting from the stabilization of the sea level and riverbeds by about 5000 BP. Increased population is also reflected in the size of sites and evidence for multiseasonal use of certain sites. An increase in burial ceremonial artifacts may provide evidence for increased sedentism and a differentiation of social ranking that usually accompanies an increase in population.

In southern New England, people of the Archaic period clearly exploited the rich deciduous forests through the harvest, storage, and consumption of acorns, hickory nuts, beechnuts, and chestnuts. They carved dugout canoes from hollowed-out trees. Evidence exists of fishing in interior rivers and lakes, shellfish collecting, the hunting of deer and other terrestrial animals, and the collecting of hundreds of plant species for food, medicinal uses, and other purposes.

Indigenous Horticulture

Evidence supports the interpretation of a diverse foraging base for New England peoples. At the beginning of the Woodland period (3000 to 400 BP), archaeological evidence across eastern North America indicates intensive exploitation of weedy plants that grow in disturbed soils, particularly goosefoot (*Chenopodium berlandieri*), sumpweed (*Iva annu*), and sunflower [*Helianthus annuus* (Smith, 1992)]. The oily, starchy seeds of these plants were boiled and made into porridge, and the species were apparently modified genetically through selective breeding, representing the earliest phases of horticulture. Supporting evidence for these crops, beginning in 3000 BP, includes soapstone bowls and ceramic pots, which made superior cooking vessels for the boiling of nuts and starchy seeds to make them palatable and digestible.

There is suggestive, although inconclusive, evidence of forest management with fire by native peoples during the Woodland period (3000 to 500 BP) (Johnson, 1996; McWeeney, 1994, 2003). One effect of burning might have been the diversification of habitats that would support a wide variety of plants and animals, including berries, grasses, birds, and land mammals (Cronon, 1983).

Mobile Farming and Maize Horticulture

During the Late Woodland period (ca. 1000 to 400 BP), important innovations arrived in New England: (1) bow-and-arrow technology and (2) tropical cultigens (maize, beans, and squash, which were originally domesticated in Mexico 5000 years BP). Although a few sites supporting maize horticulture date to 1000 BP, the earliest Accelerated Mass Spectrometry dates on maize kernels lie in the range of ca. AD 1300 to 1500 (Chilton and Doucette, 2002). Until contact with Europeans, maize was apparently only a dietary supplement to an otherwise diverse diet (Bernstein, 1999; Chilton, 1999, 2002; Chilton and Doucette, 2002; Chilton et al., 2000).

Although we know that maize horticulture was practiced by New England peoples, it may not have consumed much of their time or energy. After the planting of maize, groups would apparently disperse for 2 to 3 months as the maize ripened, to plant, hunt, and gather elsewhere (Cronon, 1983).

With regard to the interior of this region, the few postmolds located on Late Woodland sites appear to represent short-term wigwam-type structures. The overlapping pattern of these structures and other features, as well as a lack of well-defined middens, indicate repeated seasonal use of the same locations over time (e.g., Chilton et al., 2000).

The ethnohistorical literature supports an interpretation of diversity and flexibility in New England settlements. In 1674, Josselyn (1988/1674) reported: "Towns they have none, being always removing from one place to another for conveniency of food. . . . I have seen half a hundred of their Wigwams together in a piece of ground and within a day or two, or a week they have all been dispersed" (p. 91). During the second quarter of the 17th century, Johan de Laet (cited in Jameson, 1909) said of Algonquian people living in the Hudson Valley: "some of them lead a wandering life in the open aire without settled habitation. . . . Others have fixed places of abode" (pp. 105–109). Although it is clear that New England Algonquians were fairly mobile throughout their history, they were not nomadic. Rather, groups and individuals were moving within well-defined homelands and among interrelated communities.

When the English arrived, they likely misinterpreted their native pattern of mobile farming because this lifestyle and cultural practice would have been unfamiliar. It also would not have been viewed as a legitimate or proper use of land. In this light, it is easy to see the underlying justifications for the taking of Native American lands—whether implicit or explicit (see Chilton, 2005).

European Settlement History

The Rise and Fall of Agriculture, and the Fall and Rise of the Forest

The rise and fall of farming in New England, and the corresponding decline and recovery of the region's forest, is a familiar story. During the first half of the 17th century, a group of English settlers established agricultural communities on the coastal plain and in the lower Connecticut River Valley, often on former Native American village sites. With vigorous population growth, by the end of the colonial period these early towns had filled with farms, and new settlements were spreading across the interior uplands. After the Revolution, a great wave of clearing occurred throughout the region, driven by continued population growth and expanding commercial opportunities. Farmland peaked at nearly three quarters of the landscape about the middle of the 19th century, after which time forest area began to rebound as New England agriculture adjusted to the new pressures of an integrated industrializing national economy. Marginal pastures were abandoned during the late 19th century as farming intensified, and more pastures and fields were abandoned throughout the 20th century as farm contraction continued. Tree growth paralleled the late expansion of forest cover as wood harvesting diminished. However, since about 1950, urban development has sprawled outward in a spatial pattern strikingly similar to the earlier expansion of agriculture, initiating a new wave of forest fragmentation and decline that presents a new challenge to conservation.

The Crucial Role of History in Ecology and Conservation

Within this well-known story lay two sets of questions that have not been well addressed, but are crucial to conserving the landscape of New England today. One set pertains to historical ecology, whereas a parallel set concerns environmental history. The first inquires into the particular patterns of land use across the region, how those patterns changed over time, and how they shape current and future environmental conditions. The second set asks what social forces shaped these patterns and drove these transformations, and how they changed through time. These questions are crucial because history matters; we are part of it, and our modern landscape bears its legacies and imprints. The better we understand past patterns of land use and their continuing influence on the landscape, the better we can work with these evolving systems to provide desired ecological conditions and ecosystem services.

Similarly, the way land was farmed and then abandoned in New England was not a simple reflexive response to outside technological and economic signals, but a much more complicated cultural evolution and accommodation. Rural society changed its economic and ecological organization over time, and constrained its behavior in various ways. Conservation today must work across a complex political landscape of public and private landownership that is subject to evolving cultural pressures. To accomplish anything, conservationists must tell a coherent and compelling story: one that places our situation today in the midst of an ongoing history of engagement with the land, not outside it. In both cultural and ecological terms, the simplistic conservation model of defending a pristine natural landscape from human disturbance has little meaningful place in New England.

Changing Drivers of Land-Use Change

The primary driver of landscape change in New England has been the progressive integration of local land-use practices into an expanding market economy. Throughout four centuries the region has undergone a shift from production for household use and local exchange toward production for larger regional commercial markets, paralleled by a shift from consumption of local resources toward consumption of imported resources. During the course of this transformation, land use has moved from highly diversified to satisfy local needs, to highly specialized to increase cash income, to largely unused for production but valued instead for residential and commercial development or environmental amenities. The transformation has been shaped by the rise of increasingly powerful industrial technologies of extraction and transportation, and the growth of an affluent urban and suburban population. The land did not simply respond to external market forces in a mechanical fashion. Massachusetts landowners played an active role in developing new attitudes and opportunities as they engaged the market. Along the way, they obeyed or overcame various cultural constraints, and the land itself changed in the opportunities and limitations it presented. All these factors conditioned how the land was used.

All the while, the dominant tendency for land use to be determined by short-term market calculus has been restrained by subordinate but strongly held cultural values that might be defined as different historical versions of "conservation." These countertendencies are important both because they helped shape the landscape in every period and because they form the basis of the modern conservation movement. They might be summarized as agrarian, utilitarian, and romantic. Understanding their history is as important as understanding changes in the land itself.

Historical Stages

Colonial Agrarian Economies, 1600–1775

The "Great Migration" between 1620 and 1640 brought fewer than 20,000 English settlers to New England, where they established agricultural communities along the coastal plain of Massachusetts, Rhode Island, and Connecticut, and up the Connecticut River Valley. The settlers rapidly gained a strong foothold because diseases, including smallpox, reduced the Native American population from some 100,000 to about 10,000, and left them in disarray. Thereafter, the newcomers' astounding fecundity (doubling in population every 25 years), their aggressive drive for land and resources, and their view of Native Americans as heathen savages with little moral claim to the land because they failed to "improve" it ensured their expanding control of the region. Nevertheless, resisting native groups (later backed by the French from Canada) kept the northern and western frontiers of New England a dangerous place until the mid 18th century. In the meantime, the second and third generations of English settlers filled the coastal region with farms and established the basic agricultural pattern of the colonial period.

That pattern was a close adaptation of European mixed husbandry to New England soils and climate. Although it incorporated elements of the Native American ecological system—existing settlement sites and planting grounds; cultivation of pumpkins, beans, and maize; low-lying meadows of native grass; abundant fish, game, and berries; and a forest rich in white oak, hickory, and chestnut—its fundamental organization, the close integration of tilled crops and livestock, was English. The pattern of mixed husbandry revolved around producing a wide range of agricultural and artisanal goods for local economies organized at the household and community levels.

The Puritans were enterprising and not averse to participating in the market and economic improvement—as long as they did not become too "worldly." To maintain their accustomed level of material comfort, they imported textiles, metalware, sugar, rum, and tea, and engaged in a broad Atlantic trading network that focused on exports to the sugar islands of the West Indies, including fish, ship timber, pine masts, oak barrel staves, pine lumber, and cattle. Commercial production was hampered by the high cost of inland transport: It cost less to ship an iron pot from London to Boston than from Boston to Worcester, Massachusetts. New England, with its long winters and stingy, acidic soils, never developed a staple export crop like Virginia tobacco, South Carolina rice, or Pennsylvania wheat. Beyond that, New England was settled by middling families who lived in tight, religious communities and were less interested in maximizing individual wealth than in gaining a solid economic independence, providing a comfortable subsistence, and settling their many offspring successfully in the community. As a result, these yeomen produced secondarily for outside markets, but primarily for household consumption and trade with neighboring farmers and artisans such as blacksmiths, coopers, and tanners. In this way, the economies of New England towns grew mostly "within themselves," filling with farms that generated a small surplus for exchange (Donahue, 2004).

The system of husbandry that developed to fill these needs varied regionally, but included the following basic elements: tillage, mowing, pasture, orchard, and woodland. Most farmers possessed these in similar proportions, arrayed upon the landscape in similar ways.

Tillage was not the main use of agricultural land, but it was necessary for subsistence and therefore was practiced by virtually all farmers. Less than 10% of most towns was plowed in any given year: 10 acres/farm or less. Indian corn and rye were the principal bread grains. Little wheat was grown except in the Connecticut Valley, and little was consumed except by the wealthiest New Englanders. Some barley was grown for beer, although apple cider became the more common beverage. Potatoes, introduced by Scotch–Irish immigrants in about 1720, spread rapidly and became an important subsistence crop, doing well in poor acidic soils. Pumpkins and beans were grown in many corn fields, and most farmers also grew a little flax. Farm women tended gardens that produced many vegetables and herbs. The New England diet grew steadily more diverse and nutritionally complete as the generations passed (McMahon, 1985).

On the coastal plain, tillage land focused on sandy soils, which were light, easily worked, and well adapted to corn and rye. In upland regions dominated by stony till, farmers sought out patches of sandy outwash or fairly level sites from which stones could be most easily removed, often along ridgetops. The latter gave rise to distinctive double stonewalls with small stone in-fill, bounding tillage fields. The most intensive tillage was usually located close to the barn so that the corn crop could be easily manured. These factors, strong kinship bonds, and the practice of "trading work" for farm and household tasks gave rise to a pattern typical of many New England towns: small neighborhoods of farms surrounding patches of good tillage soil at intervals along the road, with extensive pastures and woodlots reaching back into the countryside. Manure was essential for corn and potatoes, whereas rye was seldom manured and was often sown after corn or in back fields broken out of pasture. Although much tillage land suffered some soil erosion, many fields likely *improved* in fertility over time. Given the thin, highly leached condition of native spodosols, the continual concentration of manure and incorporation of plant residues may have increased levels of organic matter and nutrients in many home fields. If crop yields were flat or declining in older towns by the end of the colonial period, it was probably the result not of soil exhaustion but rather of the fact that manure supplies were inadequate for increasing amounts of cultivated land.

Mowing land provided hay, which was critical to this mixed husbandry system, feeding livestock during the long winter and supplying manure to plowed land. During the colonial period, hay was largely supplied by native grasses in coastal salt marshes and wet meadows inland. These wetlands required considerable hydrological manipulation to make them productive, and they were transformed to a carefully managed resource in many towns. Extensive systems of drainage ditches, sometimes connecting for miles, rendered the meadows firm and accessible for teams during the mowing season, whereas dams, dikes, and road causeways provided hydrological control and augmented fertilization from natural flooding. Mowing, burning, and grazing, in combination with manipulation of the water table, shifted the composition of many wetlands from tree and shrub dominated to a cover of desirable grasses and sedges. The meadows returned a reliable yield of rather coarse hay, along with rich muck that was cleaned from the ditches in the fall, dried, and carted to the barnyard or plow land.

By the end of the colonial period, as the demand for hay steadily increased, native meadows were augmented by upland plantings of "English hay," consisting of red top, timothy, and red clover. This was especially true in upland regions where wet meadows were scarce.

Cattle were the principal stock in the pastoral New England economy. Many farms had a small flock of sheep (mostly for homespun wool), prosperous farmers kept a horse, and omnivorous swine ran at large in many areas. However, cattle were the economic mainstay, providing milk (converted to butter and cheese), meat, leather, locomotion, and manure. They also served as the main cash crop: Once fattened on pasture, they were driven to the ports. Farmers in most towns kept a small beef herd as a natural extension of their herd of six or eight milk cows. By the late colonial period, the Connecticut Valley, with its great surplus of hay and grain, even made a specialty of stall-fattening cattle that had been reared in surrounding hill towns (Garrison, 2003).

Pasture covered another quarter or more of most towns. Many different lands were grazed during the course of the season, including fallow plow land and mowing land after the hay was removed. But the great bulk of pasture was found on rough backland, after it emerged from the forest. During the 17th century, many older towns ran common herds on unsettled outlands, including woodlands that were being gradually harvested or cleared and burned. As regeneration was suppressed and larger trees disappeared, this created open "wood pasture," which gave way to enclosed pasture with a few lingering shade trees. These were subsequently divided into private property and settled by younger generations. European sod-forming grasses (bluegrass, bentgrass, and white clover) were sometimes planted, but often simply spread in the dung of grazing stock. As farmers pushed into upland regions, they discovered that stony till soils, although ill-suited for the plow, had good water-holding capacity and made excellent pasture. At the end of the colonial period, pasturage was increasing in most towns.

Orchards also did well on soils derived from glacial tills, and most colonial farmers grew an acre or so of apple trees, primarily for cider. Cider replaced beer as the normal daily beverage throughout New England.

Woodland was also critical to local economies as an essential source of timber, fencing, and, above all, fuel, and most towns retained significant forest—anywhere from 30% to 50% in older towns, and of course much more in the younger hill towns (Donahue, 2004). Wood products were an initial by-product of forest clearance, but as remaining forest became scarce, woodland rose in value and was husbanded. Farmers moved toward renewable woodlot management—cutting clean on rotations of several decades, which encouraged strong sprouters such as oak and chestnut, and maximized production. Because there was still a great deal of escaped fire (from its use to clear woods, brush, and stubble in fields), pitch pine continued to find a place in colonial woodlands, especially on sandy soils in southeastern Massachusetts. Woodland was concentrated on the roughest land—rocky hills, swamps, and soils too droughty for grain or grass to thrive. Certain highly valuable resources could not be quickly regenerated, such as large oak and pine timber and white cedar, and these grew scarce through time, but most towns retained adequate wood resources.

By the end of the colonial period, New England supported a mixed husbandry and woodland ecological system that was reasonably stable. It was heavily slanted toward pasture and forest, as befitted the region's soils and climate. It supported a comfortable population. A set of ecological constraints, which we might call *agrarian*, were built into this system. On a practical level, because these local economies had to supply a wide range of goods, they tended to conserve a diverse landscape that included plenty of woodland. They also lived within deeply embedded social constraints: New England yeomen thought less in terms of their own immediate profit than in terms of the long-term comfort and economic independence of their families, and they fully expected one son to occupy the land they were farming, and other daughters and sons to settle nearby. The charge that they wore down their farms and moved on to cheaper frontier land is not reflected in their behavior, which reveals the same families occupied the same farmland for generations, often prospering.

These yeomen also lived within tightly knit communities in which certain norms of behavior were expected, and some resources were managed in common—in particular, water. Drainage and flow of wet meadows by brooks and rivers were carefully regulated, and anadromous fish runs were zealously guarded and integrated with the management of mill ponds. All these factors put limits on the way individuals could readily use their land. On the other hand, there were some native ecological elements for which this agrarian society had little use, and these were in steep decline—for example, wolves, deer, beaver, pigeons, old-growth forests, and slow-growing forest trees like hemlock and beech. It was considered natural that such things should be heavily exploited or eliminated from the improved agrarian landscape. [See Cronon (1983) and Merchant (1989) for a somewhat different interpretation of New England's colonial environmental history.]

There was one aspect of this agrarian ecological system that was not at all stable: human reproduction. Family labor was extremely valuable on the land and in the house, and New Englanders reared large families. An emerging problem, of course, was how to establish grown children with farms or trades as the towns filled up and only one son could inherit the homestead. This demographic pressure led to ecological and social stress and a gradual decline in birthrates by the time of the Revolution. It also drove farmers increasingly into the cash economy as a means of giving children a start in life.

The Market Revolution, 1775–1850

As the generations after the American Revolution developed new upland towns across southern New England and pushed rapidly into the more difficult hill country of Maine, New Hampshire, and Vermont, they were part of a world undergoing powerful economic and cultural change. On the one hand, increasing population pressure in settled regions threatened many with marginalization. Farm families and new generations had to scramble to support themselves, and they had to engage deeper in the cash economy to acquire land or become established in a trade. At the same time, aspirations were changing. Far more consumer goods were available, and many rural people were eager to take part in the new economy. The maritime trading boom at the turn of the century, the rapid expansion of manufacturing along New England rivers, and an increasing urban population provided strong markets for farm produce. Improved roads and bridges, a handful of canals, and (toward the end of the period) the first railroads cut the cost of inland transportation.

New towns emerged to access water power at locations along major rivers as well as minor streams, and many existing towns that had been established on well-drained hilltops during the early agricultural period developed new industrial villages in the valleys (Fig. 2.3) (O'Keefe and Foster, 1998). The lasting landscape impacts of the mill towns followed from their infrastructural developments, particularly roads, which would shape transportation and land use in the following centuries as the urbanization process moved into the interior uplands (Kulik et al., 1982). With the building of railroads, the cost of transporting heavy, low-cost goods declined significantly.

Better farm tools and implements became available, and crop varieties and livestock breeds improved. All this crystallized by the second quarter of the 19th century in a *market revolution*. Farm families began to shift their focus to commercial cash sales. The opening of the Erie Canal in 1825, connecting

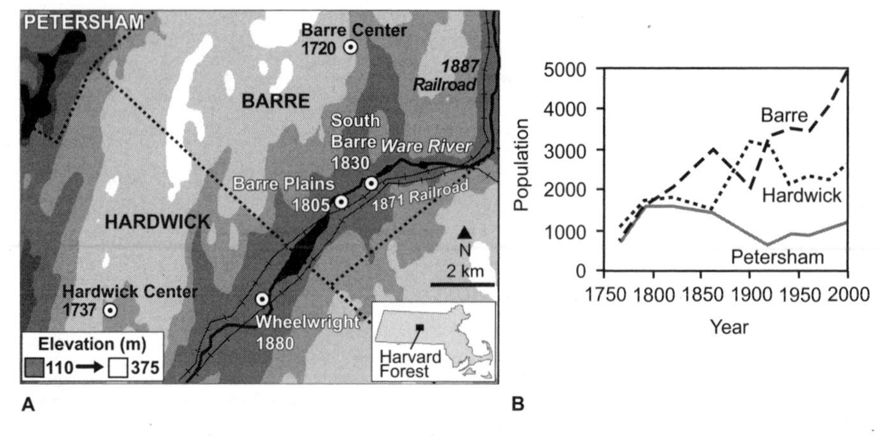

Figure 2.3 (A) Development of the towns of Barre and Hardwick, Massachusetts, showing the original agricultural town centers on high elevations during the early 18th century and subsequent creation of 19th-century industrial village centers in the valley bottoms. Because the best agricultural lands are on the broad upland ridges, and the rivers provided power for industry, the towns' population centers literally moved downhill through time. The presence of waterpower and railroads transformed towns during the 19th century. (B) Although the rural hill town of Petersham shows a population decline as agriculture waned, the industrial development of Barre and Hardwick allowed their populations to have a net increase through the mid to late 1800s (based on population data from U.S. censuses).

Lake Erie and the Hudson River, reduced the transport cost of food supplies from the Midwest to one 30th its previous cost (Van Royen, 1928). Rural communities began to meet many of their material needs through imports: wheat flour from the West replaced corn and rye meal, factory-made cotton and wool fabrics replaced homespun linsey-woolsey, shoes from Lynn replaced brogues made by a neighborhood cordwainer, coal from Pennsylvania (burned in a manufactured iron stove) augmented firewood, sawn pine two-bys (fastened with machine-cut nails) from Maine began to replace local oak and chestnut timber frames, and tea and coffee replaced hard cider as farmers became sober, calculating businessmen. The purchasing of basic commodities both allowed and required farmers to concentrate their own efforts on a few marketable crops, and the tight social and ecological constraints that had bounded the yeoman world began to dissolve.

Farmers adjusted by doing everything—from expanding their output of established crops to trying new specialty crops. Forests came under commercial pressure to supply proliferating wood industries. Almost every species spawned its own little business: spruce for the manufacture of ladders, witch hazel for liniment, and so forth. Remaining woodlands were typically cut on short rotations. However, the driving force in the commercial expansion of agriculture was pasture and hay for livestock, including the sheep boom that swept the forest from many hill towns in a generation or two. But just as important, throughout much of southern New England, was an increase in beef and dairy production.

The same transportation revolution that supported commercial production also exposed New England to the burgeoning national marketplace, bringing ruinous price pressure on traditional agricultural commodities such as meat and grain. By the 1830s, New England sheep were unable to compete with flocks raised on cheaper land in Ohio and beyond.

Ecological difficulties abounded. Rapid clearing drove forest cover to a low of perhaps no more than 25% across the region by mid century, and forest cover in some older towns fell to 10%. There were no more farms in these towns, but there was much more cleared land (Donahue, 2004; Hall et al., 2002). Deforestation was driven by the strong market for firewood and other wood products, coupled with the ability to ship in lumber, coal, and even firewood itself. The conservation of local woodlands was no longer an economic necessity. Rapid deforestation led to the population decline and even extirpation of many native wildlife species, and led (some have argued) to increased stream flooding, soil erosion, and sedimentation—all subjects worthy of investigation (Bernardos et al., 2004; Foster et al., 2002). Meanwhile, proliferating mill dams interrupted anadromous fish runs, and drowned many once-prized meadows (Cumbler, 2004; Donahue, 1997).

The expansion of upland grass proved unsustainable. Abundant English hay broke the limitations imposed by native meadows, and allowed larger livestock herds and greater manuring of plow lands. However, upland mowing ground was not recharged by annual floods, and many hayfields suffered rapid declines in yield. Pastures were also steadily drained of nutrients, and continuous grazing encouraged brushy native weeds and invasive species. Most of New England lies on acidic bedrock, which stealthily undercuts the efficacy of legumes such as red and white clover at restoring nitrogen to hayfields and pastures. It was not until the late 19th century that either the understanding or the means existed to apply lime (and other mined fertilizers such as phosphate) in sufficient quantity to counteract the downward trend. Consequently, by mid century, much of upland New England—so laboriously cleared of trees and stones—was filled with low-quality fields and scrub on its way back to forest. The land was pushed to greater productivity than during the colonial period by creating an ecologically precarious situation that was noted with alarm by many observers at the time. This was the landscape that inspired the warnings of George Perkins Marsh and Henry David Thoreau (Donahue, 1999; Foster, 2001).

Concentrated Products, 1850–1920

By the second half of the 19th century, the industrial revolution had followed the market revolution, greatly accelerating the trend toward consuming imported resources and specialized farming in New England—and in the process, reversing the trend from extensive to intensive use of the land. This was the age of rapid urban growth, and of iron, coal, and steam. New England became tightly connected to the national economy by rail and coastal steamer. The introduction of steam power, first in 1840, but more intensively in the early 1860s, fostered the rise of multiple factory complexes that filled in the industrial sections of large cities, such as Lowell and Worcester. These factories were supported by a supply of immigrant labor from Europe along with female workers from the rural hill towns (Balk, 1944). With the increases in production and population, the industrial cities of southern New England became the center of manufacturing in the country.

The shift from an agricultural to an industrial economy was complete by the beginning of the 20th century. Roads were improved, increasing labor mobility and attracting more industry (Balk, 1944). The dramatic increase in urban–industrial populations during the early 20th century, especially in Boston, necessitated the major impoundment and diversion of water from central Massachusetts. The Wachusetts and Quabbin reservoirs were opened in 1908 and 1938 respectively, totaling a combined 44 sq. mi. and associated with more than 100,000 acres of protected watershed forest (Greene, 1981). Four towns, with populations that had all declined significantly during the previous decades, were removed for the Quabbin Reservoir.

Many rural hill towns began to lose farms and population during this period, and *farm abandonment* became a concern. But this was not true regionwide. Instead, farmers focused even more narrowly on what agricultural leaders called *concentrated products*—high-value, often perishable commodities for which they enjoyed a comparative advantage in nearby urban markets. The value (and yield) of farm output actually increased—peaking in Massachusetts, for example, in about 1910 (Bell, 1989; Donahue, 1999). Farmers concentrated their production on vegetables, fruits, poultry, and hay—the last too bulky for long-distance transport, and in high demand for city horses. But above all, New England farmers produced milk—perhaps four times as much in 1900 as in 1850. They did this, paradoxically, while abandoning most of their pasture—half of it by 1900, and three quarters of it by 1920. This was accomplished by buying cheap feed grain for their cattle from the western states, turning it into milk—concentrating its value—and applying the augmented manure to corn silage and the best hay fields, while letting exhausted pastures return to forest. Again, there were not dramatically fewer farms—just much less cleared land. Agriculture had become a thoroughly businesslike profession, integrated into the national economy and increasingly divorced from household production and local exchange. Farms were surrounded by rebounding forest and a declining rural population, rather than a vigorous local economy and majority agrarian culture, a twilight atmosphere captured in the poetry of Robert Frost (Barron, 1988; Black, 1950; Wilson, 1936).

Abandoned pastures initially supported red cedar, white pine, paper and gray birch, and some red spruce at higher elevations. Because markets for wood products remained reasonably strong, these new forests and older woodlots were frequently cut. A new industry, cutting pasture pine for wood boxes and containers, provided many hill towns (including Petersham, the site of the Harvard Forest) with a major new agricultural sector by the late 19th century. By 1920, forest cover had returned to encompass nearly 50% of the landscape, where it stood at the end of the colonial period—a century and a half earlier. But it was a much younger forest, undergoing dramatic compositional changes that are still evident today.

Suburbanization and Agricultural Decline, 1920 to Present

During the 20th century, industrial capitalism evolved from the *paleotechnic* world of iron, steam, and coal into a *neotechnic* era of oil, electricity, and chemistry. The years of World War I proved to be the zenith in Massachusetts's industrial production. Between the 1930s and 1970s, the golden age of mass production for other industrial regions in the country, southern New England experienced a period of sustained economic decline (Harrison and Kluver, 1989a). Manufacturing firms closed and traditional mill-based industries relocated entirely.

By the 1970s, New England's once-powerful industrial centers had become modern-day brownfields, with more than 1,100 contaminated former industrial sites in central Massachusetts alone (Rideout and Adams, 2000). The 230 contaminated sites in Worcester occupy almost 24.3% of that city's land area, inhibiting redevelopment of much of the city (Hoover, 2001). The contamination of former industrial parcels, and the liability issues associated with their redevelopment, has made commercial/industrial and new residential development in New England's older cities extremely difficult, adding to the many factors that are pushing growth outside urban areas.

The automobile became the symbol of the age, replacing the railroad as the dominant mode of transportation. American agriculture was transformed and enhanced in productivity by new technologies: chemical fertilizers and pesticides, hybrid seeds, broad-scale irrigation, and mechanization. New England farmers could no longer compete in this world of large-scale industrial production and long-distance transport. The market for hay collapsed as cars and tractors replaced horses. Fruit and vegetable production declined with increasing land costs near cities and competition from refrigerated produce shipped in from large growers in warmer climates. Poultry production consolidated to enormous confinement facilities in other regions, and eventually—during the second half of the century—dairy farms went the same way. Consequently, the number of farms and amount of land in agricultural production declined steadily in New England through the 20th century [a mere 7% of the Massachusetts landscape is farmed today (Hall et al., 2002)]. As a result, New England farmland continued to return to forest, reaching a plateau about 1950, since which time the ongoing expansion of forest in the highlands has been roughly balanced by the loss of forest to development in the coastal lowlands (Fig. 2.4). Although older agricultural industries such as dairy farming continue to decline, recent decades have seen a resurgence in small-scale organic farming, often on land protected by the purchase of development rights. Although this trend has done little as of yet to stem the overall slow but steady loss of farmland, it does offer some hope of retaining a portion of the landscape within an economically viable agrarian tradition.

Suburban development accelerated when the Massachusetts economy experienced a resurgence in the mid 1970s. Fueled by a venture capitalist community working in collaboration with the universities and research centers in the Boston area, the economic base shifted to services and high-tech manufacturing. This economic resurgence and its associated landscape change took place in the eastern half of the state first, driving land speculation as it expanded westward.

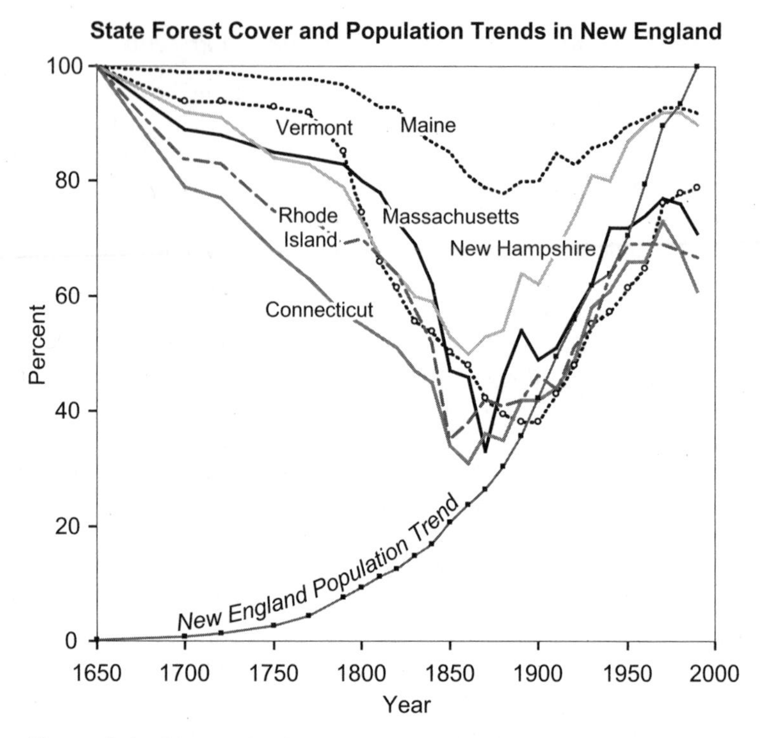

Figure 2.4 Changes in forest cover in each of the New England states and changes in population for New England as a whole.

Almost 90% of the high-tech growth in the 1980s was located in the triangle between Boston, Worcester, and Lawrence (Harrison and Kluver, 1989a). The new economy rekindled population growth, this time centered in suburban and periurban locations, filling in the spaces between old mill towns. During the 1980s, construction contracting grew twice as fast in the state as in the nation as a whole, and residential construction grew four times as fast, fueled by a high level of speculation in the housing market (Harrison and Kluver, 1989b). By the turn of the 20th century, this expansion had moved west of interstate highway 495, Boston's outer ring, making central Massachusetts one of the hottest real estate markets in the country and reawakening commuter trains from central Massachusetts to the greater Boston area.

The landscape consequences have been large. Periurban residential development, characterized by 2-acre residential lots located near major highways and junctions, is increasingly common to the east of Worcester. This pattern of development, which reflects a consumer preference for low-density housing near open space, has driven a decrease in forest cover and an increase in land fragmentation during the past few decades (Foster, 1993). From 1971 to 1999, 66,707 ha of forest and 11,648 ha of cropland were lost. During the same period, residential land use increased by 69,545 ha and commercial/industrial use increased by 12,028 ha.

The "industrial–service" economy transition of New England has had significant land-cover consequences for central Massachusetts and central New England, sustaining declines in agricultural land uses and reversing trends in forest regrowth established during the 19th- and early-20th-century industrial era. The post–World War II shift to a service–high-tech economy is linked to new "rural" pressures from expanding suburban and periurban settlement.

The past century and a half has seen a dramatic change in forest composition. Abandoned farmland is first dominated by early successional species such as white pine, cherry, birch, and red maple. Some of these species decline as the forest matures or encounters subsequent disturbances, but others persist for a long time. Meanwhile, some species common before the agricultural transformation, such as beech, have been very slow to recover. This has given rise to a forest not only very different from the frequently cut woodlands of the 19th century, but also distinct in composition from the forest of pre-European times. Suppression of fire in the landscape has favored some species, such as white pine and red maple, at the expense of others, such as pitch pine and white oak. Exotic pests such as chestnut blight and now hemlock wooly adelgid virtually eliminated some dominant trees from the forest canopy. Meanwhile, wood harvesting has declined steadily since the early 20th century, allowing the recovered forest to grow increasingly large and mature.

The practice of conservation has also evolved with the 20th century. During the 19th century, land use was largely determined by market forces and farmers' abilities to respond to those forces, as social constraints embodied in the older agrarian order dissolved. As a result, both farmland and forest were overused and degraded, in the short term. The conservation movement arose in response, with two main branches finding their roots in the figures of Henry Thoreau and George Perkins Marsh. The romantic branch emphasized the importance of a direct emotional connection with nature and appreciation of nature's wildness for itself, beyond human control or utility. The more utilitarian branch emphasized the sustainable use of resources through scientific management. These ideas have exerted considerable restraint on land use during the 20th century, beyond mere economic calculation. They have found expression in the acquisition of public forest and open space, the protection of wetlands, the promotion of sustainable practices in forestry and farming, and efforts to protect endangered species.

Today, conservation in New England faces formidable challenges in pressure for land conversion through suburban sprawl, as well as the continued influx of exotic pests, air pollution, and the likelihood of rapid climate change. But the ideals of conservation also face their own internal conflicts and contradictions, which might be alleviated by historical reflection. The romantic branch of conservation is plagued by what has been called the *illusion of preservation*, which begins with a broad misunderstanding of how "pristine" pre-European ecological systems functioned, and a simplistic wish to restore those supposedly stable conditions. A desire to "preserve" as much of New England as possible in a wild, unmanaged state is often combined with a refusal to examine seriously where the resources that support a rich material life come from. The suburban drive itself is an example of this illusion, being the impossible wish to live in an undeveloped rural landscape by means of industrial extraction, a large residential lot, and a long commute. But the utilitarian branch of conservation is equally guilty of what might be called the *illusion of management.* This can be defined as a chronically weak appreciation of the value of wild nature, combined with overconfidence in the ability of science to understand fully and control complex ecological systems, let alone the cultural forces that persist in influencing ecological development in unpredictable ways.

With an understanding of history, it is possible to imagine how these two attitudes could be combined to create the conditions for successful long-term conservation—especially if some agrarian virtues were revived as well. That is, there is room in New England for the setting aside of substantial wilderness "old-growth" preserves, the productive and sustainable management of forests that support a wide range of biodiversity, and the revival of an agrarian cultural landscape that provides habitat for open-land species, along with healthy and rewarding human engagement with the land (Foster, 2002; Foster et al., 2005). Such a landscape would be ecologically diverse and flexible, and deeply satisfying to many New Englanders. A better understanding and inculcation of our fascinating and convoluted environmental history might not only improve our ability to design and "manage" such a landscape in its ecological particulars, but also give us a better chance of inspiring the cultural ability to pull it off.

Twentieth-Century New England Forest History: Growth and Utilization

Today, Massachusetts is a forested land. These woods, though, are very different from the ones that dominated the landscape before European settlement. Although forests have displayed great resilience in the face of human impact and have reclaimed much of their former ground following agricultural abandonment, human use of the land, both historical and ongoing, has imposed strong spatial and temporal signatures on woodlands. In a dynamic fashion, people continue to react to changes in the forest, and their actions shape the forest of the future.

The Dynamic Extent of Forest

At the beginning of the 20th century, much of New England was dominated by a relatively young and rapidly growing forest as a consequence of either agricultural abandonment in southern and central parts of the region or large-scale clear-cutting during the 19th century. In many cases, the forests naturally established on old farmland in the mid 1800s were clear-cut at the onset of the 20th century. The net effect of these human activities was a thick, brushy woodland dominated by stump sprouts and rapidly growing species. Patches of more mature woodland were scattered across perhaps 20% of the landscape (Cogbill et al., 2002). A small subset of these forests had escaped direct human influence and were generally located in remote small stands of old-growth and virgin woods (Cogbill et al., 2002).

Forest area continued to increase until approximately 1960, at which point competing land uses progressively reversed this trend, although existing forest continued to grow in stature (Fig. 2.4). By the mid 1990s, despite considerable population and economic growth in much of New England, the prevailing land use remained forest (59% of Rhode Island, 60% of Connecticut, 62% of Massachusetts, 78% of Vermont, 84% of New Hampshire, and 90% of Maine). However, many of these woodlands are small and barely offer more than a shady area. Spatial analysis of all forest in Massachusetts indicates that if pieces or fragments of forest smaller than 10 acres are omitted, the total amount of forest declines by 2.5%. If pieces smaller than 50 acres are not considered, the total amount of forest declines by roughly 7%. The story is somewhat different in more heavily forested New Hampshire. Exclusion of all forest pieces smaller than 10 acres reduces the statewide area of forest by only 1%, and failure to consider pieces 50 acres or less reduces overall statewide area by only 2.5%. The Massachusetts Audubon Society estimates that the state loses 44 acres of undeveloped open space per day, resulting in an annual loss of more than 16,000 acres (Steel, 1999).

Forest conversion is only part of the story, however. Interior forest is that part of the landscape that is free of effects that penetrate in from adjacent nonforest conditions. Factors such as elevated light and temperature; invasive exotic plants; higher populations of "generalists" such as gray squirrels, raccoons, skunks, and blue jays; ranging house cats and dogs; and accumulated lawn debris and old Christmas trees all occur to a greater degree in the forest around developed lands. These "edge effects" alter predator–prey relationships and can negatively affect native plant communities. Some wildlife requires interior forest conditions free of these edge effects. In a landscape study of 19 rural towns in western Massachusetts, although total forest declined by only 2% between 1971 and 1985, the amount of interior forest declined by 6%, 12%, or 21%, depending on an edge effect distance of 400, 1,000, or 2,000 ft., respectively (Kittredge and Kittredge, 1999). Although Massachusetts is heavily forested, it also has the third highest population density in the nation. Human influences penetrate the woods and shape its size and extent.

Forest Change over Time

The structure of the woods has undergone great changes through time. In Massachusetts there has been substantial net accumulation of wood since the 1950s (Fig. 2.5). Since the late 1920s, forest structure has shifted from 80% in the seedling/sapling size class (Cook, 1929) to roughly 3% by 1998. Larger trees increasingly dominate the forest, and the brushy expanses of the early 20th century have been relegated to the history books (Hall et al., 2002). With individual forests accumulating on the order of 2 metric tons of carbon/ha/year, one globally important consequence of this forest growth is that New England (and much of the eastern United States) serves as a major sink for carbon and offsets some of the potential increase in atmospheric carbon dioxide (Barford et al., 2001; Pacala, 2001).

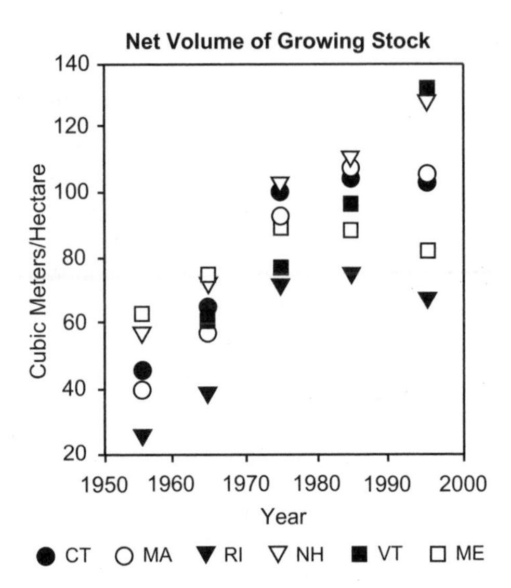

Figure 2.5 Net volume of growing stock (measured in m³/ha) of six New England states from 1950 to 2000. Based on data from USDA Forest Service (2002).

Removals/Utilization

At the beginning of the 20th century, an active timber industry produced more than one hundred million board feet (bd. ft.) of lumber annually for local secondary manufacturing (Steer, 1948). Much of the timber came from white pine stands that had established on old agricultural lands 50 to 70 years earlier. Harvesting focused on converting standing softwood timber to cash and operated without any long-term plan or silvicultural design. Lumber production peaked between 1900 and 1920, and dropped precipitously through the 1930s, when regional forests entered a period of regrowth after heavy utilization. This heavy cutting generated a shift in species composition. In southern and central New England, logging of white pine on old fields initiated a change back toward a more natural species composition of mixed hardwoods (e.g., oak, maple, birch, ash), hemlock, and white pine (Hall et al., 2002).

By the early 1970s, Forest Service inventories reported average annual growing stock removals of between 0.54 and 0.96 m^3/ha forest (Connecticut, 0.54 m^3/ha forest; Rhode Island, 0.58 m^3/ha forest; Massachusetts, 0.87 m^3/ha forest; Vermont, 0.75 m^3/ha forest; and New Hampshire, 0.96 m^3/ha forest), with Maine considerably higher at 1.69 m^3/ha forest. By the mid 1990s, annual removal rates had virtually doubled, indicating the maturation of the timber and emergence of commercially valuable diameter classes. Despite increased removal rates, the forest continues to accumulate wood. Average annual net growth exceeds removals in all New England states except Maine by a factor of between 27% (in New Hampshire) to more than 130% (in Vermont). A recent study in north-central Massachusetts indicates that from 1984 to 2001, approximately 1.5% of the forest was harvested annually, in a spatially random pattern of relatively small patches in which perhaps only 25% of the forest volume was removed (Kittredge et al.,

2003). This pattern of chronic, low-level disturbance is probably representative of harvesting across much of the Northeast. The forest-processing industries have undergone similar shifts through time

Ownership

Unlike many wooded regions of the United States that have large state and federal ownerships or extensive industrial ownerships, most of the New England forest is owned by nonindustrial private families, individuals, and nonprofit organizations.

One consequence of this ownership pattern is that management decisions concerning the forests lie collectively in the hands of hundreds of thousands of individuals. With few local to regional controls, each owner is making relatively independent and unconstrained decisions. As a result, the New England forest landscape is strongly susceptible to two transforming influences: land conversion to nonforested use, and parcelization into smaller ownership units. These processes strongly interact; as increasingly small parcels become difficult to manage, they often become prone to land conversion.

Over time there have been a growing number of individuals who are responsible for New England forest land. In Massachusetts the number of owners of nonindustrial private forestland increased from 103,900 in 1976 to 235,000 in 1985, whereas the total area of forest declined slightly. The long-term result has been a decline in average ownership size from 9.5 ha to 3.6 ha between 1976 and 1993 (Fig. 2.6). Meanwhile, the land is becoming more valuable for development (e.g., for residential use) and is increasing only slightly in value for its timber (Fig. 2.7). Only increasingly wealthy private owners may be able to resist or defy market trends and financial logic of selling, dividing, or converting their land to a developed use. As ownerships decrease in size, forest landscape functions and values such as wildlife habitat, hydrologic and nutrient cycling, and outdoor recreation potential are compromised.

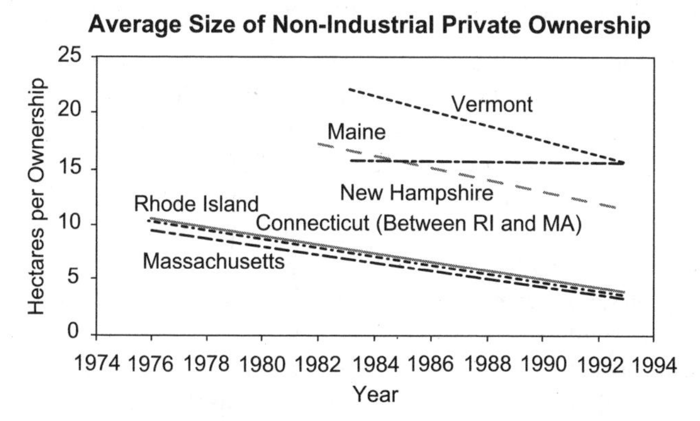

Figure 2.6 Average size of nonindustrial private ownership, by state and over time. Data modified from Kingsley (1976) and Birch (1996).

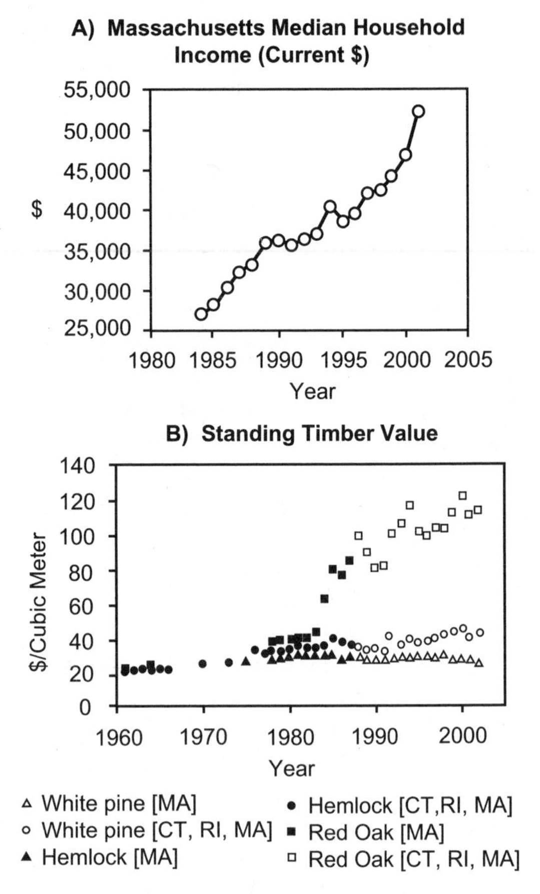

Figure 2.7 (A, B) Massachusetts median household income (A) and standing timber value (B) based on data from the U.S. Census Bureau, and SNESPR (Southern New England Stumpage Price) (2003).

Social/Human Responses to Perceived Threats and Utilization

The conservation context, major issues, and institutional setting for New England forests have evolved since the early 20th century. In the early 1900s, there were no environmental regulations in place to protect water quality or wildlife habitat, and the heavy harvest of softwood timber left large amounts of flammable material, resulting in increased fire frequency and intensity. As early successional brushfields increased on recently cut lands, public concern over forest degradation grew and manifested itself in several ways. Legislation was passed at the state and national levels to protect forests. The Weeks Act, passed in 1911, authorized the establishment of eastern national forests for purposes of watershed protection and resulted in the Green Mountain National Forest in Vermont and the White Mountain National Forest in New Hampshire and western Maine. In Massachusetts, the State Forest Commission was created in 1914 and was charged with acquiring land suitable for timber cultivation and forest reclamation (Rivers, 1998). Today, state forests cover a total of 211,000 acres. The Massachusetts

Forestry Association promoted the creation of local town forests throughout Massachusetts and other states, and in 1913 legislation was passed that authorized communities to own and manage their forests (McCullough, 1998). By the 1960s, 147 of 352 Massachusetts communities had town forests, totaling more than 43,000 acres (McCullough, 1998).

State legislation was also passed to protect forests through regulation. In 1914, the Massachusetts Slash Law required timber harvesters to leave slash in a condition that would not promote the spread of fire (Rivers, 1998). In 1945, the Forest Cutting Practices Act was passed in Massachusetts (Rivers, 1998).

In addition to public sector conservation actions, many private organizations were started in response to forest and environmental degradation. The Massachusetts Forestry Association began in 1897, in response to clear-cutting on Mount Greylock, the state's highest peak (Rivers, 1998), and similar associations were founded in other New England states (e.g., Connecticut Forestry Association in 1895, Society for the Protection of New Hampshire Forests in 1901, Forestry Association of Vermont in 1904) (King, 1998). The Massachusetts Audubon Society was founded in 1896, with an original interest in protecting birds and their habitat (Fox, 1998), and the New England Forestry Foundation was founded in 1944 to promote forestry to the thousands of nonindustrial private forest owners. The foundation sought to demonstrate good forestry practices and advocated the use of private consulting foresters to encourage sustainable management.

The early 20th century initiation of organizations and regulations to "protect forest" has evolved with newly perceived threats. During the past 20 years, distress over forest conversion to other uses has led to the strategic application of tools like easements to stem and focus the tide of widespread development and the overall loss of open space. Ironically, the perceived threat of forest destruction in the early 1900s was largely an illusion. History has shown forests to be resilient to hurricanes, fire, and harvest (Foster et al., 1997; Hall et al., 2002). In contrast, the development threat of the late 20th century is a new and different agent of change. State agencies and conservation groups have responded and are now buying private land, or the development rights to the property in the name of biodiversity (EOEA, 2001). Even the New England Forestry Foundation and conservation organizations like TNC actively pursue easements as a means to protect "working forest." Forest regulations have shifted from an emphasis on physical damage, to provision of tax breaks as incentives to landowners to retain their land in forest use (DEM, 2003). Although the New England forest has transformed greatly during the past century, so, too, have the human institutions and tools dedicated to its protection.

Future

Two parallel messages emerge from a 20th-century review of New England forest use and change: both the forest itself and corresponding human institutions have evolved in response to diverse stimuli. Perceived threats to forest have initiated public and private efforts at protection. These in turn have influenced forest trajectories. Public forests originally established to prevent deforestation are now oases from development and valued open spaces for recreation and revitalization. The large watershed of the Quabbin Reservoir was purchased to provide clean water for metropolitan Boston, but it currently serves as the largest and most intensively harvested public ownership in southern New England (Barten et al., 1998) and acts as an "accidental wilderness" (Conuel, 1990) that provides a diverse landscape for wildlife, natural processes, and human leisure. Land protection activities have curtailed the development rights on an increasing amount of private forest, and regulations have influenced harvest practices.

Future forest and societal interactions are conjectural because of the complexity of private landowner patterns and the unpredictability of social and environmental concerns. It is unlikely that governments will be able to buy a significant portion of this land, although there may be better prospects for the purchase of development rights. Furthermore, it is improbable that land-use zoning and regulation, implemented at the local level by hundreds of different volunteer boards in an environment that favors private rights, will stem the tide of forest conversion in the face of increasing population, affluence, and mobility. The future of the forest will result from the cumulative actions (or inactions) of tens of thousands of individuals, whose decisions in turn will be influenced by their relative affluence, seemingly unrelated external economic factors, and their attitudes about their land. Only if these grassroots inclinations can be marshaled into a regionwide passion for land conservation will it be possible to retain a majority of New England forest on the land (Foster et al., 2005).

Studies of private owner attitudes in New England reveal that aesthetics, wildlife habitat, recreation, privacy, and a place to live are overriding goals for ownership that far outweigh timber revenue (e.g., Belin, 2002; Birch, 1996; Kingsley, 1976; Rickenbach et al., 1998). Finley and Kittredge (2006) indicate that landowners do harvest, but at a relatively moderate intensity level. This chronic, and random disturbance may homogenize forest conditions and species composition (Cogbill et al., 2002; Fuller et al., 1998), and may create forests dominated by red maple, which commonly increases with light harvest (Abrams, 1998).

The future extent, distribution, and composition of the forest will result in large part from the decisions private owners make about their land. Decisions about land protection made by agencies and conservation organizations will also exert a lasting effect, especially on the character of large blocks of forest. The success of these groups depends in part on the way they approach private owners, and the receptivity of the message. Golodetz and Foster (1997) showed that land protection during the 20th century in central Massachusetts was haphazard and opportunistic. Conservation can have a more significant impact, and be more cost-effective, if these measures are applied in a more strategic and meaningful way (Forman 1995).

How the New England forest transforms in the coming century will depend on the pressures brought to bear by an increasing human population, a variety of external ecological factors such as climate change and invasive species, and the cumulative effects of an increasingly large and dynamic population of private owners.

The Ecological Consequences of New England's History

For more than 300 years, changing land-use activities have interacted with natural environmental and disturbance processes to generate major changes in the New England landscape. In the modern landscape, the legacies of this history shape ecological patterns and processes, condition ecosystem response to ongoing and future disturbances and stresses, underlie current dynamics in plant and animal populations, and provide the environmental context for ongoing human activities. More important, the nature of these historical effects on ecological patterns varies with spatial or ecological scale. For example, at any scale the composition and structure of modern vegetation are substantially different than they were before European arrival. However, at some spatial scales, including the site scale and subregional scale, land-use history has served to homogenize the natural patterns, whereas at a landscape scale a more patchy and heterogeneous condition has resulted.

Vegetation Response to Land-Use History

Across New England, the intensive utilization of wood products, coupled with the history of deforestation, reforestation, and other anthropogenic impacts, including burning and drainage, have produced major changes in forest composition and structure when compared with the forests first encountered by Europeans. Although forests per se have been highly resilient to repeated intense disturbances, there has been a shift toward more fast-growing, early successional, and shade-intolerant species (e.g., paper birch, red maple, white pine, and formerly chestnut) and a decline in long-lived mature-forest species (e.g., beech, hemlock, spruce, and yellow birch). Structurally, there have been accompanying declines in average stand age and an increase in the extent of even-age forests. Across the region, this substantial shift in species composition apparently has not been associated with major changes in species distributions or the geography of major forest zones. At a regional scale, major vegetation patterns are largely controlled by strong, broad-scale climatic variation, tied to latitude and elevation, and the basic patterns witnessed hundreds of years ago appear to hold in the modern landscape. Thus, in a consideration of the presettlement forests, Cogbill et al. (2002) identified a regional pattern composed of distinct distributions of northern and southern tree species that met in a fairly well-defined tension zone stretching across north-central Massachusetts. In their interpretation, this tension zone has been relatively stable through to the present despite notable shifts in temperature (+1.40 °C) and the intervening period of land-use history.

In contrast, at the subregional scale, for example across the Connecticut Valley to upland transition in central New England, it appears that the striking broad-scale variation in forest composition at the time of settlement, with oaks predominating at lower, southern localities and beech and hemlock dominating in the cooler northern higher areas, has disappeared as a consequence of land-use history. In this case, vegetation patterns were associated with fairly subtle climatic gradients. The imposition of a broadly similar land-use history across this subregion served both to shift and to homogenize forest composition. Across the region today, red maple, oak, birch, pine, and hemlock are distributed in a rather uniform pattern. Although species abundances exhibited strong, significant relationships with climate variation at the time of settlement, only a few weak relationships are detected in current distributions. One complicating, though intriguing, factor in this history is that the trend toward a more regionally homogeneous forest composition was actually initiated by climate changes approximately 550 years ago (Fuller et al., 1998). Thus, the observed shifts are evidently a consequence of the interaction between environmental change and human activity.

At a landscape scale, across hills, valleys, and township-size (i.e., 10 × 10 km) areas, land-use history has clearly created a more patchy and heterogeneous pattern of vegetation. On this scale, abrupt shifts in vegetation cover, species composition, and forest age and size are strongly tied to such factors as the specific land-use type (e.g., woodlot, pasture, tillage), the date of field abandonment or last harvesting, and the specific details of recent use of the area. The result is a complicated landscape mosaic in which transitions from hardwood forests to white pine or mature forest to sprout hardwoods are abrupt and only generally tied to edaphic conditions. At this scale, vegetation analyses indicate that land-use impacts convey an enduring effect on species distributions and community characteristics (Donohue et al., 2000; Motzkin et al. 1996, 1999a,b, 2002). Studies of soil properties demonstrate a similar lasting imprint of land use on such features as soil structure and carbon and nitrogen content (Compton and Boone, 2000, 2004; Compton et al., 1998).

There has been little work done to evaluate the consequences of land-use history at the scale of an individual stand or forest, but it is likely that within a given area of consistent vegetation, historical activities have simplified and homogenized soil conditions, vegetation patterns, and microenvironmental conditions. Although pre-European conditions would have reflected the complexities of subtle edaphic variation and prior disturbance by windthrow, animals, and other factors, most land-use activities serve to create uniform within-stand conditions. Forest clearance, tillage, pasturing, and fencing all impose fairly uniform treatments on areas frequently delimited by stone walls, other fencing systems, or property boundaries. Whether the impact was homogenization of the upper soil layers by plowing or removal of coarse woody debris by fire and decomposition, the tendency toward more consistent conditions in the individual patches within the patchwork of agrarian sites would be passed on to the modern forested landscape.

Discussion

Consequences of Land-Use History for Conservation

One of the major ecological lessons that inevitably emerges from a long-term perspective is that natural ecosystems are inherently dynamic (Davis, 1986; Whitlock and Bartlein, 1997). More important, however, a variety of studies that have assessed vegetation change over very long timescales have found that the rates of compositional change for both terrestrial and aquatic ecosystems have been, and presumably will continue to be, greater since European settlement than at any time since the last Ice Age (Foster and Zebryk, 1993; Fuller et al., 1998; Jacobson et al., 1987). In prehistory, climate and associated environmental change, as well as disturbance by pathogens, wind, ice storms, fire, and Native Americans, produced changes in vegetation composition and presumably in its structure and pattern as well. Plants and animals responded individualistically, rather than in any concerted group response to the unique combinations of environments and biotic factors that resulted. Consequently, as recently noted by Lawton (1997), although there is a relatively long fossil record of stability in the morphology of most individual plant species, the actual combinations and assemblages of species that form communities, ecosystems, and landscape patterns have no record of long-term coherency. The massive and very rapid change in land cover and land-use practices within the recent 300 to 400 years in the eastern United States have accelerated this process of natural change and recombination (Jacobson et al., 1987). The result has been a series of very transient assemblages derived from a relatively constant regional flora (Fuller et al., 1998; McLachlan et al., 1998)

One major question that concerns the reforested northeastern landscape is whether these "new" forests are similar in composition to those that occupied the same land areas at the time of European settlement (Raup, 1964; Whitney, 1994). A range of studies across New England suggest that modern plant (and animal) assemblages in upland, wetland, and lake ecosystems are historically anomalous, differing from those of four centuries earlier (Engstrom et al., 1985; Patterson and Backman, 1988). Not only do the modern groupings of species show little resemblance to their antecedents, they also show little tendency to revert in that direction as time passes and forests mature (Fuller et al., 1998). At a regional scale, for example across central Massachusetts, the forests that have formed after agricultural abandonment are remarkably more homogeneous than those of four centuries earlier, and they include more sprouting and shade-intolerant species and fewer long-lived mature-forest tree species (Foster et al., 1998b). Modern forests also exhibit much weaker relationships to regional variation in physiography, climate, and soils. At a landscape scale, the arrangement and structural and compositional characteristics of plant communities are largely the consequence of species-specific response to land-use histories and edaphic factors (Foster, 1995; Motzkin et al., 1996, 1999a). At a stand level, it has been possible to use the analysis of pollen from soils and small topographic depressions to interpret vegetation composition and disturbance histories over many centuries or even millennia, and thereby assess the extent of change (Bradshaw and Miller, 1988; Foster and Zebryk, 1993; Foster et al., 1996; McLachlan et al., 1998). Although limited in number in New England, these studies confirm that even the least disturbed sites, for example forests that were cut early during colonial history but never cleared, have been dramatically changed by human disturbance. Thus, these sites have often supported two or three distinctly different types of vegetation during the past 350 years, and the current forests generally bear little compositional resemblance to those that occupied the area when the land was first settled by Europeans.

These conclusions concerning the historical rates and types of vegetation change have many ramifications for conservation biology and the development of management policies. Primary among these is the recognition that there are no static baseline conditions that exist or have existed for comparison with current conditions or for use as a target for restoration activities. Ecologists, conservation biologists, and the public frequently use the pre-European period as a convenient benchmark for comparison with modern or historical conditions (as we have done earlier); however, this period was clearly characterized by change and flux in forest composition and structure, even if less dramatic than in the recent past. As we interpret modern landscapes or evaluate restoration and management approaches, we therefore need to recognize that forests have always been dynamic and that there is no single, ideal state to which forests should be restored. Nature changes and frequently people are a factor in this change. Thus, in our search for goals and objectives in conservation management, we should not be thinking of saving or restoring static examples of what nature is, was, or should be. These are transient entities, unreal concepts, and futile objectives (Foster et al., 1996).

The Inevitability of Future Change

The extent of human disturbance coupled with the ongoing change in global environments (climate, atmospheric composition, disturbance regimes, biota) result in the inevitability of future change in all landscapes. As a consequence, land managers of all types need to acknowledge change and anticipate future dynamics. Most New England landscapes are still in the process of recovering from past land-use activity while also responding to new changes in the physical or biotic environment, ranging from subtle stresses associated with changes in the atmospheric concentrations of nitrogen and carbon dioxide to defoliation by the gypsy moth and hemlock woolly adelgid, an insect species introduced from Asia in the 1920s (Aber, 1993; Bazzaz, 1996; Foster et al., 1997; Orwig and Foster, 1998). Hemlocks are the third most prevalent tree species in the New England forests, and they provide food for deer and protection from soil erosion along streams and rivers.

Paleoecological studies (e.g., Foster and Zebryk, 1993) and modeling approaches (Pacala et al., 1996) suggest that forest stands may take up to 500 years to recover from moderately severe disturbances such as fire, pathogens, harmful introduced species, or substantial cutting and thus we should anticipate that all vegetation, even if effectively protected from recent or future disturbance, will continue to change as it adjusts to its history of past impacts. In addition, natural disturbance and natural variation in the environment will inevitably promote future, unexpected dynamics in all ecosystems.

Wildlife Dynamics and Feedback in Perspective

Recent changes in many wildlife populations constitute one important, although often underappreciated, component of landscape change in the eastern United States that has strong implications for conservationists, natural resource managers,

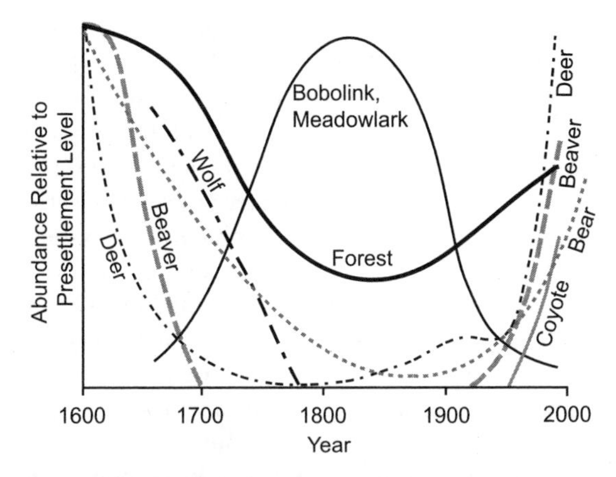

Figure 2.8 Schematic depiction of the historical changes in representative wildlife species and forest cover through time in New England. Although the wolf has been eliminated, open-field species like the bobolink and meadow lark peaked in abundance during the 19th-century period of open agriculture; the coyote is a new species in the landscape; and the deer, beaver, and bear have recovered greatly since elimination or very low historical abundance. Modified from O'Keefe and Foster (1998) and Bickford and Dymon (1990).

and many residents. As a consequence of historical variation in the relative extent and type of land cover, along with cultural and economic changes that have encouraged conservation, wildlife introductions, and regulation of hunting, the New England region is undergoing a major transformation in wildlife abundance and composition (DeGraaf and Miller, 1996; Fisher, 1929; Hosley, 1935). Although many large mammals and forest birds were eliminated during the 17th to 19th centuries and were replaced by the open-land species of meadows, fields, and shrublands, we are currently witnessing a major resurgence of native woodland species (Fig. 2.8). Some of these animals have been resident throughout the historical period and are simply expanding from small residual populations, others were locally or regionally eliminated and are immigrating from northern and western regions, whereas others have appeared as a consequence of successful programs of reintroduction. Although many of these species were important throughout the Northeast at the time of European settlement, others, such as the coyote, were originally native in other parts of the country and represent new arrivals that have been able to capitalize on changing landscape conditions and the absence of competitors or predators such as the wolf.

The recent increase in woodland species is often heralded, quite rightly, as an environmental success, but the burgeoning populations of woodland wildlife are also bringing many unexpected and occasionally undesired consequences to the landscape and to the largely suburban human population of New England. Beavers impound water creating wetlands, killing trees, and flooding roads, yards, and sewage systems while also producing important habitats for other animals and plants that utilize the resulting ponds, wetlands, and dead trees. There is evidence that the population of beavers in some regions may be surpassing presettlement levels and flooding areas such as old-growth swamp forests that have not been inundated in the previous 300 to 400 years (P. Lyons and B. Spencer, personal communication, 2000). Deer, largely unchecked by hunting or major predators in many suburban and even rural areas, may impede forest regeneration, browse ornamental and vegetable gardens, and create automobile hazards. Larger mammals such as moose and bear create even greater problems because of their potential for major impact on human safety as well as natural ecosystems. Each of these wildlife species and scenarios presents natural resource agencies and landowners with major control problems and generates ethical dilemmas for society.

The new populations of wildlife may also have unanticipated impacts on other species. Pileated woodpeckers, which have increased partly as a consequence of the greater availability of extensive forest areas and nest sites in large dead trees, create large bole cavities that shelter other animals. The swamps, ponds, and meadows that alternate through the cycle of beaver damming and abandonment provide a highly dynamic environment that presents considerable heterogeneity within a largely forested landscape. The extensive stands of dead trees produced by this flooding also provide nest sites, resulting in the large heron rookeries that have reappeared across the New England states.

Human health may also be indirectly affected by the changes in land-use, land-cover, and wildlife dynamics. The increase and spread of lyme disease has resulted, in part, from the increase in mouse and deer populations during the past several decades; a similar connection is noted between the incidence of *Giardia* (a parasitic protozoan) and beaver and other mammal populations in New England. A historical perspective on these wildlife dynamics is necessary to understand them and to anticipate how they may change in the future. Such a perspective is also extremely useful in educating a human population that is increasingly separated from nature about the changes that are occurring throughout the landscape, including their own backyard. Clearly a better understanding of these dynamics and their causes improves the ability of natural resource managers and conservationists to manage them and their consequences.

New England Is a Cultural Landscape

An evaluation of the past and current dynamics of the northeastern United States suggests that we must embrace wholeheartedly and realistically the notion that we live in a cultural landscape that is shaped in a broad pattern and controlled in fine detail in part by a history of human impacts. Recognition of this fact helps us to appreciate that humans have been, and still are, a major force and part of the functioning ecosystems that we call nature. It also helps us to shed the notion that we can somehow preserve or restore a nature independent of human history. On a regional to landscape scale, many habitats have been selectively eliminated or converted to some new status. Wetlands have been drained on a widespread basis on inland as well as coastal sites, and changes in local hydrology have left us with distinctly different habitats and vegetation cover than have occurred historically (Tiner, 1988). Across New England upland areas, sites such as the level sand plains that occupy outwash and deltaic deposits have been extensively converted to industrial and commercial activities, airfields, and landfills (Motzkin et al., 1999a). This selective habitat destruction, along with selective elimination of species, leaves us with a highly altered landscape representation of plant and animal communities. At the same time, the history of land use has increased the abundance and importance of many species and ecosystems, such as open-land and weedy taxa.

On a regional scale, the intensity and type of disturbance that has occurred is highly variable, and therefore conservation issues and priorities may vary geographically. For example, across northern New England, a history of logging, large ownerships, low population density, and relatively intact forest cover has led to a recent emphasis on the preservation of continuous, older forest; the reintroduction of large native animals; and the reestablishment of forest processes typical of large, intact ecosystems (Dobbs and Ober, 1995). In contrast, in southern and central New England, although many of these same values are embraced, the history of intense agriculture has been extremely important for the generation of a landscape of open fields and highly fragmented forests. This, in turn, has led to increased focus on rare species in limited habitats and on the maintenance of many open-habitat plants, animals, and landscapes (Dunwiddie, 1992).

Recognition of the selective creation and destruction of habitats and the tendency toward change forces us to acknowledge that the maintenance of species and habitats that were common 50 to 100 years ago will require active management either by encouraging or subsidizing historical practices such as agriculture, or by replacing them with other management regimes (Birks et al., 1988; Dunwiddie, 1992; Foster and Motzkin, 1998). In some cases, such as the conservation of open-land species of plants and animals, we may need to maintain cultural artifacts or legacies that were much less common or even absent from the landscape 300 years ago. Perhaps the best example of this phenomenon lies in the efforts to conserve grassland and shrubland habitat and species across southern and central New England and other parts of the Northeast (Foster and Motzkin, 2003). Presumably, before European arrival, few areas were large enough to maintain some of the highly restricted and rare bird species that are the current focus of major conservation efforts but that may require more than 50 ha of open grassland for the maintenance of successful populations. However, as a consequence of forest cutting and agricultural activities such as burning, plowing, planting, and grazing, extensive upland grasslands, freshwater meadows, shrublands, and heathlands were created during the 18th and 19th centuries, resulting in a dramatic increase in open-land wildlife and plants (DeGraaf and Miller, 1996; Dunwiddie, 1989). The prominence of these species is quite clear in contemporary descriptions, such as the journal notations from Henry Thoreau in which bobolinks, meadowlarks, and song sparrows are described as common (Foster, 2001). Currently, many of these taxa are in decline, which presents an interesting dilemma to land managers and conservationists who are faced with the challenge of restricting tree invasion and growth on open lands, and are confronted by the basic question of whether such cultural landscapes should be conserved.

Should these uncommon and presumably historically rare or nonnative taxa be allowed to go locally extinct as the extent of forest land increases and agriculture declines? Or should we expend increasing effort on their maintenance, based on the notion that some of these species may be native, that many have become an important or characteristic part of the landscape, and that others may be threatened elsewhere in their range? During the past decades, grassland and shrubland taxa have emerged as a major priority for conservation organizations such as TNC, Massachusetts Audubon Society, and state natural heritage programs, which are seeking to conserve such birds as the grasshopper sparrow, upland sandpiper, and meadowlark through management programs based on burning, mowing, and grazing (Scheller, 1994; Sharp, 1994). Ironically, one of the most effective protection strategies for these species has been for conservationists to work with managers of highly artificial cultural landscapes to maintain appropriate habitat. The list of top sites for open-land bird species in Massachusetts provides an indication of the precarious status of these species and the surprising nature of the remaining "prime" habitat (Jones and Vickery, 1993). Among these sites, eight are commercial airports or military air bases, one is a landfill, one is a drained cedar swamp that was converted to grassland for agricultural and industrial purposes, one is a military training ground, and only two are in a seminatural condition, albeit one that is strongly shaped by historical land use. A historical perspective reveals the landscape dynamics that enabled the development of these habitats and wildlife assemblages. It also allows us to make the conscious decision, as has been done throughout northwestern Europe, that there may be great value in maintaining diverse cultural landscapes and the aesthetic and biological qualities that they support (Birks et al., 1988; Peterken, 2003).

Consequences of the Enduring Legacies of History

Land use, like other disturbance processes, can generate legacies in terms of ecosystem structure, composition, or function that are not easily erased or changed through time or even through subsequent disturbance (Foster et al., 1998a). As a consequence, it is often erroneous to conclude that the adoption of a new management regime, even one that follows the presumed natural disturbance or environmental regime, will necessarily lead to the re-creation of "natural" conditions or the vegetation structure and composition that might have developed in the past as a result of such disturbance (Seymour et al., 2002). For example, in the study of pitch pine and scrub oak vegetation on sand plains in the Connecticut River Valley, Motzkin et al. (1996) documented that the single most important factor controlling many aspects of the modern vegetation and site conditions was the legacy of different land use across these areas. Modern soil features, such as the presence of a "plow" horizon and vegetation characteristics, including species composition and structure, reflected the prior site history even 50 to 100 years after the land-use activity ceased and despite a history of subsequent disturbance by fire. Other studies have shown a similar pattern of persistence of historical legacies in the face of hurricane impacts or other disturbances (Foster, 1993). These observations suggest that even though many management regimes are prescribed for natural areas in an effort to increase their natural character, such as prescribed burning in pine-, oak-, or grassland-dominated landscapes, the vegetation may actually be slow to respond to such disturbances or may change in unexpected and even undesirable ways (Niering and Dreyer, 1989; Patterson and Backman, 1988). The outcome of such management may not be an enhancement of "original" attributes of the area (Motzkin et al., 1999a), although it may contribute to other objectives, such as the maintenance of rare species habitat and regional biodiversity.

Conclusion and Recommendations

Much of the eastern United States has witnessed an increase in forest cover in this century as a result of a reduction in agricultural activity and natural resource extraction, presenting new opportunities and challenges for conservation planning. These changes in land-use practice have resulted from the fact that the food, energy, building materials, and natural resources for the region are no longer obtained primarily from our local landscapes, but are derived instead from highly distributed global sources. Consequently, although regions like New England are experiencing population growth and historically high levels of residential expansion, they have also reverted to a more natural condition with more extensive cover of maturing forests and more native fauna than at any time during the previous 200 years.

The rapidity and extent of change, the ongoing dynamics in the landscape resulting from recovery from prior land use as well as ongoing impacts, and the enduring legacy of past land use necessitate that historical perspectives become an essential part of all ecological study and an important basis for the development of conservation strategies. Using these perspectives we can recognize the inevitability of change and the cultural imprint on most landscapes and on many seemingly natural features. We can also recognize that many plant communities and landscapes that are of great conservation value are actually novel, highly humanized, and of recent development. As we understand the transitory and highly cultural origins of many parts of our land, we can also appreciate the relative roles of science versus social values in determining policy and management objectives. Using both historical and ecological science we can interpret and understand change, monitor and evaluate conditions and processes, and develop and inform management techniques. Ultimately, however, the decision of what we conserve or restore lies in the cultural values that we bring to this decision-making process. Thus, in New England we can retain a cultural landscape of fields and forests that support open-land and edge species, or we can allow a culturally derived forest to develop and age and harbor forest interior species. Science does not give us absolute guidelines for making these decisions, but it does inform us that either decision will produce a new landscape with a history that includes people and that is characterized by features that are not original or pristine but are constantly undergoing change.

References

Aber, J. D. 1993. "Modification of nitrogen cycling at the regional scale: The subtle effects of an atmospheric deposition," pp. 163–174. In: M. J. M. McDonnell and S. T. A. Pickett (eds.), *Humans as components of ecosystems*. New York: Springer-Verlag.

Abrams, M. D. 1998. "The red maple paradox: What explains the widespread expansion of red maple in eastern forests?" *Bioscience* 48: 355–364.

Balk, H. H. 1944. "The expansion of Worcester and its effects on surrounding towns." PhD diss., Clark University, Worcester, Mass.

Barford, C. C., S. C. Wofsy, M. L. Goulden, J. W. Munger, E. Hammond-Pyle, S. P. Urbanski, L. Hutyra, S. R. Saleska, D. Fitzjarrald, and K. Moore. 2001. "Factors controlling long- and short-term sequestration of atmospheric CO_2 in a mid-latitude forest." *Science* 294(5547): 1688–1691.

Barron, H. S. 1988. *Those who stayed behind: Rural society in nineteenth-century New England.* New York: Cambridge University Press.

Barten, P. K., T. Kyker-Snowman, P. J. Lyons, T. Mahlstedt, R. O'Connor, and B. A. Spencer. 1998. "Managing a watershed protection forest." *Journal of Forestry* 96: 10–15.

Bazzaz, F. A. 1996. *Plants in changing environments: Linking physiological, population, and community ecology.* Cambridge: Cambridge University Press.

Belin, D. 2002. "Assessing private landowner attitudes: A case study of New England NIPF owners." Masters thesis, University of Massachusetts, Amherst, Mass.

Bell, M. M. 1989. "Did New England go downhill?" *Geographical Review* 79: 451–467.

Bernardos, D., D. R. Foster, G. Motzkin, and J. Cardoza. 2004. "Wildlife dynamics in the changing New England landscape," pp. 142–168. In: D. Foster and J. Aber (eds.), *Forests in time: The environmental consequences of 1000 years of change in New England.* New Haven, Conn.: Yale University Press.

Bernstein, D. J. 1999. "Prehistoric use of plant foods on Long Island and Block Island Sounds," pp. 101–119. In: J. P. Hart (ed.), *Current northeast paleoethnobotany,* vol. 494. Albany, N.Y.: New York State Museum Bulletin.

Bickford, W., and U. J. Dymon (eds.). 1990. *An atlas of Massachusetts river systems.* Westfield, Mass.: Massachusetts Division of Fisheries, Wildlife and Environmental Law Enforcement.

Birch, T. W. 1996. "Private forestland owners of the northern United States, 1994." *USDA Forest Service Resource Bulletin,* NE-136: 293.

Birks, H. H., H. J. B. Birks, P. E. Kaland, and D. F. Moe (eds.). 1988. *The cultural landscape: Past, present and future.* Cambridge: Cambridge University Press.

Black, J. D. 1950. *The rural economy of New England: A regional study.* Cambridge, Mass.: Harvard University Press.

Bradshaw, R., and N. Miller. 1988. "Recent successional processes investigated by pollen analysis of closed-canopy forest sites." *Vegetation* 76: 45–54.

Chilton, E. S. 1999. "Mobile farmers of pre-contact southern New England: The archaeological and ethnohistoric evidence," pp. 157–176. In: John P. Hart (ed.), *Current Northeast Paleoethnobotany.* New York State Museum Bulletin, no. 494.

Chilton, E. S. 2002. " 'Towns they have none': Diverse subsistence and settlement strategies in Native New England," pp. 289–300. In: J. Hart and C. Reith (eds.), *Northeast subsistence-settlement change: A.D. 700–A.D. 1300.* New York State Museum Bulletin, no. 496.

Chilton, E. S. 2005. "Social complexity in New England: AD 1000–1600," pp. 138–160. In: Timothy Pauketat and Diana Loren (eds.), *North American archaeology.* Malden, Mass.: Blackwell Press, Studies in Global Archaeology Series.

Chilton, E. S., and D. L. Doucette. 2002. "The archaeology of coastal New England: The view from Martha's Vineyard." *Northeast Anthropology* 64: 55–66.

Chilton, E. S., T. B. Largy, and K. Curran. 2000. "Evidence for prehistoric maize horticulture at the Pine Hill Site, Deerfield, Massachusetts." *Northeast Anthropology* 59: 23–46.

Cogbill, C. V., J. Burk, and G. Motzkin. 2002. "The forests of presettlement New England, USA: Spatial and compositional patterns based on town proprietor surveys." *Journal of Biogeography* 29: 1279–1304.

Compton, J. E., and R. Boone. 2000. "Long-term impacts of agriculture on soil carbon and nitrogen in New England forests." *Ecology* 81: 2314–2330.

Compton, J., and R. Boone. 2004. "Land-use legacies on soil properties and nutrients," pp. 189–201. In: D. Foster and J. Aber (eds.), *Forests in time: The environmental consequences of 1000 years of change in New England.* New Haven, Conn.: Yale University Press.

Compton, J. E., R. D. Boone, G. Motzkin, and D. R. Foster. 1998. "Soil carbon and nitrogen in a pine-oak sand plain in central Massachusetts: Role of vegetation and land-use history." *Oecologia* 116: 536–542.

Conuel, T. 1990. *Quabbin: The accidental wilderness.* Amherst, Mass.: University of Massachusetts Press.

Cook, H. O. 1929. "A forest survey of Massachusetts." *Journal of Forestry* 27(5): 518–522.

Cronon, W. 1983. *Changes in the land: Indians, colonists, and the ecology of New England.* New York: Hill and Wang.

Cumbler, J. T. 2004. *Reasonable use: The people, the environment, and the state, New England 1790–1930.* Oxford: Oxford University Press.

Davis, M. B. 1986. "Climatic instability, time lags and community equilibrium," pp. 269–284. In: W. Case and J. Diamond (eds.), *Community ecology.* New York: Springer-Verlag.

DeGraaf, R., and R. Miller. 1996. *Conservation of faunal diversity in forested habitats.* London: Chapman Hall.

DEM. 2003. Massachusetts Department of Environmental Management. Online. Available at http://www.mass.gov/legis/laws/mgl/gl-61-toc.htm. Accessed 1/15/2008.

Dincauze, D. F. 1990. "A capsule prehistory of southern New England," pp. 19–32. In: L. M. Hauptman and J. D. Wherry (eds.), *The Pequots in southern New England*, vol. 198. The Civilization of the American Indian Series. Norman, Okla.: University of Oklahoma Press.

Dincauze, D. F., and V. Jacobson. 2001. "The birds of summer: Lakeside routes into late Pleistocene New England." *Canadian Journal of Archaeology* 25: 121–126.

Dobbs, D., and R. Ober. 1995. *The northern forest.* White River Junction, Vt.: Chelsea Green.

Donahue, B. 1997. "Damned at both ends and cursed in the middle: The 'flowage' of the Concord River meadows, 1798–1862," pp. 227–242. In: Char Miller and Hal Rothman (eds.), *Out of the woods: essays in environmental history.* Pittsburgh, Pa.: University of Pittsburgh Press.

Donahue, B. 1999. *Reclaiming the commons: Community farms and forests in a New England town.* New Haven, Conn.: Yale University Press.

Donahue, B. 2004. *The great meadow: Farmers and the land in colonial concord.* New Haven, Conn.: Yale University Press.

Donohue, K., D. R. Foster, and G. Motzkin. 2000. "Effects of the past and the present on species distributions: Land-use history and demography of wintergreen." *Journal of Ecology* 88: 14–15, 303–316.

Dunwiddie, P. W. 1989. "Forest and heath: The shaping of the vegetation on Nantucket Island." *Journal of Forest History* 33: 126–133.

Dunwiddie, P. W. 1992. *Changing landscapes: A pictorial field guide to a century of change on Nantucket.* Nantucket, Mass.: Nantucket Conservation Foundation and Nantucket Historical Association.

Engstrom, D. R., E. B. Swain, and J. C. Kingston. 1985. "A paleolimnological record of human disturbance from Harvey's Lake, Vermont: Geochemistry, pigments and diatoms." *Freshwater Biology* 15: 261–288.

EOEA. 2001. *Biomap: Guiding land conservation for biodiversity in Massachusetts.* Westboro, Mass.: Executive Office of Environmental Affairs, Division of Fisheries and Wildlife.

Finley, A. O., and D. B. Kittredge, Jr. 2006. "Thoreau, Muir, and Jane Doe: Different types of private forest owners need different kinds of forest management." *Northern Journal of Applied Forestry* 23: 27–34.

Fisher, R. T. 1929. "Our wildlife and the changing forest." *The Sportsman* March: 35–46.

Forman, R. T. T. 1995. *Land mosaics: The ecology of landscapes and regions.* Cambridge: Cambridge University Press.

Foster, D. R. 1993. "Land-use history (1730–1990) and vegetation dynamics in central New England, USA." *Journal of Ecology* 80(4): 753–771.

Foster, D. R. 1995. "Land-use history and four hundred years of vegetation change in New England," pp. 253–319. In: B. L. Turner (ed.), *Principles, patterns and processes of land use change: Some legacies of the Columbian encounter.* New York: John Wiley and Sons, SCOPE Publication.

Foster, D. R. 2001. *Thoreau's country: Journey through a transformed landscape.* Cambridge, Mass.: Harvard University Press.

Foster, D. R. 2002. "Conservation issues and approaches for dynamic cultural landscapes." *Journal of Biogeography* 29: 1533–1535.

Foster, D. R., J. D. Aber, J. M. Melillo, R. Bowden, and F. Bazzaz. 1997. "Forest response to disturbance and anthropogenic stress: Rethinking the 1938 hurricane and the impact of physical disturbance vs chemical and climate stress on forest ecosystems." *BioScience* 47: 437–445.

Foster, D., D. Kittredge, B. Donahue, G. Motzkin, D. Orwig, A. Ellison, B. Hall, B. Colburn, and A. D'Amato. 2005. *Wildlands and woodlands: A vision for the forests of Massachusetts.* Petersham, Mass.: Harvard University.

Foster, D. R., D. Knight, and J. Franklin. 1998a. "Landscape patterns and legacies resulting from large infrequent forest disturbance." *Ecosystems* 1: 497–510.

Foster, D. R., and G. Motzkin. 1998. "Ecology and conservation in the cultural landscape of New England: Lessons from nature's history." *Northeastern Naturalist* 5: 111–126.

Foster, D. R., and G. Motzkin. 2003. Interpreting and conserving the openland habitats of coastal New England: Insights from landscape history. *Forest Ecology and Management* 185: 127–150.

Foster, D. R., G. Motzkin, and B. Slater. 1998b. "Land-use history as long-term broad-scale disturbance: Regional forest dynamics in central New England." *Ecosystems* 1: 96–119.

Foster, D. R., D. A. Orwig, and J. S. McLachlan. 1996. "Ecological and conservation insights from reconstructive studies of temperate old-growth forests." *Trends in Ecology and Evolution* 11: 419–424.

Foster, D. R., F. Swanson, J. Aber, I. Burke, D. Tilman, N. Brokaw, and A. Knapp. 2002. "The importance of land-use and its legacies to ecology and environmental management." *BioScience* 53: 77–88.

Foster, D. R., and T. M. Zebryk. 1993. "Long-term vegetation dynamics and disturbance history of a *Tsuga*-dominated forest in New England." *Ecology* 74: 982–998.

Fox, S. 1998. "Massachusetts contribution to national forest conservation," p. 339. In: C. H. W. Foster (ed.), *Stepping back to look forward: A history of the Massachusetts forest.* Petersham, Mass.: Harvard University.

Fuller, J. L., D. R. Foster, J. S. McLachlan, and N. Drake. 1998. "Impact of human activity on regional forest composition and dynamics in central New England." *Ecosystems* 1: 76–95.

Garrison, J. R. 2003. *Landscape and material life in Franklin County, Massachusetts, 1770–1860.* Knoxville, Tenn.: University of Tennessee Press.

Gaudreau, D. C. 1988. "The distribution of Late Quaternary forest regions in the Northeast: Pollen data, physiography, and the prehistoric record," pp. 215–256. In: G. P. Nicholas (ed.), *Holocene human ecology in northeastern North America.* New York: Plenum.

Gaudreau, D. C., and T. Webb III. 1985. "Late Quaternary pollen stratigraphy and isochrone maps for the northeastern United States," pp. 247–280. In: V. M. Bryant Jr. and R. Holloway (eds.), *Pollen records of Late Quaternary North American sediments.* Houston, Tex.: American Association of Stratigraphic Palynologists.

Golodetz, A. D., and D. R. Foster. 1997. "History and importance of land use and protection in the north Quabbin region of Massachusetts (USA)." *Conservation Biology* 11(1): 227–235.

Greene, J. R. 1981. *The creation of the Quabbin Reservoir: The death of the Swift River Valley.* Athol, Mass.: Transcript Press.

Hall, B., G. Motzkin, D. R. Foster, M. Syfert, and J. Burk. 2002. "Three hundred years of forest and land-use change in Massachusetts, USA." *Journal of Biogeography* 29: 1319–1335.

Harrison, B., and J. Kluver. 1989a. "Deindustrialization and regional restructuring in Massachusetts," pp. 104–131. In: L. Rodwin and H. Sazanami (eds.), *Deindustrialization and regional economic transformation.* Boston: Unwin Hyman.

Harrison, B., and J. Kluver. 1989b. "Reassessing the 'Massachusetts miracle': Reindustrialization and balanced growth, or convergence to 'Manhattanization'?" *Environment and Planning* 6: 771–801.

Hoover, T. 2001. Testimony before the United States House of Representatives, Committee on Transportation and Infrastructure, Sub-committee on Water Resources and Environment. March 15.

Hosley, N. W. 1935. "The essentials for a management plan for forest wildlife in New England." *Journal of Forestry* 33: 985–989.

Jacobson, G. L., T. Webb, and E. Grimm. 1987. "Patterns and rates of vegetation change during the deglaciation of eastern North America," pp. 277–288. In: W. F. Ruddiman and H. E. Wright (eds.), *North America and adjacent oceans during the last deglaciation,* vol. K-3. *geology of North America.* Boulder, Colo.: Geological Society of America.

Jameson, J. F. 1909. *Narratives of New Netherlands, 1609–1664.* New York: Barnes and Noble.

Johnson, E. S. 1996. *Discovering the ancient past at Kampoosa Bog, Stockbridge, Massachusetts.* Amherst, Mass.: University of Massachusetts Archaeological Services.

Jones, A., and P. Vickery. 1993. Report on Massachusetts grassland bird inventory—1993, p. 8. Lincoln, Mass.: Center for Biological Conservation, Massachusetts Audubon Society.

Josselyn, J. 1988/1674. "*Two Voyages to New-England,*" pp. 1–200. In: P. J. Lindholdt (ed.), *John Josselyn, colonial traveler: A critical edition of two voyages to New-England,* Hanover, N.H.: University Press of New England.

King, W. A. 1998. "The private forestry movement in Massachusetts," p. 339. In: C. H. W. Foster (ed.), *Stepping back to look forward: A history of the Massachusetts forest.* Petersham, Mass.: Harvard University.

Kingsley, N. P. 1976. "The forestland owners of southern New England." *USDA Forest Service Resource Bulletin,* NE-41: 27.

Kittredge, D. B., A. O. Finley, and D. R. Foster. 2003. "Timber harvesting as ongoing disturbance in a landscape of diverse ownership." *Forest Ecology and Management* 180: 425–442.

Kittredge, D. B., and A. M. Kittredge. 1999. "Interior and edge: The forest in Massachusetts." *Massachusetts Wildlife* 49(3): 22–28.

Kulik, G., R. Parks, and T. Penn (eds.). 1982. *The New England mill village, 1790–1860.* Cambridge, Mass.: MIT Press.

Lawton, J. 1997. "The science and non-science of conservation biology." *Oikos* 79: 3–5.

Lindbladh, M., W. W. Oswald, D. R. Foster, E. K. Faison, J. Hou, and Y. Huang. 2007. "A late-glacial transition from Picea glauca to Picea mariana in southern New England." *Quaternary Research* 67: 502–508.

McCullough, R. L. 1998. "Town forests: The Massachusetts plan," p. 339. In: C. H. W. Foster (ed.), *Stepping back to look forward: A history of the Massachusetts forest.* Petersham, Mass.: Harvard University.

McLachlan, J., D. R. Foster, and F. Menalled. 1998. "Anthropogenic ties to late-successional structure and composition in four New England hemlock stands." *Ecology* 81: 717–733.

McMahon, S. F. 1985. "A comfortable subsistence: The changing composition of diet in rural New England, 1620–1840." *William and Mary Quarterly* 42: 26–51.

McWeeney, L. 1994. *Archaeological settlement patterns and vegetation dynamics in southern New England in the Late Quaternary.* PhD diss., Yale University, New Haven, Conn.

McWeeney, L. 2003. "Cultural and ecological continuities and discontinuities in coastal New England: Landscape manipulation." *Northeast Anthropology* 64: 75–84.

Merchant, C. 1989. *Ecological revolutions: Nature, gender, and science in New England.* Chapel Hill, N.C.: University of North Carolina Press.

Motzkin, G., R. Eberhardt, B. Hall, D. Foster, J. Harrod, and D. MacDonald. 2002. "Vegetation variation across Cape Cod, Massachusetts: Environmental and historical determinants." *Journal of Biogeography* 29: 1439–1454.

Motzkin, G., D. Foster, A. Allen, and J. Harrod. 1996. "Controlling site to evaluate history: Vegetation patterns of a New England sand plain." *Ecological Monographs* 66: 345–365.

Motzkin, G., W. A. Patterson III, and D. R. Foster. 1999a. "A historical perspective on pitch pine–scrub oak communities in the Connecticut Valley of Massachusetts." *Ecosystems* 2: 255–273.

Motzkin, G., P. Wilson, D. R. Foster, and A. Allen. 1999b. "Vegetation patterns in heterogeneous landscapes: The importance of history and environment." *Journal of Vegetation Science* 10: 903–920.

Niering, W. A., and G. D. Dreyer. 1989. "Effects of prescribed burning on *Andropogon scoparius* in postagricultural grasslands in Connecticut." *American Midland Naturalist* 122: 88–102.

O'Keefe, J., and D. Foster. 1998. "Ecological history of Massachusetts forests," pp. 19–66. In: C. H. W. Foster (ed.), *Stepping back to look forward: A history of the Massachusetts forest.* Petersham, Mass.: Harvard University.

Orwig, D. A., and D. R. Foster. 1998. "Forest response to introduced hemlock woolly adelgid in southern New England, USA." *Journal of the Torrey Botanical Society* 125: 59–72.

Pacala, S. W. 2001. "Consistent land- and atmosphere-based U.S. carbon sink estimates." *Science* 292(5525): 2316–2321.

Pacala, S. W., C. D. Canham, J. Saponara, J. Silander, R. Kobe, and E. Ribbens. 1996. "Forest models defined by field measurements: Estimation, error analysis and dynamics." *Ecological Monographs* 66: 1–43.

Patterson, W. A., III, and A. Backman. 1988. "Fire and disease history of forests," pp. 603–632. In: B. Huntley and T. Webb (eds.), *Vegetation history.* Dordrecht: Kluwer Publishing.

Peterken, G. 2003. *Woodland conservation and management.* 2nd ed. New York: Springer Publishing.

Raup, H. M. 1964. "Some problems in ecological theory and their relation to conservation." *Journal of Ecology* 52(Suppl.): 19–28.

Rickenbach, M. G., D. B. Kittredge, D. Dennis, and T. Stevens. 1998. "Ecosystem management: Capturing the concept for woodland owners." *Journal of Forestry* 96(4): 18–24.

Rideout, K., and M. Adams. November 2000. *Brownfields in central Massachusetts.* Occasional paper of the Human Environment Regional Observatory for Central Massachusetts. Worcester, Mass.: Clark University Press.

Rivers, W. H. 1998. "Massachusetts state forestry programs," p. 339. In: C. H. W. Foster (ed.), *Stepping back to look forward: A history of the Massachusetts forest.* Petersham, Mass.: Harvard University.

Scheller, W. G. 1994. "The politics of protection." *Sanctuary Magazine* November/December: 17–19.

Seymour, R. S., A. S. White, and P. H. deMaynadier. 2002. "Natural disturbance regimes in northeastern North America: Evaluating silvicultural systems using natural scales and frequencies." *Forest Ecology and Management* 155: 357–367.

Sharp, B. 1994. "New England grasslands." *Sanctuary Magazine* November/December: 12–16.

Smith, B. D. 1992. *Rivers of change: Essays on early agriculture in eastern North America.* Washington, D.C.: Smithsonian Institution Press.

SNESPR. 2003. *Southern New England stumpage price survey results fourth quarter - 2002.* Online. Available at http://forest.fnr.umass.edu/stumpage.htm. Accessed 2003.

Steel, J. 1999. *Losing ground: Analysis of recent rates and patterns of development and their effects on open space in Massachusetts*, p. 17. Lincoln, Mass.: Massachusetts Audubon Society.

Steer, H. B. 1948. *Lumber production in the United States: 1799–1946.* USDA Misc. Publication, no. 669.

Tiner, R. W. 1988. *Agricultural impacts on wetlands in the northeastern United States.* Presented at the National Symposium on Protection of Wetlands from Agricultural Impacts, Colorado State University, Fort Collins, Colo. Hadley, Mass.: U.S. Fish and Wildlife Service.

USDA Forest Service. 2002. Forest inventory and analysis tables. Online. Available at http://www.fia.fs.fed.us/documents/pdfs/2002_RPA%20FINAL%20TABLES.pdf. Accessed 1/15/2008.

Van Royen, W. 1928. *Geographic studies of population and settlement in Worcester County, MA.* PhD diss., Clark University, Worcester, Mass.

Westveld, M. V., and Committee on Silviculture, New England Section, Society of American Foresters. 1956. "Natural forest vegetation zones of New England." *Journal of Forestry* 54: 332–338.

Whitlock, C., and P. J. Bartlein. 1997. "Vegetation and climate change in northwest America during the past 125 kyr." *Nature* 388: 57–61.

Whitney, G. 1994. *From coastal wilderness to fruited plain: A history of environmental change in temperate North America 1500 to the present.* Cambridge: Cambridge University Press.

Wilson, H. F. 1936. *The hill country of northern New England: Its social and economic history, 1790–1930.* New York: Columbia University Press.

3

Agricultural Transformation of Southern Appalachia

Ted L. Gragson
Paul V. Bolstad
Meredith Welch-Devine

Humans impose patterns on the earth through purposeful as well as inadvertent land use, and these patterns affect local, regional, and global ecological processes. The effects ultimately influence the sustainability of biophysical and cultural landscapes, as well as the quality of life. The challenge is to understand biophysical *and* cultural landscapes as the result of integrated socioeconomic and ecological dynamics playing out across potentially vast scales of space, time, and organizational complexity (Levin, 1999; Turner, 1990; Vitousek et al., 1997). Certainly the intrinsic temporal rhythms and spatial arrangement of ecological systems bear the signature of human activities and institutions (Carpenter and Gunderson, 2001; Pyne, 1997; Scheffer et al., 2001; Turner et al., 2002). However, temporal rhythms and spatial arrangements of human activities and institutions are in turn shaped and influenced by the ecological systems in which they are embedded (Berkes and Folke, 1998; Cronon, 1983; Diamond, 1997; Dove and Kammen 1997; Ostrom et al., 1999). The reciprocal "imprinting" between socioeconomic and biophysical systems means that artificially separating landscapes into these two component systems will fail to improve understanding of the sustainability of either landscapes or the quality of life.

Our objective is to understand the agrarian transformation of southern Appalachia. We are specifically interested in what happens when humans impose their signature on ecological systems, how they must then respond to the systems they helped create, and how the reciprocity between the two realms results in a coupled socioeconomic–biophysical system that has a unique dynamic across space and time. The general outline of our approach is to examine three aspects of the agrarian transformation of southern Appalachia:

1. How human activities influenced the spatial and temporal structure of agrarian landscapes
2. What the ecological and environmental consequences of this structure are
3. How humans responded to landscape transformation in the realm of conservation

Our examination of the agricultural transformation of southern Appalachia follows the approach of narrative positivism (Abbott, 2001). Rather than being a synonym for *discourse* or the telling of a story composed of a succession of events, our narrative strives for an account of regularities in the socioeconomic–biophysical process of agrarian transformation. It focuses on the cumulative effect of human activities in southern Appalachia at moments in time that constrain the opportunities of current and future generations. The duration of stages results in a coercive narrative in the sense of implying a certain result, thus subsuming the relatively smooth, directional *trajectory* implied by duration as well as the *turning points* that mark relatively abrupt, diversionary moments separating trajectories. The approach is appropriate for examining the agricultural transformation of southern Appalachia as discrete and categorical rather than continuous and numerical (Abbott, 2001; Isaac and Griffin, 1989; Isaac and Leight, 1997; Isaac et al., 1994). The objective is to reveal the pattern of transformation in anticipation of ultimately defining general processes and relationships.

The introduction, spread, and abandonment of agriculture represents the most pervasive alteration of Earth's environment during the past 10,000 years (Farina, 2000; Matson et al., 1997; Vitousek et al., 1986). Most places on the earth bear the signature of agricultural use even though the past 100 years of U.S. history have been characterized by the progressive transition of agrarian landscapes and lifeways to other uses. A significant portion of former agricultural lands is in the process of being developed and incorporated into the urban–suburban sprawl that characterizes contemporary U.S. landscapes. Despite the reforestation that has followed agricultural abandonment in southern Appalachia and that seems like a rewilding of the landscape, the signature of past agricultural use is still present in diverse scalar elements of the contemporary landscape (Fraterrigo et al., 2005). Beyond describing how a shift in land use has occurred, there is a growing need to understand the underlying mechanisms for the transition and the legacies associated with prior states so that possible future conditions can be anticipated (Bennett et al., 2003; Groves et al., 2002; Nilsson et al., 2003; Wollenberg et al., 2000). This is where history will cease being retrospective and begin contributing to current practice. In our view, conservation is not only the preservation of natural patches in a complex landscape matrix dominated by human use, but also must include the preservation of desirable cultural qualities associated with production on *working landscapes.*

Background

By virtue of its geographical position, southern Appalachia serves as a natural laboratory for evaluating across diverse gradients the proportional contribution of socioeconomic and biophysical processes to the structure and function of ecosystems. Physical environmental forces exert strong influences on the organization of southern Appalachian ecosystems. Much previous research on ecosystem responses to disturbance in southern Appalachia has focused on a subset of important forces acting on large scales and/or short time intervals. For example, the pattern and magnitude of wind damage from Hurricane Opal was substantially controlled by local physiography (Hunter and Forkner, 1999; Wright and Coleman, 2002). However, direct human disturbances such as farming, logging, mining, and road construction have altered more than 98% of the southern Appalachian landscape. The introduction of the chestnut blight (*Cryphonectria parasitica*) and the balsam wooly adelgid (*Adelges piceae*), two indirect cases of human disturbance, caused or are in the process of causing profound changes in everything from the composition of vegetative communities to the flux of nitrogen in streams.

We still know relatively little about the diversity and magnitude of the human contribution to southern Appalachian disturbance. For example, the shift from Native American to European American dominance in southern Appalachia entailed important changes well-known to historians in the ways both peoples organized their lives. However, there has been little research to address the fundamental reorganization of biophysical systems in response to social, political, and economic forces. The typical approach has been to consider the pressure exerted by external political and economic forces on the organization of southern Appalachian society with the environment relegated to the role of scenery (e.g., Dunaway, 1996). The point is to build explanations of cause for the patterned relationships observed in the agricultural transformation of southern Appalachia that go beyond vague statements that disturbance is driven by technology, socioeconomic organization, level of economic development, and culture.

Although the vast majority of research on environmental change has been directed at global scale processes, estimates at the scale of kilometers or hectares are ultimately more important. This is because such estimates give recognition to the fact that policymakers, resource managers, and the public at large make decisions in response to local and regional conditions more so than to global conditions. For example, residents of the southwestern United States are more likely to care about changes in water availability and fire frequency during the coming decades than residents of the southeastern United States, who are more likely to be concerned about how changes in forest cover might affect their recreational opportunities and ability to produce timber. Determining the regional-scale consequences of environmental change rests on a fundamental ecological understanding of how the population dynamics of plants, animals, and microbes are linked to biogeochemical processes. However, it also depends on understanding how pressures such as urban expansion, gentrification, and industrial diversification shape ecosystems and landscapes.

Building regional-scale estimates for such processes will ultimately assist decision makers' responses to local and regional conditions. By understanding the reciprocal influences between socioeconomic and biophysical systems, we can address how the conjunction of events in both domains during previous time periods both constrains and structures future environmental and societal opportunities.

Regions, in effect, constitute a multitude of distinctive, self-organized landscapes. The periodization of distinct regimes and the long view of history can move understanding from the initial recognition of pattern to the determination of process. By this means, it becomes possible to dispel popular and sometimes scholarly scenarios for southern Appalachia of early settlers patiently chopping their way out of the dark woods into the sunlight or timber barons slashing and burning their way across the landscape. In short, focusing on the agricultural transformation of southern Appalachia presents the opportunity to address fundamental issues in the historical ecology of the region that can help bridge the longstanding parochialism of southern history (Kolchin, 2003).

Agricultural Regimes

We rely on the concept of an agricultural regime to organize our analysis of the agricultural transformation of southern Appalachia. We define an agricultural regime as a unique configuration of factors—crops, livestock, humans, and management technologies—applied to a landscape for the purpose of achieving a return on their investment. We use the concept qualitatively to facilitate the recognition of patterns in the data, although it can be used quantitatively to define the specific configuration of relations among independent and dependent variables (Campbell and Allen, 2001; Isaac et al., 1994). A brief evaluation of how the term *regime* has been variously used in the physical and social sciences suggests a widespread if sometimes ambiguous use of the concept.

Geomorphology uses *regime theory* to describe how streams reach a balance between creating part of their boundaries from the transported load they carry and creating part of the transported load they carry from their boundaries. A *regime waterway* (i.e., stream, river, or canal) is one that has achieved an average equilibrium between the two processes of deposition and scouring (Blench, 1957; Poff et al., 1997). A *fire regime* is the natural fire equilibria on a landscape in the absence of human mechanical intervention (Agee, 1993; Brown, 1995). It is a function of the average number of years between fires, combined with the severity of the fire on the dominant overstory vegetation. In political science, democracy or authoritarianism are referred to as *political regimes* that are differentiated by the types of rules and the distribution of political resources within them that enable actors to exercise authority over their constituents (Kitschelt, 1992).

An *agricultural regime* refers to the practices used by individuals to draw on common pool resources—soil, climate, technology, and labor—to achieve production success (Agrawal, 2001; Young, 1982; Zimmerer, 1999). It minimally includes

- *Resources*—the spatial variability and temporal unpredictability of biophysical factors along with the adaptive dynamics and environmental tolerances of crop types
- *Groups*—the sociospatial incentives and disincentives affecting the capacity of individuals to coordinate land use and to create cohesion at the level of the landscape (e.g., size, levels of wealth, and income)

- *Relationships*—the network and hierarchy relations between resource systems and locally situated groups vis-à-vis circumstances beyond their immediate control (i.e., uneven development, governmental policy, consumer tastes, extralocal forces)

In the final analysis, an *agricultural regime* is a human artifact that has no existence or meaning apart from the behavior of individuals within the regime (Young, 1982). The "rugged individualism" of the American way tends to ignore the fact that agriculture is a common pool resource system that depends on participants (voluntarily) adopting policies and interacting to foster credible commitments and to facilitate recurrent transactions among themselves. The properties of such a regime serve as the constraints to human interaction that derive from shared understandings about internalizing costs to forestall or attenuate conflict. As a qualitative description of the agricultural transformation of southern Appalachia, a regime refers to the patterns of behavior around which participant expectations converge. It differs from previous approaches to agricultural transformation of southern Appalachia that tended to rely on contemporary statistics or invoke cultural determinism.

The agricultural transformation of southern Appalachia is a process that began nearly 5,000 years ago, for which contemporary statistics alone provide little insight. More important, contemporary statistics fail to capture the major turning points in the trajectory of change other than to reveal that agriculture is no longer very important. As of AD 2000, less than 2% of the population of southern Appalachia listed agriculture as their primary occupation, and less than 3% of households were classed as rural–farm. Both measures reflect the strong proximity effect of the major cities in and surrounding southern Appalachia that structure the economy of the region: Asheville, Atlanta, Birmingham, Roanoke, Winston-Salem, and Greenville. The current rates of agricultural dependence and the distribution of the population reflect social, political, and economic forces during the past 50 years.

The understanding of agricultural transformation of southern Appalachia is further confounded by a longstanding "local color" narrative tradition (Anglin, 2002). Emerging at the turn of the 20th century, the tradition creates images of Appalachia that characterize the people as independent, religious fundamentalists with strong family ties who live in harmony with nature yet are traditional and fatalistic in their outlook (Philliber, 1994). A recent article in the *Atlanta Journal–Constitution,* the largest daily newspaper in the region, began as follows: "The ancient, misty mountains that surround the 'hollers' of southern Appalachia remain a wall between the region's proud, melancholy people and American prosperity" (Burritt, 1997, p. A1). Within this narrative tradition, argument tends to center on whether the current problems are the result of long-term geographical isolation or the product of outsiders plundering the region's natural and social endowments (Caudill, 1963; Eller, 1982; Moore, 1994; Rothblatt, 1971; Salstrom, 1994). It evaluates the agricultural transformation of southern Appalachia by reference to victimization rather than in terms of reciprocity between the component elements of the system.

Agroecology of Southern Appalachia

Political definitions of Appalachia abound, but we define southern Appalachia as extending across the Blue Ridge Province of the southern Appalachian highlands (Fig. 3.1), which numerous authors recognize as a distinct biophysical and cultural region of the continental United States (Bailey, 1996; Markusen, 1987; Whittaker, 1966). The area is also conterminous with the project area of the Coweeta LTER site within which our study of agricultural transformation takes place. The Blue Ridge Province of southern Appalachia comprises the unfolded highlands from north Georgia through western Virginia. The original Paleozoic land surface has been severely eroded and dissected by numerous streams and rivers into a series of deep, steep-sided valleys separated by narrow ridges. The Tennessee–North Carolina state line follows the ridgeline of the Appalachian mountain chain and contains the highest peaks in the entire province. Rainfall and temperature vary widely as a result of differences in topography, elevation, and thermal belts, creating a highly varied agroecological landscape. Temperatures decrease as we go from low to high elevation, and average summer temperatures on the higher peaks are more similar to those in central New England 1,400 km to the north than they are to the lower Piedmont only 150 km to the southeast. Precipitation is abundant, averaging more than 1,600 mm a year, although it typically increases from low to high elevation and exhibits local mountain effects in the form of wet zones and rain shadows. In addition, microclimate and related soil properties vary considerably on short spatial scales from ridgetops to streamside bottoms.

Temperate deciduous forests are the dominant vegetation of southern Appalachia. Because it is both cooler and wetter than the adjacent provinces to

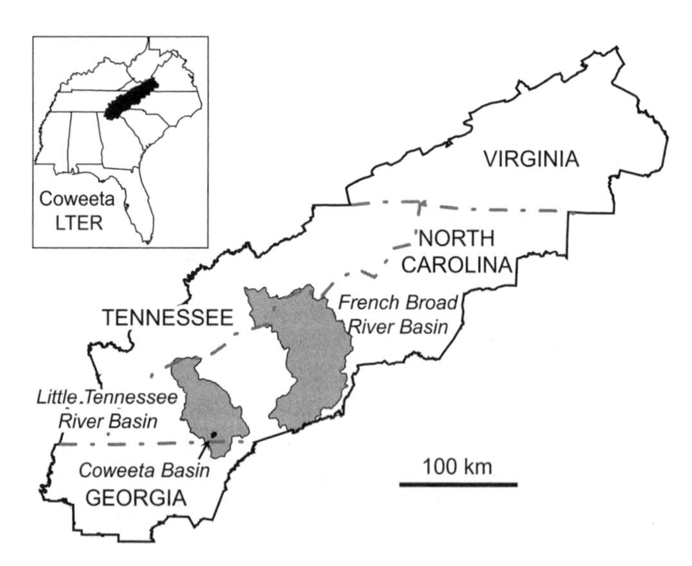

Figure 3.1 Southern Appalachian study region with the location of state lines and basins where research efforts are concentrated.

the west (i.e., Ridge and Valley) and the east (i.e., Piedmont Plateau), southern Appalachia is a refugium for "northern" taxa that reached here from the last glaciation (Barnes, 1991; Braun, 1950), and in intermixing with "southern" taxa give rise to one of the most biodiverse regions of North America. Despite the contemporary vigor and beauty of the forest, the ecosystems in the region harbor the "ghost of land use past" (Harding et al., 1998) and reveal that use of the land, albeit changing, has been constant throughout time. Southern history focusing on the agricultural use of southern Appalachia has concentrated on the period of the so-called *plain folk*. These were the free southern whites living outside the plantation economy during the Antebellum period (1750–1850) (Kretzschmar et al., 1993; Otto, 1989; Owsley, 1949; Salstrom, 1994). However, southern Appalachia has been occupied since at least 8000 BP.

The earliest human occupants were mobile, transient, and dependent on naturally available animal and plant food resources (Perkinson, 1973; Ward and Davis, 1999), but the earliest evidence for manipulation of plants goes back to approximately 2300 BP (Chapman and Shea, 1981; Fritz, 2000; Yarnell and Black, 1985). Larger, more widely distributed settlements had developed by AD 800 in valleys, coves, and adjacent uplands, and southern Appalachian societies were actively trading with native populations both north and south of the region. The first major agricultural transformation of the region took place before European contact during the Mississippian period that flourished between AD 1200 and 1450. Forest clearing to convert land to agricultural use during the 13th and 14th centuries, and again in the 19th and 20th centuries, led to the direct removal of many native species from large portions of the landscape for extended periods of time. Although the rationales and practices differed, the outcomes were similar in many ways. In the following section we trace the major periods of the transformation of southern Appalachia.

Agri-*Cultural* History of Southern Appalachia

Time prior to European contact is generally divided into nine major archaeological periods, and after contact, into an additional eight periods. The names for periods as well as the boundaries between them vary among authors and are a function of whether they are archaeologically or historically determined. By the year 1700, documentary evidence for the Blue Ridge proper, including maps, diaries, and reports from which to discern disturbance, existed. Studies of events prior to that time rely more heavily on physical evidence and analogy—as might be available from archaeology. When outlining the agri-*cultural* history of southern Appalachia we focus extensively on the work of others before 1700 and draw from our own research subsequent to that date. From 1850 forward we also incorporate information from the U.S. agricultural census using data for three pivotal dates—1850, 1900, and 1950—to characterize the distribution of production across the study region spatially and temporally. These dates respectively represent (1) the peak of antebellum agricultural production, (2) the point of recovery after Reconstruction yet prior to the radical expansion of the timber industry, and (3) the end of agriculture as a way of life in Appalachia.

Despite the 17 conventional historical periods, we collapse time into four great periods in line with the objectives of the AgTrans Project and our use of agricultural regimes. We divide time prior to European contact into a Premanipulation period and a Hearth of Domestication period. Because these two periods fall outside our direct research experience and because the archaeology of southern Appalachia is poorly understood, we cover these two periods only cursorily. We also divide time after European contact into two periods. The Columbian Revolution period covers the protohistoric period through the Revolutionary War of 1776. The Nationhood period subsumes all the major transitions characteristic of the Southeast, from establishment of the federal government through the demise of agriculture as a way of life.

Premanipulation Period

The earliest human presence in southern Appalachia probably dates to the early Holocene between 11,000 and 12,000 years ago (Anderson and Sassaman, 1996; Walker, 2002). The evidence largely consists of widely scattered, isolated surface finds of fluted Clovis- and Folsom-like spear points. The spruce–fir boreal forest interspersed with open parkland was giving way to a continuous mesic oak–hickory hardwood forest, a vegetative transition that was completed 9,000 years ago (Anderson, 1995; Delcourt and Delcourt, 1981, 1983). Dust Cave, located just outside our study region at approximately 34° 46′ N, 85° 00′ W, is one of the few stratified Paleo-Indian or Archaic sites in the Southeast (Meltzer, 1988; Walker, 2002). The interpretation of the faunal remains from this site suggests that the human populations from the Paleo-Indian through the Middle Archaic periods relied on a diverse array of aquatic and terrestrial species with little change over time (Davis, 1990; Ward and Davis, 1999).

Many of the ideas about Paleo-Indian and Early Archaic adaptations in southern Appalachia are little more than speculation, and most of the research has so far focused on developing and refining chronologies rather than reconstructing lifeways (Ward and Davis, 1999). The simple yet ubiquitous tool assemblages from the Middle Archaic parallel those left by small, kin-based foraging groups moving as a unit from place to place. The drier and warmer Altithermal (approximately 6000 BC to 2000 BC) created a patchy and less predictable environment, and the archaeological remains have generally been interpreted as reflecting flexible subsistence strategies (Ward and Davis, 1999; Wendland and Bryson, 1974). Overall, the evidence indicates that humans from their first entry into southern Appalachia 11,000 to 12,000 years ago to approximately 3000 BC practiced an extractive rather than a manipulative use of natural resources.

The Hearth of Eastern Domestication

After the climate shift of the Altithermal, the Late Archaic through Mississippian periods are characterized by dramatic increases in human population, the beginning of pottery making, and the gradual shift toward sedentary villages. Beginning about 2500 BC, several locally and distantly domesticated species converged to

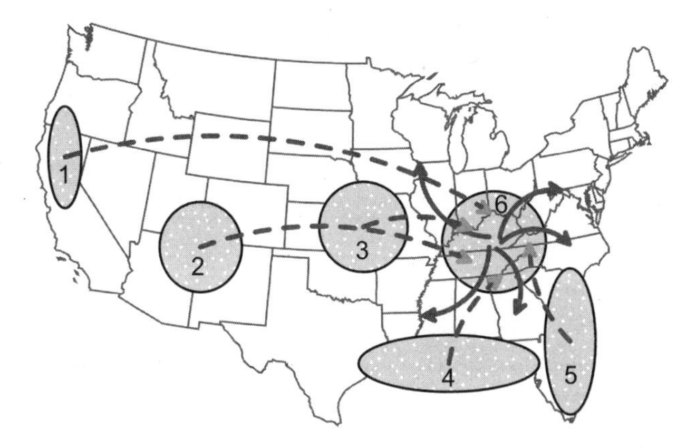

Figure 3.2 Early domesticate influx to and dispersal from the southeastern agricultural hearth. 1, Tobacco (*Nicotiana sp.*); 2, maize (*Zea mays*), garden bean (*Phaseolus vulgaris*), cushaw squash (*Cucurbita argyrosperma*); 3, sunflower (*Helianthus annuus*), little barley (*Hordeum pusillum*); 4, eastern gourd (*C. pepo*); 5, bottle gourd (*Lagenaria siceraria*); 6, sumpweed/marshelder (*Iva annua*), chenopod/goosefoot (*Chenopodium berlandieri*), maygrass (*Phalaris caroliniana*), knotweed (*Polygonum erectum*).

form a true garden complex denoting the transition from extraction to manipulation (Fritz, 1993). The species include "starchy" (high-carbohydrate) species such as chenopod (*Chenopodium berlandieri* ssp. *jonesianum*), little barley (*Hordeum pusillum*), maygrass (*Phalaris caroliniana*), and erect knotweed (*Polygonum erectum*) as well as "oily" (high-lipid) varieties such as gourd/squash (*Cucurbita argyrosperma, C. pepo, Lagenaria siceraria*), sunflower (*Helianthus annuus* var. *macrocarpus*), and sumpweed (*Iva annua* var. *macrocarpa*). By approximately AD 1, tobacco (*Nicotiana quadrivalvis* or *N. multivalvis*) was introduced from the West Coast and maize (*Zea mays*) from the Southwest (Anderson and Mainfort, 2002; Chapman and Shea, 1981; Crites, 1993; Fritz, 2000; Yarnell and Black, 1985).

The evidence indicates the larger Central Mississippi Valley was part of the Hearth of Eastern Domestication (Fig. 3.2) (Gremillion, 2002). From this area it spread outward in all directions, with the first evidence of domesticates in southern Appalachia proper being small, charred fragments of squash rind dated to 2440 BC (Chapman, 1994; Chapman and Shea, 1981). The potential yield of the species in the garden complex was as high as 1,000 kg/ha, which is comparable with what can be achieved with maize. However, there is little evidence to suggest the populations of the lower Southeast were fully committed to food production until after AD 800. This corresponds to the appearance of larger sites containing multiple features and structures, and covering several acres, indicating an increased occupation intensity relative to the first half of the period (Anderson and Mainfort, 2002; Chapman, 1994; Davis, 1990; Keel, 1972; Smith, 1992; Ward and Davis, 1999). After AD 1000, there were villages with several hundred inhabitants and massive ceremonial mound centers, indicating the emergence of highly stratified societies. Maize, first introduced to the region about AD 1, became the staple, and all other native crops with the exception of sunflower began to decrease in relative importance (Johannessen, 1993).

By AD 1400, a simplified, maize-dominated farming system was fully entrenched and supported the Mississippian chiefdoms (Gremillion, 1989; Ward and Davis, 1999; Yarnell, 1976). The form and structure of Mississippian villages clearly suggest a shift toward a hierarchical form of sociopolitical organization centered on a class of hereditary elite rulers (Anderson, 1994; DePratter, 1983). The effect of increased political integration on farming strategies could have included allocation of fields and surplus production to tribute (Rose et al., 1991; Scarry, 1993). However, the chiefdom-level Mississippian polities were notoriously unstable, and few lasted more than a hundred years (Hally, 1996). As a consequence, no locality in the region was subject to continuous and intensive human impact for more than a few centuries.

Columbian Revolution

Time of first contact varied widely across the Southeast. In 1525, an expeditionary force led by Pedro de Quejo sailed along the Atlantic coast from Andrews Sound in south Georgia to Delaware Bay, making landfall at various places in between (Hoffman, 1994). The effect on interior groups is unknown, but numerous authors have long speculated on the biological consequences of this and other early contacts (Thornton et al., 1992). Hernando de Soto was the first to visit the interior, embarking from Florida in 1539 and reaching what is now west-central North Carolina in the spring of 1540. He traveled west over the Blue Ridge into Tennessee and finally south into Georgia, where he visited the paramount chiefdom of Coosa (Hudson et al., 1985). Juan Pardo retraced de Soto's route in two separate expeditions in 1566 and 1568 (Hudson, 1990). Nevertheless, there is no significant evidence that 16th-century Spanish expeditions had a lasting impact on aboriginal groups in southern Appalachia.

The English settled Jamestown in 1607, and by 1670 there was a steady stream of traders and packhorses making their way to the eastern edge of the Blue Ridge, bringing tools, weapons, ornaments, and disease. By 1690, the Cherokee were the principal aboriginal group occupying southern Appalachia when sustained contact with English traders from Charles Town was established. The Cherokee had the largest population of the many groups in the southern Appalachian region, with a population of approximately 12,000 and a claim to 322,600 km^2 that included most of the current states of Kentucky and Tennessee, and large sections of Georgia, South Carolina, North Carolina, Virginia, and West Virginia (Mooney, 1995; Royce 1975).

Cattle were the last European production element adopted by the Cherokee. Martin Schneider noted during his travels in Cherokee country in 1783 and 1784 that "every family has its own field... they have not fences about their fields, on which account no cattle are kept except by traders" (Williams, 1928, p. 261). The acceptance of cattle marked the Cherokee conversion from a mixed agriculture and hunting subsistence to full-time farming. The federal Indian policy of directed culture change further encouraged and accelerated the transformation in land-use practices and culture (Newman, 1979). By the early 19th century, according to Wilms (1991), the Cherokee "transformed their aboriginal landscape into a new cultural landscape that resembled and perhaps sometimes surpassed their white frontier neighbors" (p. 1).

White settlement in southern Appalachia was really set in motion when the British crown assumed control of the Carolina colony in 1744 and bought out seven of the original eight proprietors. Lord Granville, the heir of the eighth proprietor, refused to sell and received as compensation the northern half of North Carolina. He offered lands in return for modest rents to attract settlers from the northern colonies, in particular Germans from Pennsylvania. After 1761, the early Pennsylvanian settlers were joined by Scotch–Irish and German immigrants from Europe who were encouraged by South Carolina's offer of bounties to white settlers (Bridenbaugh, 1971; Fischer, 1989; Merriwether, 1940). By 1775, the southern Appalachian backcountry was exporting an array of commodities, including flax seed, wheat, indigo, livestock, and poultry (Otto, 1989).

The backcountry frontier was different from other period frontiers in that it lacked a major cash crop, there were relatively few slaves, and the majority (approximately 52%) of white inhabitants were from northern England, Scotland, and Ireland rather than from southern England (Bridenbaugh, 1971; Fischer, 1989; Otto, 1989). These differences played into a pivotal event called the *Regulation rebellion*, which was in response to the comprehensive political reform agenda with social and racial overtones. It was suppressed in 1771 at the battle of Alamance, North Carolina (Hatley, 1993; Otto, 1989). "Regulators" fled to frontier lands beyond the crest of the Appalachian Mountains, which marked the Proclamation Line of 1763 that separated Native American from American colonial lands. Many of the Regulators arrived at the Watauga settlements in what is now eastern Tennessee and that had been established by Virginians in approximately 1769 (Arthur, 1914; Dixon, 1976; Hatley, 1993; Summers, 1903). They would form an important nucleus to the early post–Native American settlement of southern Appalachia.

Nationhood: 1776 to Present

After the American Revolution, before lands could be opened to European American agricultural settlement, the federal government had to extinguish Native American claims. In southern Appalachia the process involved a progressive series of quit-claim treaties between the United States and the Cherokee. The final Treaty of New Echota, signed May 23, 1836, confirmed that the Cherokee Nation ceded to the United States all its claims to lands east of the Mississippi River (Otto, 1989; Royce, 1975). The availability of cheap public lands made possible by these treaties attracted thousands of agriculturalists from the southern seaboard. They first settled the piedmont areas of North Carolina, Georgia, Tennessee, and Alabama, and gradually converged on the interior highlands, particularly the Little Tennessee and French Broad River basins, which offered attractive farming opportunities (Inscoe, 1989). Between 1790 and 1850, the population of the study region grew at an average annualized rate of 3.4%, peaking at 7.6% between 1790 and 1800 (Fig. 3.3).

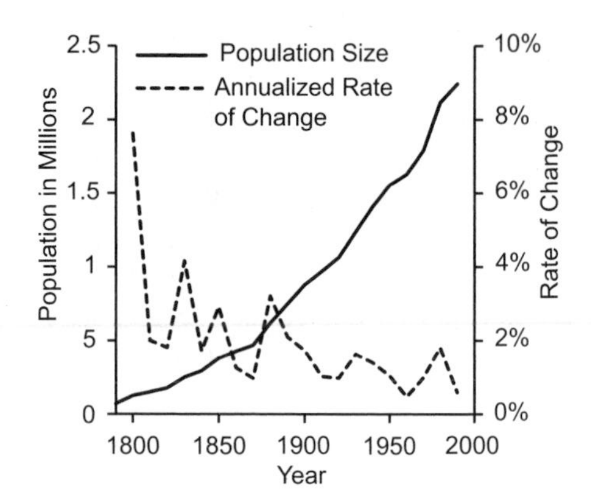

Figure 3.3 Relative and absolute population growth of the southern Appalachian study region, 1790–1990.

Throughout the course of the 19th century, the population of Appalachia became enmeshed in the emerging global system, drawn in by the plantation economy of the lower south and the commodity chains that spread outward from the Atlantic Rim to the Caribbean and the urban European centers (Dunaway, 1996; Salstrom, 1994). The economic ties to the lower south in particular led to a shift of production priorities in southern Appalachia from subsistence and shelter to marketing agricultural surpluses. The singular focus of coastal plain plantations on cash crops such as cotton and rice limited their ability to produce the foodstuffs necessary to sustain their enslaved agricultural labor. Southern Appalachia met this need through the export of subsistence commodities produced on small- and medium-size family-operated farms (Groover, 2003). Cattle and other livestock were critical components of the overall production system in southern Appalachia; hogs were fed on mast, and cattle were allowed to wild forage on unfenced lands (Davis, 2000; Inscoe, 1989; Otto, 1989).

The level of export, although not known with precision prior to 1850 when the first agricultural census was carried out, is believed to have been significant. Slaves represented, on average, 46% of the population in the seven Deep South states (South Carolina, Mississippi, Louisiana, Alabama, Florida, Georgia, and Texas). These were the states that seceded from the Union prior to the battle of Fort Sumter (April 12–14, 1861), marking the beginning of the Civil War (Kolchin, 2003). In South Carolina alone, slaves composed 57% of the total population, which was the largest of any state in the Union. In the aggregate, slaveholding in the states of the Deep South set the region apart from the 18 northern states, which had no slaves at all. However, it also distinguished the Deep South from southern Appalachia, where the distribution of slaves was more variable and ranged by area from 2% to 25% (Inscoe, 1989).

Although the Civil War led to the emancipation of black slaves, there were many things that changed little in the South. Appalachia continued to supply subsistence products to the lower south, where cotton farming and the tenancy system (slavery of a different kind) undermined rural sufficiency and diversified farming (Fite, 1984; Groover, 2003; Wright, 1986). Appalachia was characterized by a different type of inequality that had been carried forward from colonial times, and was first manifested in the battle of Alamance mentioned earlier. Southern Appalachia was characterized by a few large absentee owners, a small class of yeomanry, and many landless families (Dunaway, 1996; Fischer, 1989; Salstrom, 1994). By the last decade of the 18th century, Gini ratios for total wealth in four Tennessee counties in the upper northwest corner of the study area ranged from 64 to 75. The top decile of wealth holders owned between 47% and 73% of land and slaves, whereas between 28% and 39% of the population had neither. Gini ratios from Cocke and Johnson counties (respectively, 62 and 71) indicate the pattern of landownership changed little well into the 20th century (Fischer, 1989; Geisler 1983; Winters, 1987).

There is some support for a cyclical expansion and contraction with a 50-year periodicity of the rural economy of Appalachia during the 18th, 19th, and 20th centuries conforming to a so-called *Kondratieff wave* (Kondratieff, 1979). Economic stagnation provided the impetus for frontier expansion and colonialism, which led to the capture of new resources that revived the stagnating economy (Dunaway, 1996; Groover, 2003; Salstrom, 1994). The takeoff for the first cycle in southern Appalachia occurred in approximately 1820. This corresponded to the first major territorial quit claim between the Cherokee and the federal government, in which the Cherokee relinquished some 650,000 acres of western North Carolina lands. Returns were favorable through the 1830s with cheap transportation and favorable prices. A banking crisis in 1837 shook the American economy, and commodity prices fell, but prices recovered quickly and per-farm production peaked during 1850 to 1860 (North, 1961; Otto, 1989).

The agricultural prosperity of the 1850s stimulated construction of both interstate and local transportation lines, in particular the laying of train tracks (Cotterill, 1924; Stover, 1978). In 1850, the southern states had only 3,219 km of railroad—largely confined to the Atlantic seaboard states. By 1860, the southern states had about 16,093 km of railroad tracks. The improved transportation network helped reorient settlement away from waterways in the coastal plain and piedmont areas, both Atlantic and interior. Agriculturalists in the rugged highlands in between, including the Little Tennessee and French Broad drainages, however, continued to depend on canoes, large flatboats or keelboats, and wagons for transportation (Black, 1952; Groover, 2003; Stover, 1978; Taylor, 1951). There was a movement after the Civil War toward increased marketing of surplus grains and livestock (Salstrom, 1994; Weingartner et al., 1989), although the backward benefits of the outward flow of commodities was rather ephemeral.

The decade of the 1850s not only marked a watershed for agricultural production in southern Appalachia, but it also reflected the onset of a comprehensive political reform agenda by the United States. The federal government began imposing banking, landownership, and agricultural reforms that, combined with the central government's increasing effectiveness at ensuring policy compliance, eventually resulted in a dramatic shift in the agricultural regime of the United States. The year 1850 also marked the beginning of the regular collection of national agricultural census information as well as other types of information that reflect a state strategy of increasing legibility and control over the national territory and its resident population (Scott, 1998).

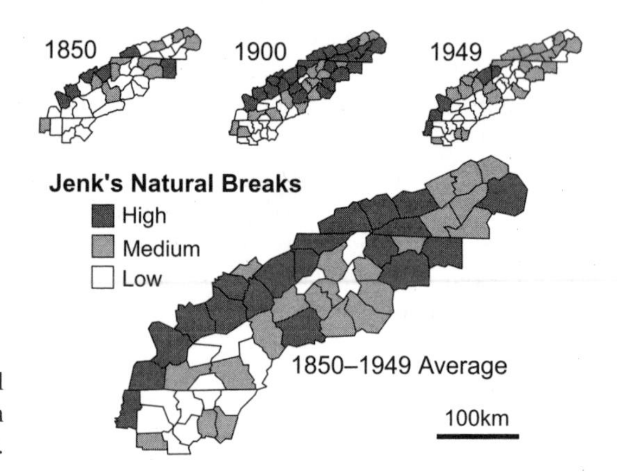

Figure 3.4 Standardized total county agricultural production (1850, 1900, and 1949).

The agroecology of Appalachia limits the production of crops such as cotton and tobacco to the warmer and lower areas, whereas fruits that require an extended cold season to set them (e.g., apples, peaches) are concentrated in the cooler and higher areas. The highest production in 1850, 1900, and 1949 corresponded roughly with the area of median to maximum agroecological suitability (Fig. 3.4). Furthermore, there were shifts throughout time in the relative production of different commodity classes. Grain and forage accounted for 94% of total production in 1850; grain, forage, and fruit accounted for 93% of total production in 1900; and forage, fiber, and grain accounted for 97% of production in 1949. Access to transportation was the most significant force at work in determining county-level production. By 1855, the counties in eastern Tennessee and southwestern Virginia were linked to Atlanta, Charleston, and the eastern seaboard via the East Tennessee and Georgia Railroad. Much of western North Carolina and north Georgia, however, had only limited roadways until after World War II (Black, 1952; Cotterill, 1924; Groover, 2003; Otto, 1989). In the early 1960s, the Appalachian Regional Commission identified "building transportation routes" as the first step to opening the region to the opportunity for economic development.

During the 19th century, southern Appalachia was considered one of the most self-sufficient regions in the country (Salstrom, 1994), with farmers growing a wide variety of fiber, forage, fruit, grain, and other products. However, the claim about self-sufficiency is not without challenge when subsistence is defined as the level at which the farm as a production unit is able to reproduce itself. Evidence from Beech Creek is highly suggestive in this regard. Beech Creek is a neighborhood in the hills of eastern Kentucky (i.e., central Appalachia) where, by 1880, up to 35% of households may have been producing below subsistence level, contributing to poverty and malnutrition (Weingartner et al., 1989). Family and kin relationships probably made possible the reproduction of marginal and below-subsistence-level farms through interhousehold strategies of survival (Halperin, 1990).

Examined globally, the average farm size for the study region in 1850 was 131 ha (323 acres). Land clearing was carried out annually by the vast majority of mountain farmers, with the general cropping procedure being to girdle trees, burn the area to kill pests and further clear the land, then plant corn between the remaining stumps (Otto, 1989; Otto and Anderson, 1982). It was necessary to clear land on an annual basis because the environment was not capable of supporting continuous cultivation, and only the wealthiest could afford to purchase fertilizers that would allow them to plant the same area every year (Otto and Anderson, 1982). By one estimate, it cost more to lime 1 acre of land in southern Appalachia than to buy 3 acres of land in the western United States (Lebergott, 1985). The principal grain across time was corn, despite an overall decrease from 60% to 13% of total production between 1850 and 1949. Corn was used to feed livestock, but it was also converted to grain alcohol for storage, transportation, and sale as "moonshine" (Davis, 2000; Otto, 1989).

Over time, the number of farms and the population increased, yet slash-and-burn agriculture remained the general practice well into the 20th century. Between 1850 and 1900, total farm area increased 19%, but the number of farms increased 275% and the average farm area decreased by 66%. The main driver behind this dramatic shift was the ever-increasing population of Appalachia, which had the highest reproductive rate of any region in America (DeJong, 1968). Between 1900 and 1949, total farm area decreased by 25%, whereas the number of farms increased by 14% and the average farm area decreased by 36%. Population growth, combined with the practice of partible inheritance, triggered the transition in Appalachia from frontier expansion to infilling during the second half of the 19th century (Salstrom, 1994). The ever-decreasing farm size meant more land needed to be cleared each year, which made slash and burn less and less desirable as a production strategy. This further undermined the viability of southern Appalachian farms. Although phosphate mining and the expansion of the rail system made fertilizer more widely available, its cost was still prohibitive for most mountain farmers (Otto and Anderson, 1982).

An additional constraint on production was the widespread introduction of fencing after the Civil War and, by 1880, the rise of the timber industry. Farmers with small holdings could no longer forage hogs and cattle on common lands. Furthermore, cattle and hog holdings as well as the human population were increasing during the latter half of the 19th century. More animals per unit area had two clear consequences. More animals meant an increase in soil compaction, which affected the capacity of the soil to retain moisture, which increased runoff and soil erosion. In addition, cattle grazed on saplings, which slowed the rate of reforestation (Otto and Anderson, 1982). By the end of the 19th century, the overall viability of the small farm in Appalachia was in jeopardy, and many full-time farmers had to seek part-time wage employment in mining and timbering to ensure their families' survival (Dunaway, 1996; Groover, 2003; Otto, 1989; Salstrom, 1994). By 1920, erosion was resulting in serious soil loss (Salstrom, 1994).

The mounting problems with agricultural production within Appalachia between 1880 and 1950 corresponded in time with the rise and consolidation of the state and federal institutional framework regulating agriculture. For example, the Hatch Act of 1887 created a national network of agricultural experiment stations, whereas the Smith-Lever Act of 1914 created the U.S. Cooperative Extension Service. However, the most far-reaching instrument was the Agricultural Adjustment Act. Its initial effect in southern Appalachia was to reduce the production of corn and hogs—the region's two largest outputs. Many mountain families did not participate in the program because it meant they would then need to spend money they no longer had to buy the corn and pork they no longer produced. The acreage cutbacks had little effect on large landholders because they could increase productivity per acre by using pesticides, fertilizers, herbicides, and mechanization that remained collectively, as before, cost prohibitive for small producers (Salstrom, 1994).

Modifications to the Agricultural Adjustment Program after 1930 allowed for regional and farm-to-farm differences, which began to improve conditions materially for small-scale farmers. In addition, the federal government began distributing phosphate freely and encouraging its use to promote an increase in groundcover—using legumes to decrease soil erosion (Salstrom, 1994). Despite such programs, livelihood remained precarious and there was a net outmigration of 1.8 million people between 1940 and 1960 (DeJong, 1968; Salstrom, 1994). The event known as the *Great Out Migration* greatly affected the subsequent history of the region.

The push-and-pull factors linked to this mass movement of population are complex. They include the increased economic opportunities in the upper Midwest created by the industrial–military complex that emerged out of World War II, as well as the previously described cycle of land and household impoverishment. By the early 1960s, Appalachia was described as "an island of distress in a sea of affluence" (Moore, 1994, p. 319, quoting Hansen, 1990, p. 133), and some authors characterized the living conditions as analogous to those found in many Third World countries (Falk and Lyson, 1988). The situation led to the creation, in the early 1960s, of the Appalachian Regional Commission, with the aim of resolving the economic disparity between Appalachia and the rest of the United States. The strategy of the commission was to build highways between population centers to further economic development and thus improve local access to educational, health, recreational, commercial, and industrial facilities. Changes were dramatic. Between 1970 and 1991, certain economic sectors—particularly tourism and service—grew as much as 600% in the 13 states composing "political" Appalachia. By reference to a matched control group of counties elsewhere in the United States, the fast-growing Appalachian counties showed superior economic, social, and public health gains (Isserman and Rephann, 1995).

As transportation networks developed, many Appalachian families abandoned a difficult, meager, and uncertain agricultural livelihood and moved to the rural areas surrounding regional cities. In doing this they took their cultural traditions with them (Halperin, 1990). For example, the strong resistance to zoning and other land-use restrictions in southern Appalachia has been related to the strong tradition of individual and family independence (Cho et al., 2003; Falk and Lyson, 1988). At the same time, large numbers of individuals from other parts of the country began migrating to southern Appalachia, reversing the effects of the Great Out Migration. The situation was set for the transformation of the Old South to the New South. The transformation of Appalachia during the past 30 years is just as dramatic, even though it has been less well studied than the transformation of the "Wild" West into the "New" West (Hansen et al., 2002).

Settlement in southern Appalachia through the 1960s was concentrated in low-lying areas on large flats or near the confluence of rivers. The contemporary trend favors individual dwellings, dispersed in loose clusters across the landscape—particularly on steep slopes and upland ridges (Wear and Bolstad, 1998). The new inhabitants, rather than striving for the proximity of kin, are seeking the relative isolation and amenity of distant views afforded by houses built high on forested slopes. Previously farmed land near streams is now reverting to forest as agricultural production is supplanted by service and recreation activities. In short, an aging local population is not being replaced reproductively or economically by its descendants and is subdividing former agroforestry lands into recreational properties for sale to in-migrants.

Appalachia is gentrifying. Newcomers (on average) are wealthier, are more highly educated, and have more urban interests than traditional southern Appalachian inhabitants (Falk and Lyson, 1988; Gragson et al., n.d.). They are also older, and most future changes in land use will be based on the need to house and serve this older, wealthier population. There is currently little evidence, however, of state-level initiatives to respond either to the future needs of this population or to the consequences of this shift in age structure and values (Cho et al., 2003).

Ecological Consequences of Agricultural Transformation

In different locations around the world, and for various reasons, land once dedicated to agriculture has been abandoned and, thus, forest and woody vegetation has expanded. Southern Appalachia is undergoing this very change with the demise of agriculture as a way of life. Our understanding of what this means for regional environments is still limited, because we have yet to link systematically the underlying mechanisms to their various long-term ecological consequences. Much of our Coweeta LTER research has been directed at reducing the similar lack of understanding of the cumulative temporal and spatial effects of multiple land uses and land-use change on water quantity, quality, and biota. The current failure to link the biophysical and socioeconomic realms seriously constrains efforts to forecast future ecosystem responses or to execute management strategies that anticipate the most likely outcomes of change trajectories. As a first step in correcting this situation, the following section summarizes what we have learned from Coweeta LTER research on the general regional consequences of agricultural land use on terrestrial and aquatic ecosystems.

We preface this discussion by emphasizing the concentrated nature of agricultural transitions to near-stream regions. Flat or gently sloping land composes less than 5% of the southern Appalachian landscape. Near-stream flat lands are among the most desirable for agriculture. The proximity to water, and hence potential habitation, is also easier, and colluvial deposits and downslope locations typically have increased fertility and moisture. Near-stream areas were the first to be farmed, and had the longest tenure and the most intense manipulations. Early surveys and aerial photographs show the highest valued and most intensively cultivated row crops, such as tobacco, were planted near streams. Grazing, farming, and living were more common in these near-stream areas, and the resultant modification of riparian vegetation and changes in flow and sediment regimes have cascaded down through the embedded aquatic environments.

These near-stream areas bear the heaviest human footprint, which continues to the present. Early transportation networks were easiest to place in the gently sloping near-stream areas, and also served the concentration of agriculture there. Cherokee villages served by footpaths gave way to farm villages and county centers connected by narrow roads. These have been replaced by local economic centers connecting urban refugees and retirees. With a few notable exceptions, the current settlement patterns and impacts are driven by the history of agricultural land use.

Terrestrial Ecosystems

In the mesic forests of southern Appalachia, past agricultural land use is associated with a decrease in herbaceous species richness and total herbaceous cover. Where past land use was intense, the cover of liliaceous, old-growth, and mesophytic forest herbs is reduced and that of weedy species increased (Fig. 3.5). However, life-history characteristics interact strongly with landscape pattern to determine final species distribution in disturbed forests (Pearson et al., 1998). Native mesophytic species are less abundant in small patches because they lacked adaptations for long-range dispersal by wind or animals; native species with adaptations for long-range dispersal are equally abundant in small and large patches. Modeling studies based on long-term seed dispersal data demonstrate large differences among tree species in the rates at which they colonize abandoned agricultural land (Clark et al., 1998). Even in closed stands, only a subset of species predictably disperses seed to open sites. Dispersal limitations appear to be a major obstacle to the rate of spread. Species producing large quantities of well-dispersed seed, including *Betula*, *Acer*, and *Liriodendron*, have the advantage (Clark et al., 1999).

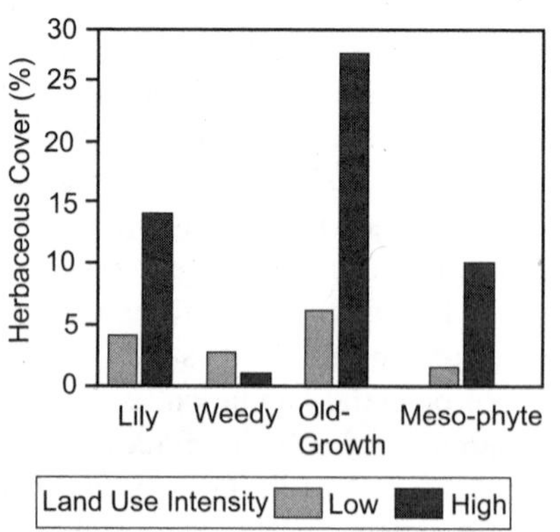

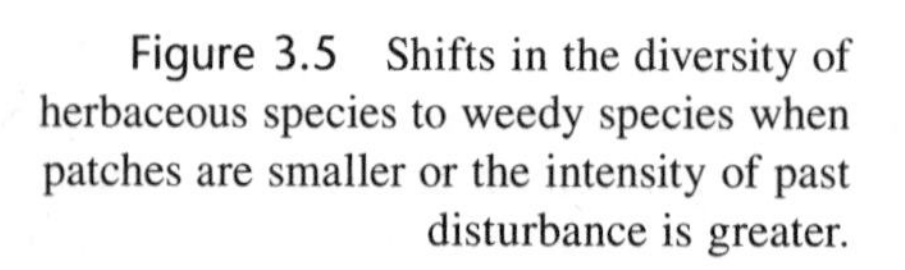
Figure 3.5 Shifts in the diversity of herbaceous species to weedy species when patches are smaller or the intensity of past disturbance is greater.

There were undoubtedly indirect effects of agricultural transformation on ecosystems, and converse indirect effects of ecosystems on agricultural transformation. Some of these effects are well supported whereas others await investigation. For example, a number of studies have documented an increase in the brown-headed cowbird (*Molothrus ater*), a nest parasite of many forest birds, resulting from increased agriculture (Brittingham and Temple, 1996; Ortega, 1998). Cowbirds forage primarily in open areas, often on invertebrates disturbed by the movements of cattle, and more frequently parasitize birds nesting near forest edges. Increases in agricultural edges and livestock have resulted in substantial declines in many forest bird species, although impacts from the Precontact through the Columbian Revolution periods are speculative.

Agricultural transitions also affected large herbivore populations. White-tailed deer (*Odocoileus virginianus*) and eastern elk (*Cervus elaphus canadensis*) are often the most common species found in paleo-Indian middens (Guilday and Tanner, 1965; McMichael, 1963), and were a primary food source prior to agriculture. Both species thrive in a forest/open mosaic, and early journals note extensive habitat manipulation by Native Americans to improve deer habitat (McCabe and McCabe, 1984). The area, frequency, and intensity of deliberate burning is still in dispute (Pyne, 2001). Ungulate populations were undoubtedly affected by agricultural expansion and contraction during the Mississippian period, but in an unknown and probably unknowable way. Hunting pressures may have decreased as a result of a shift toward agricultural expansion, or increased because of a net human population increase.

Journal entries and economic records provide some estimate of human impacts on these large-game species during the second agricultural expansion that took place during the 1700s and early 1800s, a period straddling the Columbian Revolution and early Nationhood periods. Subsistence requirements prior to contact drove the deer and elk harvest, and needs for leather may have been more a factor behind harvest rates than protein or caloric requirements (Driver, 1969; McCabe and McCabe, 1984). Early contact harvests are generally estimated to be between 3 and 10 deer and a fraction of an elk per person annually—rates that are substantially below levels that would control ungulate populations. Harvest rates increased substantially because hides were the primary commodity through which Native Americans could obtain axes, hoes, pots, and other desired manufactured goods. Woodland harvest reached unsustainable proportions during the early Nation Building period. Agricultural development in the coastal region supported a local European American population, the development of trading centers, and overharvest in the as-yet uncolonized interior regions. As agriculture and colonization advanced toward the uplands, hide and market hunting by both Native Americans and European Americans increased. The pattern continued after Native Americans moved out, and eventually led to the extirpation of eastern elk and substantial reduction of eastern deer populations.

Agricultural transitions also affect mass, energy, and elemental cycling through ecosystems. As forests were converted to agricultural fields, carbon stored in standing wood and forest floor biomass was lost; live biomass and necromass contain a majority of the labile carbon in southern Appalachian forests. Agricultural

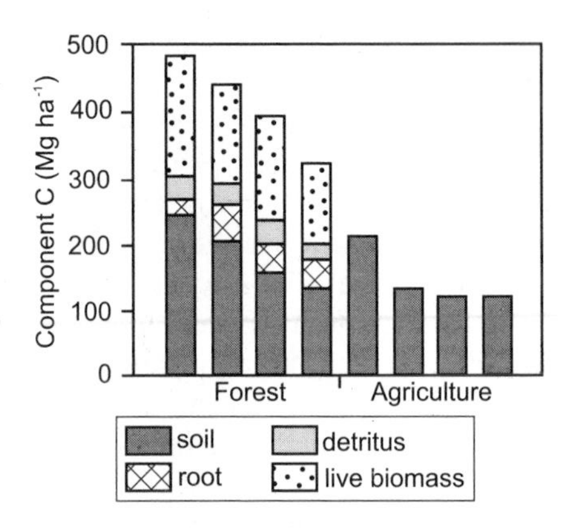

Figure 3.6 Carbon loss resulting from conversion of forest to agriculture in system components.

production prior to the introduction of commercial mineral fertilizers in the early 1900s (Davidson and Ackerman, 1993), measured as annual dry biomass accumulation, is only a small fraction of forest production. Aboveground net primary production in southern Appalachian cove and lowland forest sites typically ranges from 10 to 12 Mg biomass/ha/year, whereas nonfertilized agricultural plots typically produce less than 2 Mg biomass/ha/year. Productivity on fertilized agricultural sites is higher, but rarely reaches 50% of that observed on forested sites (Fig. 3.6).

Valley and cove forests potentially have the highest aboveground carbon pools in southern Appalachia. Cove forests, in particular, carry up to 230 Mg carbon/ha (Whittaker, 1966). However, valley and cove forest sites are also the most affected by human land use across all time periods, save perhaps the last 30 years (Bolstad et al., 1998; Wear and Bolstad, 1998). Comparative measurements on such sites in southern Appalachia indicate that aboveground carbon pools on row-crop and pasture sites average approximately 2 Mg carbon/ha, which is only a tiny fraction of the 151 Mg carbon/ha found on comparable sites with mature, once-harvested forests (Vose and Bolstad, 2007).

Belowground soil carbon loss depends on many factors, and across southern Appalachia ranges from low (Kalisz, 1986) to substantial (Vose and Bolstad, 2007). Most of the belowground carbon pool on forest sites is contained in total coarse root and stump carbon (Harris et al., 1977). The majority of this pool is lost in the first few decades after such sites are converted to agricultural use. On agricultural sites, carbon losses often increase relative to carbon inputs resulting from the increase in soil temperature, reduced soil carbon inputs, and the addition of nitrogen fertilizers (Davidson and Ackerman, 1993). When agricultural use of a site ends and forest is allowed to regrow, much of the aboveground live biomass recovers within the first century.

Aboveground necromass recovers more slowly than biomass, and soil carbon recovers even more slowly. It can take from several decades to several centuries to return soil carbon to preforest clearing levels (Schlesinger, 1990). In general, regional carbon stocks decreased in southern Appalachia in concert with increases in agricultural land use from AD 800 to the early 1900s (Delcourt and Harris, 1980). Since 1900, carbon stocks have increased as farms were abandoned, fires were suppressed, and regional manufacturing and service economies replaced agriculture. Soil carbon changes resulting from changes in land use also affect the habitat quality for understory forest herbs. This results in lower forest herb diversity and abundance (Pearson et al., 1998). The change in soil quality is a consequence of increased organic matter with greater water retention capacity, improved aeration and tilth, and enhanced supplies of plant nutrients (Coleman and Crossley, 1996). In summary, land-use changes have been most complete, intense, and persistent in floodplain and cove sites. This creates numerous potential pathways for impacts to occur on aquatic ecosystems. For example, conversion of forest to agriculture removes trees that both shade and deliver substantial quantities of matter and energy (as leaves and woody litterfall) to streams (Wallace et al., 1999).

Aquatic Ecosystems

The aquatic ecosystems of southern Appalachia are among the world's most diverse for a number of reasons. The mountains encompass a broad range of elevation and hence climactic regimes, providing both cool, shaded uplands and warmer reaches that border the piedmont. Higher elevations are geographically isolated, particularly for cool-adapted species, as thermal barriers downstream often prevent or reduce interbasin migration. The region has never been glaciated and many of the river systems drain to the south, which allowed a range shift to southern refugia during the preceding glaciation. Fish diversity is high, and mussel diversity is among the highest in the world, but it is particularly imperiled. Damage is the result primarily of increased sediment from farming and development, changes in a suite of characteristics resulting from dam building, and the introduction of exotic organisms.

Places like southern Appalachia where substantial portions of the landscape were in agriculture for decades and then abandoned and reforested raise important questions about how much stream ecosystems reflect past versus current land use. A series of recent Coweeta LTER studies have focused specifically on the legacy of past land use evident in contemporary aquatic systems (Harding et al., 1998; Jones et al., 1999). Forest streams generally have higher diversity and abundance of clean-water benthic macroinvertebrates than streams in agricultural land. However, forest streams also have lower fish diversity and abundance, largely consisting of introduced rainbow or brown trout that presumably ate or displaced most other species. Agricultural streams do not contain trout, but rather have a mixture of native and introduced species that tolerate high levels of fine sediment and higher water temperatures. The best indicator of 1990s stream biodiversity in southern Appalachia is land use in 1950 as well as certain measures of current stream water quality (Groves et al., 2002; Scott, 2001). Most streams on land forested in 1950 had higher biodiversity than streams on agricultural land at the same time, irrespective of land use in 1990.

The widespread abandonment of farms across the region means that much of the landscape that is currently forested was farmed in 1950. However, land-use change has been largely unidirectional. Many farms were converted to forests,

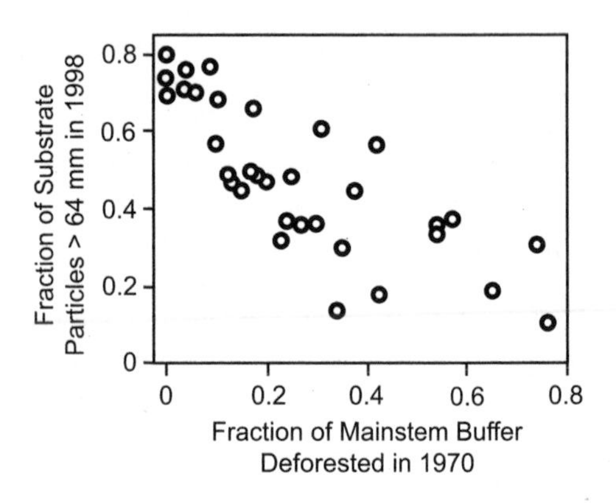

Figure 3.7 Fine sediment input to the substrate of small southern Appalachian rivers as a function of land use.

but relatively few new farms were established in forest-covered land (Scott, 2001). Nevertheless, multivariate analyses of stream faunal communities were used to distinguish sites linked to land in agriculture in 1950 and sites linked to land in forest in 1950 (Harding et al., 1998). Moreover, these groupings cut across current land use. Streams on currently forested land that was farmed within the past 50 years have fish and invertebrate communities comparable with streams on land currently in agricultural use. They are not comparable with streams on forested land that was not cleared within the past 50 years. Despite their appearance, currently forested sites harbor the "ghost of land use past" (Harding et al., 1998). The critical determinant is stream substrate, which is important at many life-history stages for both vertebrate and invertebrate organisms. Therefore, the quality of stream substrate is most strongly related to past, not present, land use (Fig. 3.7) (Scott, 2001).

Riparian corridor width has long been considered the most important determinant of the impact of land use on streams. Recent Coweeta LTER research, however, indicates that riparian corridor length may be as important as corridor width (Jones et al., 1999). Fish diversity and abundance in sampled streams were most strongly related to the length of unbroken forest immediately upstream from the sampling location. Invasive and sediment-tolerant species were most common where agricultural clearing extended more than 1 km upstream from the sampling location. They were least common on stream reaches where forest cover prevailed. Linear agricultural patches paralleling streams are associated with higher inputs of stream sediments. As the area and length of upstream agricultural patches increased, fish species that nest on the bottom and do not clean sediment from their nests decreased in abundance whereas species that keep their nests free of silt increased in abundance. In summary, our findings suggest that nearly 50 years of forest regrowth failed to return southern Appalachian stream biota to that characteristic of forested streams.

Although agricultural practices and transitions have substantially affected the biotic integrity of aquatic ecosystems, these impacts do little to constrain or affect land use and human actions directly. Although fish were a mainstay of native diets during long intervals in the Precontact and European American periods, few material goods were harvested from aquatic ecosystems during the post-European periods. The biotic consequences of agricultural transitions after colonization are in some sense unidirectional, in that the activities have directly changed the functioning of aquatic biota, and feedback on human action is light and indirect. The largest impacts are within a context of a cultural appreciation of biodiversity and wild spaces, and attempts to manage public and private land to maintain aquatic biodiversity.

Although agricultural impacts on the aquatic biota are great, they appear largely unidirectional, and impacts on physical resources derived from aquatic ecosystems are pervasive and profound. Agriculture altered stream physiography, structure, and transport capacity. Stream bottoms are more finely textured, banks are more incised, and floodplains expanded as a result of past farming practices. Floods are higher, more frequent, and occur more quickly than in times past, thereby suggesting more stringent limits to building in floodplains and other flood-prone areas.

Conservation in Southern Appalachia

The gentrification of southern Appalachia in combination with the legacy of past land-use practices on contemporary terrestrial and aquatic ecosystems has important implications for regional conservation efforts. Are they reactive or anticipatory? Do they focus on integrative systems or on rescuing biophysical systems to the exclusion of socioeconomic systems? To begin answering these questions, we concentrate in this section on characterizing the nature of contemporary conservation efforts in southern Appalachia. We do this based largely on an assessment of Web information presented by the 81 conservation organizations currently active in southern Appalachia.

These organizations range from national in scope, with and without local chapters, to site-based organizations: national, 19 (7.1%); regional, 14 (5.2%); state, 15 (5.6%); local, 30 (11.2%); and site, 3 (1.1%). In total, they target some 267 conservation-related issues that can be classified into 15 overarching classes. When the dominant scope of the organization (national to site) is related to issue classes, clear priorities emerge. Growth is somewhat of an exception (e.g., urban sprawl, development) because it is addressed by organizations of both national and local scope.

Local groups often serve as watchdog organizations targeting disturbance activities such as asphalt plants, chip mills, power lines, solid waste incinerators, and landfills. They may also advocate improved planning of forests and wilderness and recreation areas. Most are isolated from any regional or national umbrella organization and are typically run by unpaid volunteers. Once the issue that galvanized the formation of the group is resolved, they may cease to operate, either by decision or for lack of interest. Although this can mean a high turnover through time in the presence and activity of local groups, some local groups build a lasting organization that reflects regional visions and foments public involvement. Such organizational transformation is evident in watershed organizations such as the Chattooga Conservancy and the New River Community Partners. Although these are local groups, they nevertheless remain active by putting forth a vision for the future of their target watersheds. They do this by prioritizing their efforts to fend off incompatible land uses and taking concrete steps toward achieving their vision.

Regional organizations tend to fall into one of three categories: forest issue groups, air quality groups, and support service groups. Forestry issue groups are well organized, active, and interconnected. They include organizations such as the Southern Appalachian Biodiversity Project, the Southern Appalachian Forest Coalition, and the Dogwood Alliance. These three organizations work closely together, using grassroots strategies to monitor timber sales, the designation of wilderness areas, the listing of endangered species, and forest product marketing campaigns. Appalachian Voices is a membership organization focused on many of the same issues, and is also the most visible group in the Blue Ridge fighting mountaintop removal. Katuah Earth First! is a regional network of activists most active in Tennessee, with members who focus on forest and nuclear issues and who use highly visible and often illegal tactics.

The current Bush administration's changes to the Clean Air Act have recently focused conservation efforts on air quality issues in southern Appalachia. The New Source Review regulations of the Clean Air Act require power companies and other industrial facilities to install advanced emission controls when undertaking major modifications that would significantly increase their emissions. The Southern Alliance for Clean Energy and the Canary Coalition are regional groups focused on "green power" legislation and coal-fired power plants at the state level, and New Source Review and climate change at the federal level. The effectiveness of such organizations and their proposals on air quality are highly variable. This is a function of the inherent uncertainty of the issues, the ambiguity of the regulations themselves, and the internal and external scientific expertise on which these organizations can draw.

National organizations active in southern Appalachia tend to be those with local chapters or local affiliates working with grassroots volunteers. The Sierra Club is the paradigm for such a group because it straddles the boundary between a local grassroots and a large national organization. The same is true of The National Forest Protection Alliance. Both organizations work on forestry, air quality, and education issues. Staff-based national advocacy organizations are much less active in the region. Although their websites may place southern Appalachia within an ecoregional panorama, local participation is generally limited to supporters sending money. Exceptions are organizations focused on national parks such as the National Parks Conservation Association working in the Great Smoky Mountains National Park.

There are two types of site-based organizations operating in the region: eco-villages and land trusts. Eco-villages are intentional communities of conservation-minded people living a sustainable lifestyle, such as Earthaven Eco-Village and Narrow Ridge Earth Literacy Center. Only a minute percentage of the population of the region live in eco-villages, but the approach of such groups is notable. Instead of activism, residents of these centers seek to build a sustainable village that others may join or mimic. They also offer educational programs in permaculture, sustainable design, and folk arts.

Land trusts are supported by a different clientele than other conservation organizations, but are united in their strategy of conservation through preservation of real property. Land trusts need money, and people like to see and live near where their money is used. For example, the greatest concentration of land trust organizations in the region is in and around Highlands, North Carolina, a historically wealthy pocket in southern Appalachia. Nevertheless, there is a strong support network for land trusts in southern Appalachia and there are local-, state-, regional-, and national-level organizations. The Conservation Trust for North Carolina is a service organization helping to form and operate local land trusts.

Forestry is the most significant concern of conservation organizations in southern Appalachia. The origin of this interest is the move within the last 15 years of the forest industry from the Pacific Northwest to the Southeast. The identified conservation issues revolve around fighting chip mills, invasive species, timber sales on public lands, logging, road building, the Bush administration's "Healthy Forest Initiative," and Revised Statute 2477 (a loophole allowing counties to claim nearly anything as an abandoned right-of-way where a road can be built). Some organizations focus on positive mechanisms. They concentrate on designating wilderness areas, protecting roadless areas, promoting sustainable private forestry, designating critical habitats for endangered species, reforming the U.S. Forest Service, and developing national recreation areas. All these issues coalesce in the 114-page document drafted by the Southern Appalachian Forest Coalition entitled *Return the Great Forest* (Irwin et al., 2002), which advocates establishing an interconnected network of wildlands covering 2.8 million acres in southern Appalachia. The plan is endorsed by more than 200 stakeholders and identifies concrete steps to be taken by funders, legislators, and the public.

There is a notable absence of organizations in southern Appalachia concerned with issues related to urban sprawl and smart growth. Although some groups have organized to stop road construction, such as the North Shore Road in Great Smoky Mountains National Park, such activities have been primarily carried out in the context of a focused effort at combating forest fragmentation. Virginians for Appropriate Roads (a chapter of the Blue Ridge Environmental Defense League) concentrates its efforts on stopping two proposed highways in southwestern Virginia. Local groups of the Sierra Club have included land use and transportation in their efforts, but these are tangential to their other activities and are also typically restricted in their area of influence. In general, there is a true lack of attention to the systematic problem of sprawling development now occurring throughout the region and taking place not only adjacent to cities but also in what are officially "rural" areas. Smart Growth Partners of Western North Carolina is one of the few organizations with a specific urban outlook, although it is restricted to the Asheville area.

Conclusion

This overview of the agricultural transformation of southern Appalachia clearly reveals how humans have insinuated themselves into regional biophysical systems at all levels. The spatial and temporal patterns of ecological systems bear the signature of human activities and institutions; however, it is also true that human activities and institutions have been shaped by the ecological systems in which they are embedded. This reciprocal imprinting means that the artificial separation of the two components will fail to improve sustainability of either the landscape or the quality of life. The general linear reality view of how and why events occur in either realm continues to underpin regional historiography, environmental studies, and conservation efforts.

Our multiscalar, historical, and comparative approach to the agricultural transformation of southern Appalachia is the beginning of a reconciliation of the socioeconomic and biophysical realms. This pattern-based assessment within the framework of narrative positivism is the first step toward developing knowledge of general processes and relationships. Our next step is more analytical, and focuses on the network and hierarchy of specific commodity classes at finer temporal and spatial scales (Bolstad and Gragson, 2008; Gragson and Bolstad, 2007). The objective of this subsequent study is to establish the trajectory and turning points that we can use to identify transformation signatures from which to formulate high probability forecasts of future ecosystem and socioeconomic responses. Decision makers will then be in the position of defining and executing management decisions that truly anticipate the most likely outcomes of change.

References

Abbott, A. 2001. *Time matters: On theory and method.* Chicago, Ill.: University of Chicago Press.

Agee, J. K. 1993. *Fire ecology of Pacific Northwest forests.* Washington, D.C.: Island Press.

Agrawal, A. 2001. "Common property institutions and sustainable governance of resources." *World Development* 29: 1649–1672.

Anderson, D. G. 1994. *The Savannah River chiefdoms: Political change in the late prehistoric southeast.* Tuscaloosa, Ala.: University of Alabama Press.

Anderson, D. G. 1995. "Paleoindian interaction networks in the eastern woodlands," pp. 3–26. In: M. S. Nassaney and K. E. Sassaman (eds.), *Native American interactions: Multiscalar analyses and interpretations in the eastern woodlands.* Knoxville, Tenn.: University of Tennessee Press.

Anderson, D. G., and R. C. Mainfort Jr. 2002. "An introduction to woodland archaeology in the Southeast," pp. 1–19. In: D. G. Anderson and R. C. Mainfort Jr. (eds.), *The woodland Southeast.* Tuscaloosa, Ala.: University of Alabama Press.

Anderson, D. G., and K. E. Sassaman. 1996. *The Paleoindian and Early Archaic Southeast.* Tuscaloosa, Ala.: University of Alabama Press.

Anglin, M. 2002. *Women, power, and dissent in the hills of Carolina.* Urbana, Ill.: University of Illinois Press.

Arthur, J. P. 1914. *Western North Carolina: A history from 1730 to 1913.* Raleigh, N.C.: Edwards and Broughton Printing.

Bailey, R. G. 1996. *Ecosystem geography.* New York: Springer-Verlag.

Barnes, B. V. 1991. "Deciduous forests of North America," p. xx. In: E. Röhrig and B. Ulrich (eds.), *Ecosystems of the world: 7 temperate deciduous forests.* Amsterdam: Elsevier.

Bennett, E. M., S. R. Carpenter, G. D. Peterson, G. S. Cumming, M. Zurek, and P. Pingali. 2003. "Why global scenarios need ecology." *Frontiers Ecology and Environment* 1: 322–329.

Berkes, F., and C. Folke (eds.). 1998. *Linking social and ecological systems.* London: Cambridge University Press.

Black, R. C., III. 1952. *The railroads of the Confederacy.* Chapel Hill, N.C.: University of North Carolina Press.

Blench, T. 1957. *Regime behavior of canals and rivers.* London: Butterworths Scientific Publications.

Bolstad, P. V., and T. L. Gragson. 2008. "Resource abundance as constraints on early post-contact Cherokee populations." *Journal of Archaeological Research* 20: 1–14.

Bolstad, P. V., L. Swift, F. Collins, and J. Regniere. 1998. "Measured and predicted air temperatures at basin to regional scales in the southern Appalachian Mountains." *Agricultural and Forest Meteorology* 91(3–4): 161–176.

Braun, E. L. 1950. *Deciduous forests of eastern North America.* New York: McGraw-Hill.

Bridenbaugh, C. 1971. *Myths and realities: Societies of the colonial South.* New York: Atheneum.

Brittingham, M. C., and S. A. Temple. 1996. "Vegetation around parasitized and non-parasitized nests within deciduous forest." *Journal of Field Ornithology* 67: 406–413.

Brown, J. K. 1995. "Fire regimes and their relevance to ecosystem management," pp. 171–178. In: *Proceedings of Society of American Foresters National Convention, Anchorage, AK, 18–22 September, 1994.* Washington, D.C.: Society of American Foresters.

Burritt, C. 1997. "Habitat comes to help in the hills." *Atlanta Journal–Constitution* June 15: A1.

Campbell, J. L., and M. P. Allen. 2001. "Identifying shifts in policy regimes: Cluster and interrupted time-series analyses of U.S. income taxes." *Social Science History* 25: 187–216.

Carpenter, S. R., and L. H. Gunderson. 2001. "Coping with collapse: Ecological and social dynamics in ecosystem management." *BioScience* 51: 451–457.

Caudill, H. 1963. *Night comes to the Cumberlands.* Boston: Little, Brown.

Chapman, J. 1994. *Tellico archaeology: 12,000 years of Native American history. Report of investigations.* Vol. 43. Knoxville, Tenn.: University of Tennessee Press.

Chapman, J., and A. B. Shea. 1981. "The archaeobotanical record: Early Archaic period to contact in the Lower Little Tennessee River Valley." *Tennessee Anthropologist* 6: 61–84.

Cho, S.- H., D. H. Newman, and D. N. Wear. 2003. "Impacts of second home development on housing prices in the southern Appalachian highlands." *Review of Urban and Regional Development Studies* 15(3): 208–225.

Clark, J. S., E. Macklin, and L. Wood. 1998. "Stages and spatial scales of recruitment limitation in southern Appalachian forests." *Ecological Monographs* 68(2): 213–235.

Clark, J. S., M. Silman, R. Kern, E. Macklin, and J. Hille Ris Lambers. 1999. "Seed dispersal near and far: Patterns across temperate and tropical forests." *Ecology* 80: 1475–1494.

Coleman, D. C., and D. A. Crossley Jr. 1996. *Fundamentals of soil ecology.* San Diego, Calif.: Academic Press.

Cotterill, R. S. 1924. "Southern railroads, 1850–1860." *Mississippi Valley Historical Review* 16: 396–405.

Crites, G. D. 1993. "Domesticated sunflower in fifth millennium BP temporal context: New evidence from middle Tennessee." *American Antiquity* 58: 146–148.

Cronon, W. 1983. *Changes in the land: Indians, colonists, and the ecology of New England.* New York: Hill and Wang.

Davidson, E. A., and I. L. Ackerman. 1993. "Changes in soil carbon inventories following cultivation of previously untilled soils." *Biogeochemistry* 20: 161–193.

Davis, D. E. 2000. *Where there are mountains: An environmental history of the southern Appalachians.* Athens, Ga.: University of Georgia Press.

Davis, R. P. S., Jr. 1990. *Aboriginal settlement patterns in the Little Tennessee River Valley.* Vol. 54, Tennessee Valley Authority publications in anthropology. Knoxville, Tenn.: University of Tennessee.

DeJong, G. 1968. *Appalachian fertility decline: A demographic and sociological analysis.* Lexington, Ky.: University Press of Kentucky.

Delcourt, H. R., and P. A. Delcourt. 1981. "Vegetation maps for eastern North America: 40,000 years ago B.P. to the present," pp. 123–165. In: R. C. Romans (ed.), *Geobotany II.* New York: Plenum.

Delcourt, H. Z., and W. F. Harris. 1980. "Carbon budgets of the southeastern US biota: Analysis of historical change in trend from source to sink." *Science* 210: 321–323.

Delcourt, P. A., and H. R. Delcourt. 1983. "Late Quaternary vegetational dynamics and community stability reconsidered." *Quaternary Research* 13: 111–132.

DePratter, C. B. 1983. *Late prehistoric and early historic chiefdoms in the southeastern United States.* PhD diss., University of Georgia, Athens, Ga.

Diamond, J. 1997. *Guns, germs, and steel: The fates of human societies.* New York: W. W. Norton.

Dixon, M. 1976. *The Wataugans.* Nashville, Tenn.: Tennessee American Revolution Bicentennial Commission.

Dove, M., and D. Kammen. 1997. "The epistemology of sustainable resource use: Managing forest products, swidden, and high-yielding variety crops." *Human Organization* 56(1): 91–101.

Driver, H. E. 1969. *Indians of North America.* Chicago, Ill.: University of Chicago Press.

Dunaway, W. A. 1996. *The first American frontier: Transition to capitalism in southern Appalachia, 1700–1860.* Chapel Hill, N.C.: University of North Carolina Press.

Eller, R. 1982. *Miners, millhands, and mountaineers: Industrialization of the Appalachian South, 1880–1930.* Knoxville, Tenn.: University of Tennessee Press.

Falk, W. W., and T. A. Lyson. 1988. *High tech, low tech, no tech: Recent industrial and occupational change in the South.* Albany, N.Y.: State University of New York Press.

Farina, A. 2000. "The cultural landscape as a model for the integration of ecology and economics." *BioScience* 50: 313–320.

Fischer, D. H. 1989. *Albion's seed: Four British folkways in America.* New York: Oxford University Press.

Fite, G. C. 1984. *Cotton fields no more: Southern agriculture, 1865–1980.* Lexington, Ky.: University Press of Kentucky.

Fraterrigo, J. M., M. G. Turner, S. M. Pearson, and P. Dixon. 2005. "Effects of past land use on spatial heterogeneity of soil nutrients in southern Appalachian forests." *Ecological Monographs* 75: 215–230.

Fritz, G. J. 1993. "Early and Middle Woodland Period paleoethnobotany," pp. 39–56. In: C. M. Scarry (ed.), *Foraging and farming in the eastern woodlands.* Gainesville, Fla.: University Press of Florida.

Fritz, G. J. 2000. "Native farming systems and ecosystems in the Mississippi River Valley," pp. 225–249. In: D. L. Lentz (ed.), *Imperfect balance: Landscape transformation in the precolumbian Americas.* New York: Columbia University Press.

Geisler, C. C. 1983. *Who owns Appalachia? Landownership and its impact.* Lexington, Ky.: University Press of Kentucky.

Goodwin, G. C. 1977. *Cherokees in transition: A study of changing culture and environment prior to 1775.* University of Chicago, Department of Geography, research paper, vol. 181. Chicago, Ill.: University of Chicago Press.

Gragson, T. L., and P. V. Bolstad. 2007. "A local analysis of early 18th century Cherokee settlement." *Social Science History* 31: 435–468.

Gragson, T. L., D. Newman, and C. Espy. n.d. *Private land management practice and intent in north Georgia.* (Forthcoming.)

Gremillion, K. J. 1989. *Late Prehistoric and Historic period paleoethnobotany of the North Carolina piedmont.* PhD diss., University of North Carolina, Chapel Hill, N.C.

Gremillion, K. J. 2002. "The development and dispersal of agricultural systems in the Woodland Period Southeast," pp. 483–501. In: D. G. Anderson and R. C. Mainfort Jr. (eds.), *The woodland Southeast.* Tuscaloosa, Ala.: University of Alabama Press.

Groover, M. D. 2003. *An archaeological study of rural capitalism and material life: The Gibbs farmstead in southern Appalachia, 1790–1920.* Contributions to global historical archaeology. New York: Kluwer Academic/Plenum Publishers.

Groves, C. R., D. B. Jensen, L. L. Valutis, K. H. Redford, M. L. Shaffer, J. M. Scott, J. V. Baumgartner, J. V. Higgins, M. W. Beck, and M. G. Anderson. 2002. "Planning for biodiversity conservation: Putting conservation science into practice." *BioScience* 52: 512.

Guilday, J. E., and D. P. Tanner. 1965. "Vertebrate remains from the Mount Carbon Site (46 Fa 7), Fayette County, West Virginia." *West Virginia Archeology* 18: 1–14.

Hally, D. J. 1996. "Platform–mound construction and the instability of Mississippian chiefdoms," pp. 92–127. In: J. F. Scarry (ed.), *Political structure and change in the prehistoric southeastern United States.* Gainesville, Fla.: University Press of Florida.

Halperin, R. 1990. *The livelihood of kin: Making ends meet the "Kentucky Way."* Austin, Tex.: University of Texas Press.

Hansen, A. J., R. Rasker, B. Maxwell, J. J. Rotella, J. D. Johnson, A. W. Parmenter, U. Langner, W. B. Cohen, R. L. Lawrence, and M. P. Kraska. 2002. "Ecological causes and consequences of demographic change in the New West." *BioScience* 52: 151–162.

Harding, J. S., E. F. Benfield, P. V. Bolstad, G. S. Helfman, and E. B. D. Jones III. 1998. "Stream biodiversity: The ghost of land use past." *Proceedings of the National Academy of Sciences* 95(25): 14843–14847.

Harris, W. F., R. S. Kinerson, and N. T. Edwards. 1977. "Comparison of belowground biomass of natural deciduous forest and loblolly pine plantations." *Pedobiologia* 17: 369–381.

Hatley, T. 1993. *The dividing paths: Cherokees and South Carolinians through the era of revolution.* New York: Oxford University Press.

Hoffman, P. E. 1994. "Lucas Vázquez de Ayllón's discovery and colony," pp. 36–49. In: C. H. Hudson and C. C. Tesser (eds.), *The forgotten centuries: Indians and Europeans in the American South, 1521–1704.* Athens, Ga.: University of Georgia Press.

Hudson, C. M. 1990. *The Juan Pardo expeditions: Spanish explorers and the Indians of the Carolinas and Tennessee, 1566–1568.* Washington, D.C.: Smithsonian Institution Press.

Hudson, C., M. T. Smith, D. J. Hally, R. Polhemus, and C. B. DePratter. 1985. "Coosa: A chiefdom in the sixteenth century United States." *American Antiquity* 50: 723–737.

Hunter, M. D., and R. E. Forkner. 1999. "Hurricane damage influences foliar polyphenolics and subsequent herbivory on surviving trees." *Ecology* 80: 2676–2682.

Inscoe, J. C. 1989. *Mountain master: Slavery and the sectional crisis in western North Carolina*. Knoxville, Tenn.: University of Tennessee Press.

Irwin, H., S. Andrew, and T. Bouts. 2002. *Return the Great Forest: A conservation vision for the southern Appalachian region*. Asheville, N.C.: Southern Appalachian Forest Coalition.

Isaac, I. W., S. M. Carlson, and M. P. Mathis. 1994. "Quality of quantity in comparative/historical analysis: Temporally changing wage labor regimes in the United States and Sweden," pp. 54–92. In: T. Janoski and A. M. Hicks (eds.), *The comparative political economy of the welfare state*. New York: Cambridge University Press.

Isaac, I. W., and L. Griffin. 1989. "A historicism in time-series analyses of historical process: Critique, redirection, and illustrations from U.S. labor history." *American Sociological Review* 54: 873–890.

Isaac, I. W., and K. T. Leight. 1997. "Regimes of power and the power of analytic regimes: Explaining U.S. military procurement Meynesianism as historical process." *Historical Sociology* 30: 28–45.

Isserman, A., and T. Rephann. 1995. "The economic effects of the Appalachian Regional Commission: An empirical assessment of 26 years of regional development." *Planning Journal of the American Planning Association* 61: 345–364.

Johannessen, S. 1993. "Farmers of the Late Woodland," pp. 57–77." In: C. M. Scarry (ed.), *Foraging and farming in the Eastern woodlands*. Gainesville, Fla.: University Press of Florida.

Jones, E. B. D., III, G. S. Helfman, J. O. Harper, and P. V. Bolstad. 1999. "The effects of riparian deforestation on fish assemblages in southern Appalachian streams." *Conservation Biology* 13: 1454–1465.

Kalisz, P. J. 1986. "Soil properties of steep Appalachian old fields." *Ecology* 67: 1011–1023.

Keel, B. C. 1972. *Woodland phases of the Appalachian summit area*. PhD diss., Washington State University, Pullman, Wash.

Kitschelt, H. 1992. "Political regime change: Structure and process-driven explanations?" *American Political Science Review* 86: 1028–1034.

Kolchin, P. 2003. *A sphinx on the American land: The nineteenth-century South in comparative perspective*. Baton Rouge, La.: Louisiana State University Press.

Kondratieff, N. D. 1979. "The long waves in economic life." *Review of Economic Statistics* 17(6) 1935: 105–115. Full English translation in *Review* 2(4) 979: 519–562.

Kretzschmar, W. A., Jr., V. G. McDavid, T. K. Lerud, and E. Johnson. 1993. "Settlement history," pp. 154–164. In: W. A. Kretzschmar, V. G. McDavid, T. K. Lerud, and E. Johnson (eds.), *Handbook of the linguistic atlas of the Middle and South Atlantic states*. Chicago, Ill.: University of Chicago Press.

Lebergott, S. 1985. "The demand for land: The United States, 1820–1860." *Journal of Economic History* 65: 181–212.

Levin, S. A. 1999. *Fragile dominion: Complexity and the commons*. Reading, Mass.: Perseus Books.

Markusen, A. R. 1987. *Regions: The economics and politics of territory*. Totowa, N.J.: Rowman and Littlefield.

Matson, P. A., W. J. Parton, A. G. Power, and M. J. Swift. 1997. "Agricultural intensification and ecosystem properties." *Science* 277: 504–509.

McCabe, R. E., and T. R. McCabe. 1984. "Of slings and arrows: An historical perspective," pp. 19–72. In: L. K. Halls (ed.), *White tailed deer ecology and management*. Washington, D.C.: Wildlife Management Institute.

McMichael, E. V. 1963. "1963 excavations at the Buffalo Site, 46 PU 31." *West Virginia Archeology* 16: 12–23.

Meltzer, D. J. 1988. "Late Pleistocene human adaptations in eastern North America." *Journal of World Prehistory* 2: 1–52.

Merriwether, R. L. 1940. *The expansion of South Carolina 1729–1765*. Kingsport, Tenn.: Southern Publishers.

Mooney, J. 1995. *Myths of the Cherokees*. New York: Dover Publications.

Moore, T. G. 1994. "Core-periphery models, regional-planning theory, and Appalachian development." *Professional Geographer* 46(3): 316–331.

Newman, R. D. 1979. "The acceptance of European domestic animals by eighteenth century Cherokee." *Tennessee Anthropologist* 4: 101–107.

Nilsson, C., J. E. Pizzuto, G. E. Moglen, M. A. Palmer, E. H. Stanley, N. E. Bockstael, and L. C. Thompson. 2003. "Ecological forecasting and the urbanization of stream ecosystems: Challenges for economists, hydrologists, geomorphologists, and ecologists." *Ecosystems* 6: 659–674.

North, D. C. 1961. *The economic growth of the United States 1790–1860*. Englewood Cliffs, N.J.: Prentice-Hall.

Ortega, C. P. 1998. *Cowbirds and other brood parasites*. Tucson, Ariz.: University of Arizona Press.

Ostrom, E., J. Burger, C. B. Field, R. B. Norgaard, and D. Policansky. 1999. "Revisiting the commons: Local lessons, global challenges." *Science* 284: 278–282.

Otto, J. S. 1989. *The southern frontiers, 1607–1860: The agricultural evolution of the colonial and antebellum South*. New York: Greenwood Press.

Otto, J. S., and N. E. Anderson. 1982. "Slash-and-burn cultivation in the highlands South: A problem in comparative agricultural history." *Society for Comparative Study of Society and History* 24: 131–147.

Owsley, F. L. 1949. *Plain folk of the Old South*. Baton Rouge, La.: Louisiana State University Press.

Pearson, S. M., A. B. Smith, and M. G. Turner. 1998. "Forest patch size land use and mesic forest herbs in the French Broad River Basin, North Carolina." *Castanea* 63: 382–395.

Perkinson, P. H. 1973. "North Carolina fluted projectile points: Survey report number two." *Southern Indian Studies* 23: 3–60.

Philliber, W. W. 1994. "Introduction: Appalachia and the study of regionalism," pp. xv–xviii. In: P. J. Obermiller and W. W. Philliber (eds.), *Appalachia in an international context: Cross-national comparisons of developing regions*. Westport, Conn.: Praeger.

Poff, N. L., J. D. Allan, M. B. Bain, J. R. Karr, K. L. Prestegaard, B. D. Richter, R. E. Sparks, and J. C. Stromberg. 1997. "The natural flow regime: A paradigm for river conservation and restoration." *BioScience* 47: 769–784.

Pyne, S. J. 1997. *Fire in America: A cultural history of wildland and rural fire*. Seattle, Wash.: University of Washington Press.

Pyne, S. J. 2001. *Fire: A brief history*. Seattle, Wash.: University of Washington Press.

Rose, J. C., M. K. Marks, and L. L. Tieszen. 1991. "Bioarchaeology and subsistence in the central and lower portions of the Mississippi Valley," pp. 7–21. In: M. L. Powell, P. S. Bridges, and A. M. Wagner Mires (eds.), *What mean these bones? Studies in Southeastern bioarchaeology*. Tuscaloosa, Ala.: University of Alabama Press.

Rothblatt, D. N. 1971. *Regional planning: The Appalachian experience*. Lexington, Mass.: Heath Lexington Books.

Royce, C. C. 1975. *The Cherokee Nation of Indians*. Chicago, Ill.: Aldine Publishing.

Salstrom, P. 1994. *Appalachia's path to dependency: Rethinking a region's economic history, 1730–1940*. Knoxville, Tenn.: University of Tennessee Press.

Scarry, C. M. 1993. "Variability in Mississippian crop production strategies," pp. 78–90. In: C. M. Scarry (ed.), *Foraging and farming in the eastern woodlands.* Gainesville, Fla.: University Press of Florida.

Scheffer, M., S. Carpenter, J. A. Foley, C. Folke, and B. Walker. 2001. "Catastrophic shifts in ecosystems." *Nature* 413: 591–596.

Schlesinger, W. 1990. "Evidence from chronosequence studies for a low carbon-storage potential of soils." *Nature* 348: 232–234.

Scott, J. C. 1998. *Seeing like a state: How certain schemes to improve the human condition have failed.* New Haven, Conn.: Yale University Press.

Scott, M. C. 2001. *Integrating the stream and its valley: Land use change, aquatic habitat, and fish assemblages.* PhD diss., University of Georgia, Athens, Ga.

Sirmans, M. E. 1966. *Colonial South Carolina: A political history, 1663–1763.* Chapel Hill, N.C.: University of North Carolina Press.

Smith, B. D. 1992. *Rivers of change: Essays on early agriculture in eastern North America.* Washington, D.C.: Smithsonian Institution Press.

Stover, J. F. 1978. *Iron road to the west: American railroads in the 1850s.* New York: Columbia University Press.

Summers, L. P. 1903. *History of southwest Virginia, 1746–1786, Washington County, 1777–1870.* J. L. Richmond, Va.: Hill Printing.

Taylor, G. R. 1951. *The transportation revolution 1815–1860.* New York: Holt, Rinehart, and Co.

Thornton, R., J. Warren, and T. Miller. 1992. "Depopulation in the Southeast after 1492," pp. 187–195. In: J. W. Verano and D. H. Ubelaker (eds.), *Disease and demography in the Americas.* Washington, D.C.: Smithsonian Institution Press.

Turner, B. L., D. R. Foster, and J. Geoghegan (eds.). 2002. *Land change science and tropical deforestation. The final frontier in southern Yucatan.* New York: Oxford University Press.

Turner, M. 1990. Landscape changes in nine rural counties in Georgia. *Photogrammetric Engineering and Remote Sensing* 56: 379–386.

Vitousek, P. M., P. R. Ehrlich, A. H. Ehrlich, and P. A. Matson. 1986. "Human appropriation of the products of photosynthesis." *BioScience* 36: 368–373.

Vitousek, P. M., H. A. Mooney, J. Lubchenco, and J. Melillo. 1997. "Human domination of Earth's ecosystem." *Science* 277: 494–499.

Vose, J. M., and P. V. Bolstad. 2007. "Biotic and abiotic factors regulating forest floor CO_2 flux across a range of forest age classes in the southern Appalachians." *Pedobiologia* 50: 577–587.

Walker, R. B. 2002. "Early Holocene ecological adaptations in north Alabama," pp. 21–41. In: B. J. Howell (ed.), *Culture, environment, and conservation in the Appalachian South.* Urbana, Ill.: University of Illinois Press.

Wallace, J. B., S. L. Eggert, J. L. Meyer, and J. R. Webster. 1999. "Effects of resource limitation on a detrital-based ecosystem." *Ecological Monographs* 69: 409–442.

Ward, H. T., and R. P. S. Davis Jr. 1999. *Time before history: The archaeology of North Carolina.* Chapel Hill, N.C.: University of North Carolina Press.

Wear, D. N., and P. V. Bolstad. 1998. "Land use changes in southern Appalachian landscapes: Spatial analyses and forecast evaluation." *Ecosystems* 1: 575–594.

Weingartner, P. J., D. Billings, and K. Blee. 1989. "Agriculture in preindustrial Appalachia: Subsistence farming in Beech Creek, 1850–1880." *Journal of the Appalachian Studies Association* 1: 70–80.

Wendland, W. M., and R. A. Bryson. 1974. "Dating climatic episodes of the Holocene." *Quaternary Research* 4: 9–24.

Whittaker, R. H. 1966. "Forest dimensions and production in the Great Smoky Mountains." *Ecology* 47: 103–121.

Williams, S. C. 1928. *Early travels in the Tennessee Country, 1540–1800.* Johnson City, Tenn.: The Watauga Press.

Wilms, D. C. 1973. *Cherokee Indian land use in Georgia, 1800–1838.* Athens, Ga.: University of Georgia Press.

Wilms, D. C. 1991. "Cherokee land use in Georgia before removal," pp. 1–28. In: W. L. Anderson (ed.), *Cherokee removal: Before and after.* Athens, Ga.: University of Georgia Press.

Winters, D. L. 1987. "'Plain folk' of the Old South reexamined: Economic democracy in Tennessee." *Journal of Southern History* 53: 565–586.

Wollenberg, E., D. Edmunds, and L. Buck. 2000. "Using scenarios to make decisions about the future: Anticipatory learning for the adaptive co-management of community forests." *Landscape and Urban Planning* 47: 65–77.

Wright, C. J., and D. C. Coleman. 2002. "Responses of soil microbial biomass, nematode trophic groups, N-mineralization and litter decomposition to disturbance events in the southern Appalachians." *Soil Biology and Biochemistry* 34: 13–25.

Wright, G. 1986. *Old South, New South: Revolutions in the southern economy since the Civil War.* New York: Basic Books.

Yarnell, R. A. 1976. "Plant remains from the Warren Wilson site," pp. 217–224. In: R. S. Dickens (ed.), *Cherokee prehistory: The Pisgah phase in the Appalachian Summit Region.* Knoxville, Tenn.: University of Tennessee Press.

Yarnell, R. A., and M. J. Black. 1985. "Temporal trends indicated by a survey of Archaic and Woodland plant remains from southeastern North America." *Southeastern Archaeology* 4: 93–106.

Young, O. R. 1982. *Resource regimes: Natural resources and social institutions.* Berkeley, Calif.: University of California Press.

Zimmerer, K. S. 1999. "Overlapping patchworks of mountain agriculture in Peru and Bolivia: Toward a regional–global landscape model." *Human Ecology* 27: 135–165.

4

Dustbowl Legacies

Long-Term Change and Resilience in the Shortgrass Steppe

Kenneth M. Sylvester
Myron P. Gutmann

For centuries, European observers perceived the western High Plains as a desert. Few signs remained, on the surface, of the Pawnee villages that once grew maize and bean crops and built earthen homes in the bottomlands of the South Platte and Republican River basins. Decades-long drought during the 13th century had forced an eastward retreat of the first agricultural peoples on the High Plains. In the centuries that followed, humans were mainly visitors to a shortgrass steppe dominated by bison, drought, and fire. The landscape that emerged favored resilient shortgrass species, like buffalograss and blue grama, which thrived on natural disturbance. The longevity of these forces only began to unravel when Europeans introduced horses and firearms into plains ecology. The near demise of the bison in the 19th century, and the expansion of agriculture that followed, had dramatic consequences for the shortgrass, reducing its biomass and diversity. But unlike the tallgrass prairie, human intervention was limited by the interaction between the environment and the very institutions that promoted wholesale change farther east, thereby preserving sizeable grassland corridors in the process.

The shortgrass thus represents an important counterpoint to the larger narrative of the grasslands' declension into the Dust Bowl. During and since the 1930s, many questioned the wisdom of allowing agricultural land use to continue in the High Plains. Several officials instrumental in creating the Soil Conservation Service in Washington believed that careless farming methods were to blame (Bennett, 1947). Generally, historians have shared this negative assessment, seeing poor land-use practice as a major cause of the soil erosion and subsequent dust storms that blanketed the plains (Opie, 1987, 2000; Worster 1979, 1992).

However, drought has likely become too large a part of our mental model of human impacts in the region. Climate variation was characteristic of the region long before agriculture was a significant factor in its ecology. Many studies indicate that drought was far more severe in plains prehistory, with sand dunes reaching their maximum height or range 5,000 to 6,000 years ago (Forman et al., 1995, 2001). Human impacts have certainly added stress in modern times; but, over time, signs of resilience have also become clearer. Overstocked ranges in the 1880s recovered, and homesteaders from the 1890s to the 1910s moved cautiously into the region, seeking out landscapes with rich soils and water access, but not pushing beyond them. These local patterns of persistence and adaptation continue to frame the ecology of the High Plains.

To some extent, early ecosystem science has stayed within the dust bowl paradigm, while working toward a better understanding of the presettlement functioning of the shortgrass. Early research in the shortgrass began with a grassland recovery effort initiated by the U.S. Forest Service in the 1930s. Charged with the care of the Pawnee National Grasslands (PNG) in Weld County, Colorado, the Forest Service began to study grassland management practices in 1939. At a 15,500-acre site in the northwest corner of the PNG, 13 km northeast of Nunn, Colorado, the Forest Service focused its scientific effort on understanding the diversity and functioning of native plants in the shortgrass, the pace of recovery on abandoned cropland, and techniques for measuring plant responses to grazing by cattle. Known as the Central Plains Experimental Range (CPER), the Forest Service site eventually shifted in 1968 and the service began working jointly with Colorado State University's Natural Resources Ecology Lab on studies funded under the initiatives of the NSF, which included the International Biological Program (IBP) and the LTER. During IBP, research examined ecosystem interactions and grassland productivity, and pioneered approaches for modeling the complexity of ecosystem processes. Since 1982, research projects have focused on monitoring primary production, organic matter accumulation, inorganic inputs and transport, disturbances, and populations (Stafford et al., 2002). The area chosen for this chapter embraces the northernmost extent of the shortgrass, first delimited in a rigorous way by vegetation scientist A. W. Küchler (1964), in his map of potential natural vegetation zones in the conterminous United States (Fig. 4.1). Twenty-six counties surrounding the PNG, which is located in Weld County, Colorado, embrace two major river basins in the region, the South Platte and Republican, and experienced very different settlement patterns in historic times. The South Platte basin was favored early during American territorial expansion. It quickly became the scene of agricultural settlement schemes designed to capture mountain runoff in stream flow irrigation systems and reservoirs. But in the Republican River basin, the high rolling tablelands of northeastern Colorado remained little more than "coal lands" in the minds of early scientific observers like Ferdinand Hayden (Hayden and United States, Geological and Geographical Survey of the Territories, 1877). Without water from the Rocky Mountains, it was difficult for contemporaries to imagine how farms might rely on rainfall less than 450 mm (18 in.) a season (Fig. 4.2) or the intermittent flows that gathered in the tablelands and occasional springs to form the many lesser creeks that became the

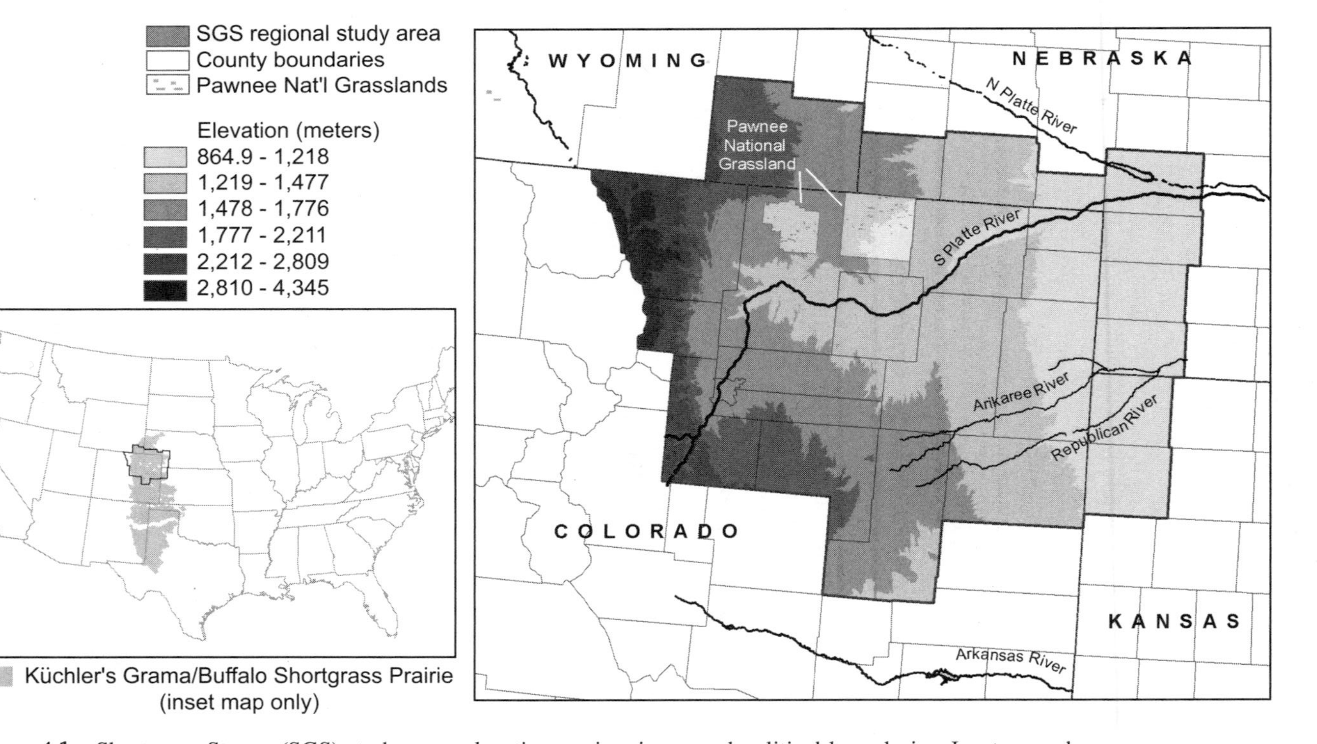

Figure 4.1 Shortgrass Steppe (SGS) study area: elevation, major rivers, and political boundaries. Inset map shows potential natural vegetation of the conterminous United States. Digital vector data (digitized from 1:3,186,000 scale map) on an Albers Equal Area Conic polygon network in ARC/INFO format. Corvallis, OR: US-EPA Environmental Research Laboratory. Based on data from Küchler (1993).

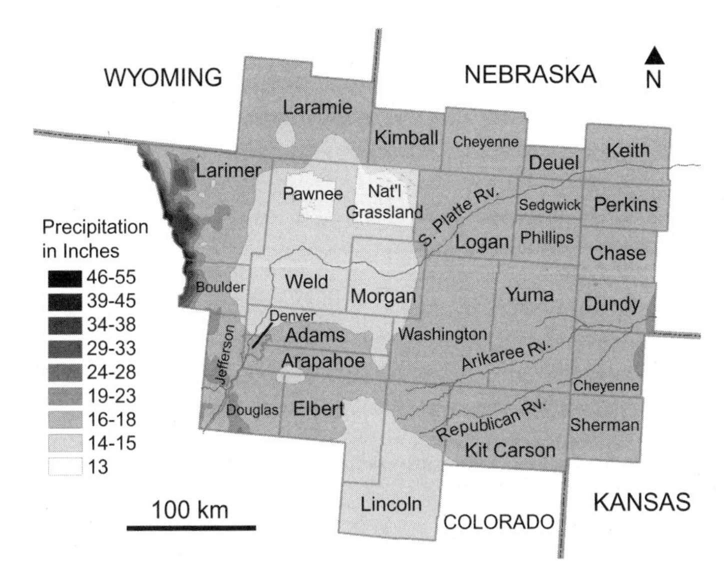

Figure 4.2 Climate normal precipitation (1961–1990) in Shortgrass Steppe counties. Based on PRISM (2008).

Republican and Arikaree rivers. Although the shallow valleys created by these streams afforded relief from broad, fertile plains, the tablelands were occasionally punctuated with narrow, steep-walled canyons and dune-covered sand hills with poor drainage and no flowing streams (Wedel, 1986). Environmentally cautious patterns of land use by European American settlers continue to define the extent to which human management has changed the ecosystem of the Republican River basin.

Geophysical, Environmental, and Biological Setting

Grasses have dominated the Shortgrass Steppe region of North America for millennia. Twelve thousand years ago, in what are now the High Plains of Texas, New Mexico, Oklahoma, Kansas, Nebraska, and eastern Colorado, a diverse landscape of unbroken grasslands was punctuated by small forests, typically in river valleys and canyons. Woodland almost certainly began a steady decline with the arrival of a more arid climate beginning around 11,000 BP, but recent work has questioned the presence of extensive forest cover in the plains during the late Pleistocene era. Investigations in Texas and Kansas have concluded that spruce and fir pollen were only a small part of otherwise high percentages of grass and composites sampled from glacial lake clays (Fredlund, 1995; Hall and Valastro, 1995). Studies of Holocene dune activity in eastern Colorado indicate repeated activation of eolian dunes during the past 20,000 years. Radiocarbon ages on organic matter from buried soils suggest periods of aridity, visible in parabolic dunes straddling the South Platte (Forman et al., 1995; Madole, 1995; Muhs et al., 1999). The botanical and climate signatures point to a grassland vegetation regime extending over a wide geography for much of the past 20,000 years.

The Shortgrass Steppe region is part of the larger Great Plains physiographic province in North America, extending roughly 1.16 million km^2 from southern Canada to the Gulf of Mexico (Bailey and Ropes, 1998; Commission for Environmental Cooperation, 1997; Hunt, 1967). Although never under ice, the Shortgrass Steppe's topography and geology were a by-product of continental glaciation. Outwash from the melting of mountain glaciers is evident in alluvium deposited in landscapes across the region. Soils are largely Mollisols, characterized by a thick, dark surface horizon, resulting from the long-term addition of organic materials from plant roots; and Aridisols, derived from volcanic activity and characterized by their dryness, with subsurface horizons in which clays, calcium carbonate, silica, salts, and gypsum have accumulated. Mollisols typically occupy middle latitudes and are extensive in grassland regions like the Great Plains. In the United States, they are the most extensive soil order, accounting for roughly 21.5% of the nation's land area, and are among the most important and productive agricultural soils in the world (Quandt et al., 1998). Soils that are coarse in texture (sandy) generally have high infiltration rates and store most of their water beyond the influence of evaporation, and they increase soil fertility (Lauenroth and Milchunas, 1992).

The major environmental gradient in the Shortgrass Steppe is an increase in annual precipitation from west to east. From the front range of the Rockies to the eastern margin of the shortgrass region, annual precipitation ranges from less than 300 mm to more than 550 mm, with a discernible rain shadow effect from the mountains that carries east of Denver. Most precipitation occurs between May and September, with a dry period from December through February. Much of the rainfall occurs as torrents or light, ineffective showers. Often hot weather and high winds further reduce this effectiveness, especially during drought periods. Variations in precipitation, however, drive most processes in the region, including the abundance of small mammals, vegetation structure, and the availability of mineral nutrients in the soil. Distribution of plant populations is very closely related to available water, with very little evidence to suggest that differences in the supply of mineral nutrients greatly alter local composition (Lauenroth and Milchunas, 1992). Indeed the plains ecosystem as a whole is unlike most similar grasslands in the world in the relative smoothness of its environmental gradients.

Grasses, shrubs, forbs, and succulents are the major components of the vegetation structure of the Shortgrass Steppe. Vegetation scientist Küchler (1964) identified the chief native plant populations as the blue grama (*Bouteloua gracilis*) and buffalograss (*Buchloe dactyloides*) species, both of which are tolerant to grazing and are well suited to the semiarid climate. In upland sites like the CPER, plant communities tend to be dominated by blue grama, mixed with plains prickly pear (*Opuntia polyacantha*) in half-shrub communities. Lowlands are dominated by either blue grama or buffalograss, and plains prickly pear is less common. In saline–alkali soils, western wheatgrass (*Agropyron smithii*) and inland saltgrass (*Distichlis stricta*) predominate. Blue grama was particularly well adapted to the pre-European ecology of the region, because it responded so well to grazing activity from large bison herds and could withstand variations in temperature that are known to range from 48 °C to –34°C (Savage and Costello, 1948, p. 504). Blue grama is extremely resilient and can recover from almost total destruction resulting from drought, plowing, or overgrazing. Most studies of grassland succession have found that blue grama requires a period of at least 40 years to regenerate after use for cropland agriculture. Costello (1944) described a successional sequence for abandoned farmland in north-central Colorado where the initial stage is dominated by annual plants, like Russian thistle (*Salsola kali*); and perennial forbs that appeared 3 to 5 years after abandonment, like foxtail barley (*Hordeum jubatum*), tumble grass (*Schedonnardus paniculatus*), squirrel tail (*Sitanion hystrix*), and spike dropseed (*Sporobolus cryptandrus*). After 20 years, wiregrass or red threeawn (*Aristida longiseta*) came to dominate. During this interim stage of succession, wiregrass could account for as much as 50% of grass density in fields where as much as 80% of the vegetation had recovered. Wheatgrass and blue grama usually returned to dominance after 40 to 50 years.

Pre-European Context and Dynamics

It is believed that the Central Plains was the first region in North America to be inhabited by humans at the end of the last Ice Age. Descendants of the Clovis, the first peoples in North America, made a home in what was a much wetter plains ecosystem—one punctuated by lakes, rain-fed ponds, and hardwood groves (Dixon, 1999; Smith, 2003; West, 1998; Wyckoff, 1999). These hunters preyed on the extraordinary abundance of game during the Pleistocene era: wild horses, camels, pronghorns, deer, peccaries, *Bison antiquus* (a much larger predecessor of the plains bison), and Columbian mammoths. When climate change and human predation led to the disappearance of many of these species around 11,000 BP, the Clovis focused on the one mammal that was able to survive and proliferate in the central grasslands: *B. antiquus*. In turn, even this plains ecology broke down during the middle prehistoric period, around 7000 BP, with a sustained drought. During a dry epoch lasting for centuries, humans retreated from the plains, and *B. antiquus* disappeared and was replaced by a smaller modern cousin: *B. bison*.

Very little evidence can be found of human presence in the plains again until around 3000 BP, when the woodland peoples of the Mississippi River basin began to expand westward. Agriculture came to the Central Plains along the lower Republican, when former Mississippi peoples brought traditions of mound building and maize production to the tallgrass and mixed-grass plains. Expansion farther west was assisted by still more climate change, as the grasslands entered into one of the wettest periods of its prehistory. Moist tropical air circulated much farther north beginning in about 1200 BP, and the tallgrass and mixed-grass flora moved westward with increased rainfall, pushing the shortgrass perhaps 200 mi. closer to the Rocky Mountains (West, 1998). By 1000 BP, small communities of hunters and horticulturalists had spread out along the tributaries of the upper Republican, and in the valleys of the Loup and Platte Rivers (Calloway, 2003;

White, 1983). Most of these peoples made their fields beside streams, planting in the spring, harvesting in the fall, and moving west during the winter and summer to hunt buffalo.

The Pawnee were among the first High Plains peoples to leave a substantial archaeological record west of the 100th meridian (Parks, 1989; Parks and Wedel, 1985; Wedel, 1986). Pawnee men would clear the fields prior to planting, and the women would plant the seeds in clusters of six or seven kernels after digging small holes with sticks. Like elsewhere in the Americas, corn, squash, and beans were grown in the same fields. Unfortunately, the Pawnee proved unequal to the return of extended drought in the High Plains. Tree-ring analysis from western Nebraska shows that in the 97 years after 1220, all but 34 were years of drought. Just as the wet period had drawn hunter-farmers westward, the return of drought pushed them eastward in the centuries that followed.

When Francisco Vasquez de Coronado headed north into Kansas in 1541, the only substantial settlements he encountered were at a place known to his guide as Quivira, where several villages of conical grass huts sheltered a few hundred inhabitants (West, 1998). Coronado's chronicler recorded in his diary that "in some villages there are as many as two hundred houses; they have corn and beans and melons; they do not have cotton or fowls, nor make bread which is cooked, except under the ashes. Francisco Vazquez went twenty-five leagues [105 km or 65 mi.] through these settlements, to where he obtained an account of what was beyond, and they said that the plains came to an end, and that down the river there are people who do not plant, but live wholly by hunting" (Thomas, 1935, pp. 5–6).

Later, Spanish explorers came across horticultural societies in southern Colorado, along the Purgatoire River, and as far north as Sand Creek, where Juan de Ulibarri reported that he found a village composed of several *rancheria* in 1706. Everywhere Ulibarri journeyed into southeastern Colorado that year he remarked on the horticultural activities of the Apache, who then dominated the shortgrass plains. On the Purgatoire, the Penxayes Apache planted "much land to corn, frijoles, and pumpkins," he wrote in his diary on Sunday, July 25, 1706, and when Ulibarri reached his destination, El Cuartelejo, up Sand Creek from the Arkansas River, the Apache chiefs "brought us buffalo meat, roasting ears of corn, tamales, plums, and other things to eat" (Thomas, 1935, pp. 64, 68).

When explorer Etienne Véniard de Bourgmont made the first official expedition of the French crown west of the Missouri in 1724, he arrived at very extensive village settlements of Apache in central Kansas, probably very near the settlements Coronado had seen 183 years earlier in present-day Macpherson County, along the Arkansas River. Unlike the Apache in Cuartelejo, the Apache in central Kansas, whom the French referred to as the Padoucas, were further along the transition to nomadism. In his journal of the expedition, deposited in the *Archives de la Marine* in Paris, he described the manner of living in the chief Padouca settlement:

> In the villages of this tribe that are far way from the Spaniards, all subsist solely on the hunt, in winter as in summer. However, they are not entirely nomadic, for they

> have large villages with sizable dwellings. They go on the hunt in bands of 50 to 80, sometimes even 100 households together; when they return to their permanent villages, those who had stayed at home leave at once, while those returning bring with them provisions of dried meat, either bison meat or venison, killed not far from their villages. When they travel from their villages as far as five or six days' journey, they find herds of bison in great numbers, and they kill as many as they want. . . .
>
> This tribe sows hardly any maize; however, it does sow a little and a few pumpkins. They grow no tobacco; nevertheless, they all smoke when they have it. The Spaniards bring them some when they come to trade, and they also bring them horses. The Padoucas trade to them dressed bison skins, as well as bison skins dressed in the hair, which are used as blankets. (Bourgmont, 1724, as cited in Norall, 1988)

Bourgmont estimated the territory of the Apache at more than 200 leagues (837 km or 520 mi.) in breadth, extending all the way to the Spanish settlements in New Mexico. In the village where he made a peace pact with the Apache, he estimated that there were 140 dwellings, 800 warriors, 1,500 women, and about 2,000 children.

It was in many ways the sheer physical extent of the bison ecology developed by various plains peoples that made the deepest impression on European chroniclers. The Apache transition to the bison hunt was one of many that caught their attention and reinforced an image of the west as a wilderness inhabited only by nomadic peoples (Norall, 1988; Thomas, 1935; West, 1998). However, the weight of accumulated archeological and ethnographic evidence indicates that nomadism was experimental for the plains peoples. It was experimental both because it was really only possible on the scale seen during the 18th century because of the horse and gun revolution, and because of the migration of displaced Native Americans into the region. Although it is difficult to sequence the transition for specific peoples, during the pedestrian era plains people remained closer to their riverside homes, used fire to regenerate the soil and reshape the landscape, and were increasingly drawn out onto the plains in response to the spread of horses, guns, and disease. Some peoples never left the bottomlands and canyons in search of the highlands and streams where bison gathered during rutting season. But like the Pawnee along the Platte River basin in Nebraska, those who never abandoned their permanent villages were also the ones that suffered the most severe population losses in the face of smallpox (Binnema, 2001; Isenberg, 2000).

Most accounts of precontact bison ecology have scaled back the estimates of early 20th century naturalists. Zoologist Tom McHugh (McHugh and Hobson, 1972) estimated the number of bison using a metric based on the direct observation of bison in Yellowstone Park. He found that, with adequate rainfall, one bison needed 25 acres of grassland to be adequately fed each year. McHugh therefore concluded that the carrying capacity of the plains lay close to 32 million bison. Other naturalists have estimated that this number may still be a few million too high.

These populations were known to fluctuate even further than these maxima, given the considerable climactic and predatory variation in the region. Bison as a species were particularly susceptible to ecological disturbance, as reduced forage from grass fires, drought, or severe cold in winter could delay puberty in females. Wolf predation likely reduced the number of calves who survived by an additional one third (Meagher, 1973). Human turmoil in the region added a dramatic new dimension to the ecological pressures that regulated the size of the bison herd and its range across the plains.

North America was very much a managed and evolving landscape at the time of European contact, the product of as many as 350 generations of human presence. Wildlife biologists have increasingly acknowledged the precontact dimensions of this management, noting that wildlife was more abundant in buffer zones between hostile neighbors (Flores, 2001; Martin and Szuter, 1999; White, 1983). Paul Martin and Cristine Szuter conclude that part of the reason why Lewis and Clark reported such differences in wildlife east and west of the Continental Divide was that the Blackfeet had imposed a "war zone" on the upper Missouri, whereas the relative peace among peoples of the Columbia basin was responsible for the game sink the famed explorers experienced there (Martin and Szuter, 1999). Over time, these game sinks became more common as displaced peoples from the east and north made their way to the bison range in the Shortgrass Steppe. From as far away as eastern Minnesota, the Cheyenne slowly made their way to the Black Hills in South Dakota and then southward, reaching Bent's Fort on the Arkansas River. The Comanche also migrated southward into the shortgrass in search of diminishing bison herds. None of these peoples had built the same mental maps as the Apache of where water was available, and of the many tributaries of major rivers and buffalo wallows where the High Plains aquifer spilled out onto the surface. However, peoples who shifted more fully to a nomadic existence escaped the full onslaught of European disease.

Drivers of Agriculture and Land-Use Change

The final eclipse of bison ecology came in the 1860s and 1870s with the expansion of American settlement and the sanctioning of a commercial hunt that nearly exterminated the bison. Signs of decline were visible earlier in the century, particularly during a sustained drought between 1845 and 1856 (Isenberg, 2000).

The effects of the drought on bison populations were well-known by contemporaries. Native American tribes often harvested 100 bison in a single hunt, according to Jacob Fowler, who traveled amongst the Kiowa, Comanche, Cheyenne, and Arapahos in the 1820s. This suggests by various calculations an annual kill of between 360,000 and 450,000 bison—a harvest that could quickly exceed the natural increase in times of ecological distress (Isenberg, 2000, p. 83). The migration of California-bound gold-seekers, bringing half a million cattle and sheep through the Platte River basin after 1849, only added to the ecological pressure. Like the Arkansas River valley to the south, the Platte provided critical winter refuge for bison. But during the early 1850s, the bison herds were greeted by overgrazed bottomlands (West, 1998).

Typically, the grasslands were resilient enough to recover from these kinds of stresses, and the bison themselves often adjusted by migrating farther in all directions in search of forage. However, as the 19th century progressed, the vice was closing. The political bottleneck that had slowed American expansion was removed at the conclusion of the Civil War, and the full weight of European American expansion was brought to bear. For its part, the U.S. government did not impede the acceleration of a commercial hunt of the bison, begun before the Civil War to exploit a growing trade in buffalo hides, but deferred instead to interests in Congress, which favored the use of western plains as open rangeland and homestead land grants, and those that simply wanted to secure routes for proposed transcontinental rail lines. The scorched-earth effects of federal inaction were well understood, particularly by military officers who welcomed the hunt's potential to ease their task in subjugating the western tribes. In 1867, the Treaties of Medicine Lodge ended 3 years of hostilities with the Comanche, Kiowa, Southern Cheyenne, and Southern Arapaho by promising to forbid agricultural settlement south of the Arkansas River. A year later, the Treaty of Fort Laramie secured peace with the Sioux, Northern Cheyenne, and Northern Arapaho by guaranteeing Native American hunting rights north of the North Platte and west of the Missouri. However, with little or no attention given to the bisons' survival, the terms of the treaties were Pyrrhic at best (Isenberg, 2000).

In the near term, the disruption of bison herbivory in the shortgrass was alleviated to some extent by the introduction of open-range cattle. Ranchers in southern Texas had bred an animal, the Texas longhorn, that was able to thrive on grass. During the Civil War, with their southern markets cut off, the ranchers began to range northward, driving cattle toward markets centered in St. Louis. After the war, the trend continued and the Texas ranchers, urged by overseas investors from Great Britain and encouraged by a period of wet weather in the High Plains, drove cattle north in increasing numbers. Between 1866 and 1884, it is estimated that five million head of cattle were driven north and west onto the plains (Jordan, 1993).

As it turned out, the longhorn was not well adapted to the colder climate in the High Plains. The number of calves surviving into adulthood decreased 20% from what ranchers could expect in Texas, because the cattle did not know how to react to snow, standing and starving rather than clearing a hole in the snow with their heads to permit them to eat the dry grass that lay underneath. In some areas in the winter of 1871 to 1872, ranchers lost up to half their herds on the open range because of the severe conditions in Kansas and Nebraska. Without an infrastructure of local feed supplies, huge numbers of cattle died in the severe winters of 1879–1880, 1884–1885, and 1886–1887. There was clear evidence that ranchers had stocked the key areas in the western plains with too many cattle by the mid 1880s, as annual grasses started to replace perennials in some overstocked ranges, and undernourished animals faced gale force winds and unusually high snow cover.

Drought returned to the plains with a vengeance in 1887, putting the government's land policies in some doubt. Despite warnings from the scientific surveyors of the west like John Wesley Powell and Ferdinand Hayden, the federal government decided that the shortgrass plains between the Arkansas and the North Platte would be developed in the same way as farmland east of the 99th meridian.

In his *Report on the Arid Lands of the United States*, Powell, soon to be installed as director of the U.S. Geological Survey (USGS), urged Congress to adopt a different land-granting framework for western land alienation than the one used for the American interior west of the Ohio River (Powell and Geographical and Geological Survey of the Rocky Mountain Region, 1879). Powell predicted that less than 3% of the western two fifths of the United States could be farmed in an ordinary way (Opie, 1998; Worster, 2001). He recommended a system of watershed commonwealths, with small public land parcels close to rivers to provide wider access to water, and larger ones out on the plains to allow for better rotation of crop and grazing land. However, these intended resource regimes remained more of an inspiration for future land-use planners than the basis of western land law. The federal government duly noted his recommendations to sell land in 2- to 4-sq. mi. sections, totaling 1,280 to 2,560 acres, but did not allow his advice to slow the work of the General Land Office (Pisani, 2002; Worster, 2001). In the post–Civil War climate, too many interests were arrayed on the left and right in favor of continuing the distribution of public lands in the form and at the pace they were proceeding, even if scientific knowledge of the potential uses of the semiarid region was distinctly lacking. Moving west of the Kansas and Nebraska borders in the mid 1880s, the timing of the land surveys proved unfortunate for many. The 5 years of residence required for a successful homestead claim fell in the middle of a severe drought. Some, but not all, of the abandonment foreshadowed by Powell was realized. Local populations swelled to take advantage of the initial grants of free land, and generally found the semiarid climate unforgiving of inexperience and single-mindedness.

In eastern Colorado, the effects of expansion were dramatic. More than a half dozen new counties were created between 1885 and 1890 that focused on dryland farming in response to the campaigns of the Kansas Pacific and Burlington Railroads (Wyckoff, 1999). Towns sprung up along the tracks with names like Yuma, Akron, Burlington, and Cheyenne Wells, serving as key market towns. Phillips, Yuma, Washington, and Kit Carson counties experienced the largest growth, reaching populations of around 2,500 residents by 1890. Although the railways cynically promoted these counties as the "Rain Belt" of eastern Colorado, farmers soon discovered how tenuous grain farming was in the region. Corn was quickly abandoned in dryland settings in favor of wheat, rye, oat, and sorghum, and feed crops like millet and hay became commercial crops. The reality of raising crops in semiarid conditions, without the benefit of irrigation, quickly became known as *dryland farming*—a vernacular that farm families understood to mean the blend of cropping and ranching that even the smallest of farms was required to pursue to make the best of changing conditions. Families also cultivated small, hand-irrigated gardens of potatoes, lettuce, peas, beans, and melons. In the easternmost counties of the state, the census reported more than 10,000 cattle in 1890. With the return of drought in 1893 through 1897, many settlers finally packed up and left. By 1900, some counties had lost 30% of the settlers who had arrived in the 1880s.

By 1906, the conditions were set for further expansion of dryland farming. Most cattle in the High Plains were raised on individual farms, and their feed increasingly came from farm production. State extension agencies were also convinced that a more scientific approach to agriculture would mitigate the seasonal hazards of farming on the High Plains. To control surface evaporation, agronomists recommended deep plowing after harvests, frequent disking after rains to work moisture into the ground, packing the subsoil and covering it with mulch, and alternating crops with summer fallows to preserve moisture. A state-funded dryland experiment station at Cheyenne Wells led the call for more contour plowing and diversified farming in the region. A federally sponsored station opened at Akron in 1907 to promote further dryland farming research. Then, finally, in a symbolic acknowledgment of Powell's argument, the federal government passed the Enlarged Homestead Act in 1909 to allow a claim of 320 acres of dryland in areas removed from known sources of irrigation (Wyckoff, 1999). As further inducement, in 1912 the federal government shortened the proving-up period (during which land claimants promised to fulfill duties that included plowing a minimum number of acres) from 5 years to 3 years. As a result, many of the counties that had lost population during the 1890s began to grow again.

The incentives proved necessary because sizeable portions of the High Plains had still not been claimed at the beginning of the 20th century. Ironically, the federal government's determination to pursue homestead grants in the High Plains ultimately slowed the transfer of land to private hands. In the High Plains, homestead-size grants situated in a township grid designed to promote dense settlement forced entrants to be more, not less, selective. The size of the grants excluded ranchers, who needed much larger tracts of land with access to pasture and water organized by watershed, and it obliged small farmers to locate parcels with fertile soils close to groundwater or rivers and streams. Because these resources were not integrated evenly across the region, as Powell suggested, early land selection slowed noticeably after the best locations and endowments were spoken for. Eventually, nearly all public land in eastern Colorado would pass into private hands. However, the tentative nature of early settlement suggests that settlers proceeded with a keen awareness of the limitations that faced them in the High Plains. Irrespective of the temptations to convert grassland to cropland at various points during the 20th century, particularly with the development of groundwater irrigation technology in the 1950s, cropland has remained largely within limits established early during the settlement process.

An important baseline of information about these environmental choices survives in a series of maps commissioned by the Bureau of Reclamation (United States Reclamation Service, 1902) for its first annual report. Wanting to create a stunning visual reference of the public lands still vacant in Nebraska, Wyoming, and Colorado (to shape public debate in favor of its proposed land management), the new federal agency commissioned a detailed cartographic inventory of entry information from public land records. Digitizing the parcel-level information in those historical maps, by classifying a georeferenced image of the historic map against a public land survey Geographic Information System (GIS) layer, subdivided into homestead units (160-acre/68-ha square parcels), unearths fascinating comparisons with recent environmental and land-cover information (Vogelmann et al., 2001). In Figure 4.3, the darker shading represents land parcels that were

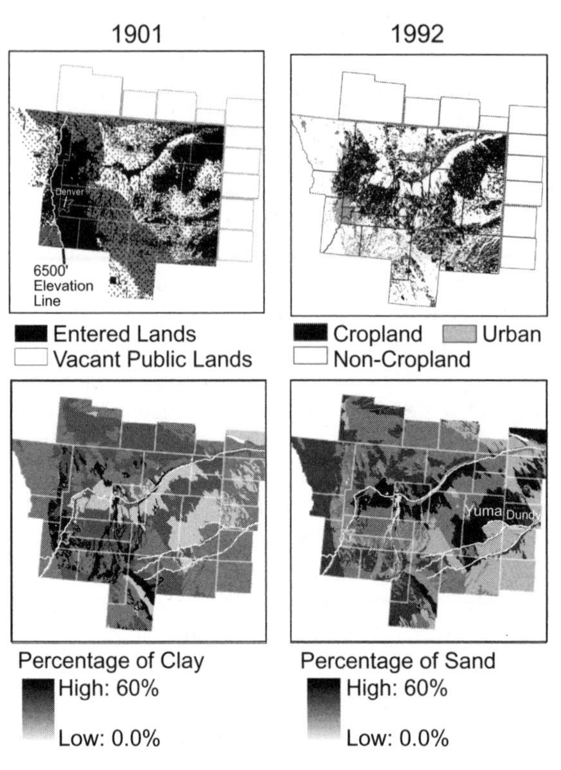

Figure 4.3 Land claimed in 1901 and land use in 1992. Sources: Entered lands derived from U.S. Bureau of Reclamation (1902) and land cover for 1992 is derived from U.S. Geological Survey (1992). Soil texture derived from the USDA's Natural Resources Conservation STATSGO soils database, available online at www.ncgc.nrcs.usda.gov/products/datasets/statsgo/.

claimed by 1901 and the white areas represent parcels still in the public domain in northeastern Colorado. It is not surprising that most early land claims focused on parcels along the front range of the Rocky Mountains, with its proximity to Denver, streams fed by runoff from mountain snowmelt, and access to proposed rail lines (seen in the corridors of alternately claimed and unclaimed land running north and south of Denver, next to land claimed along the front range). Elsewhere the caution was more obvious. Settlers focused early land selection in the eastern approaches to Denver on land close to major rivers (the South Platte, Arikaree, and Republican) and a broad plain below the sand hills that lay all along the southern banks of the South Platte. That high plain, we now know from 20th-century soil surveys, is dominated by permeable soils, with dark surface horizons. In American soil taxonomy, which standardized its classification schema relatively late in the 20th century (Smith, 1986), these are known as *Mollisols*. In the lower left corner of Figure 4.3, the elevated percentage of clay in the soils in that high plain, parallel to the northeasterly course of the South Platte River, illustrates the location of these rich grassland soils. The opposite effect is illustrated in the map of sand content in the region's soils in the lower right corner of Figure 4.3. Settlers largely avoided claims where sand content was high. Few early claims were made to the sand hills below the South Platte River prior to 1901, nor to the sandy plain that cuts across Yuma County, Colorado, into Dundy County, Nebraska.

Whatever advantages they may have had as grazing land, the restrictions placed on the size of land grants effectively excluded their integration into early land claims. Therefore there is a remarkable continuity in the magnitude and spatial dimensions of early land claims and cropland agriculture today. This is illustrated in the upper right corner of the panel in Figure 4.3 in a very simplified representation of the National Land Cover Data (NLCD), a national land-cover data set developed using Landsat Thematic Mapper satellite imagery from 1992. Reducing the 21 Anderson Level II land-cover classifications in the NLCD to just three categories (cropland, noncropland, and urban), we see a striking similarity between the land in crops in 1992 and land that was selected from the public domain by 1901. This is the case despite the much higher spatial resolution of the NLCD, derived from satellite imagery at a 30-m scale, and the inclusion of small grains, row crops, pasture, hay, and fallow in a combined representation of cropland in 1992.

The continuity is also evident from mid century at more refined scales. In the 1930s, federal researchers working in eastern Colorado worried that ownership levels had fallen too rapidly during the Depression and environmental stewardship had suffered. Too many tenant farmers, small and large, responded to short-term incentives and ignored management practices then emerging to improve soil conservation. However, when soil scientists began working individually with local farmers in demonstration projects, they often found that established practices were not far removed from the conservation measures they promoted. Although adding some innovations like contour plowing and novel rotation plans to the routine of High Plains farmers, scientists discovered patterns of mixed livestock raising and winter wheat cultivation that demanded careful use of pasture, cropland, and fallow rotations already in place. In one land-use study made between 1936 and 1938, the USDA's Bureau of Agricultural Economics (BAE) conducted a field-level survey of land use in Kit Carson County, Colorado, and mapped its findings in a report sent to Washington (Watenpaugh et al., 1941). The BAE field staff generalized their observations to ownership parcels across the entire county in a map digitized here to examine land-use change over time. By compressing the 1992 NLCD classifications into categories that mirrored the survey of BAE field staff in the 1930s, we were able to generate the comparison seen in Figure 4.4. The digitized parcels are not an exact match for the 30-m data derived from Landsat imagery, but they point to obvious continuities. Despite the coarser grain of analysis in 1938, the comparison shows that cropland remains concentrated in tablelands between the streams that define Kit Carson's topography. The bottomlands along the South Fork of the Republican River (and its feeder streams running from the southwest to the northeast of the county) have never really been converted from grassland to cropland. Some farmers were obviously experimenting with cropping closer to streams in the 1930s, but the retreat from them was well entrenched by

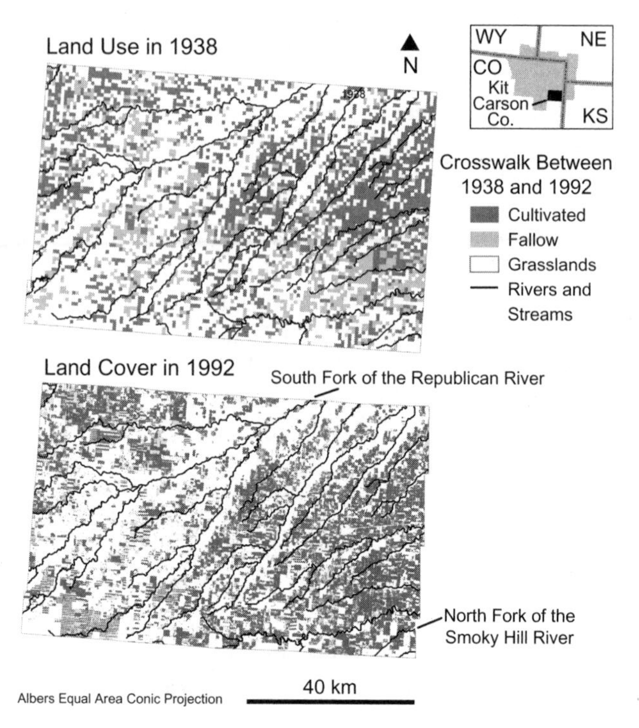

Figure 4.4 Land use in 1938 and land cover in 1992, Kit Carson County, Colorado. Based on data from 1938 land use derived from a parcel-level map contained in Watenpaugh et al. (1941). Land cover for 1992 is derived from U.S. Geological Survey (1992).

the 1990s. The biggest difference in land-use practice may be a visible increase in pasture and reduction in fallow cropland by the early 1990s. This suggests a more continuous and spatially concentrated pattern of cropping, with larger field sizes associated with the more challenging landscapes in the western half of the county, and smaller fields planted to make more intensive use of the better cropland in the eastern half of the county.

This kind of historical mapping indicates that land use in northeast Colorado has respected biophysical limits to a far greater degree than generally acknowledged in the wider historical literature (Opie, 1998; Steinberg, 2002; Worster, 1979). Ecologists have been aware of the differences in conservation outlook for some time. The issues facing the tallgrass prairie, which is estimated to have shrunk to less than 5% of its presettlement range (Knapp et al., 1999; Samson and Knopf, 1994), are very different from ones in the shortgrass, which has shrunk to only 45% of its presettlement extent (Lauenroth et al., 1999). The legacies of prior land use are simply not the same in the western High Plains as they are elsewhere.

Although cropland has not expanded beyond broad maxima since the Depression (Cunfer, 2005), in the Shortgrass Steppe study area, county-level returns from the U.S. agricultural census indicate that cropland did not peak until the 1950s—just as the population census shows that urbanization in the metropolitan Denver area began to sever the tight relationship between population and the extent of cropland that had existed in the region since the 1880s. In Figure 4.5A we see this illustrated by the similar trends in cropland plotted on the left axis, which reached a stable maximum in the 1959 agricultural census at 8,374,566 acres across the region, just as urbanization allowed population growth to accelerate in the 26 counties (plotted on the right axis). The advent of groundwater irrigation technology in the 1950s helps to explain the modest expansion of cropland in the High Plains during that decade, but as we will see, this did not dramatically reshape the spatial distribution of cropland in the Shortgrass Steppe region. Cropland also spiked higher in the 1970s and 1980s as wheat growers responded to the lifting of trade embargoes against China and Russia, settling back to historic maxima as those export markets declined in the 1990s. Cropland has remained stable throughout eastern Colorado since 1945 (Parton et al., 2003).

Figure 4.5B shows that wheat has dominated the crop regime in the shortgrass since the boom in wheat prices that preceded World War I. Early settlers stubbornly planted corn west of what the instrumental record showed to be a 500 mm (20 in.) rainfall line near the 101st meridian (Wedel, 1986). Wheat did briefly overtake corn by the 1900 census, because poor-yielding midwestern corn varieties were no match for conditions west of the 30-in. rainfall line near the 98th meridian, nor for farming outside the alluvial bottomlands where Native Americans had grown drought-resistant varieties of corn in centuries past (Wedel, 1986). Nevertheless, the acreage devoted to corn grew steadily from the 1900s to the 1930s, mainly because of the development of stream flow irrigation networks in the South Platte River basin. According to the 1920 U.S. Census, across the study area irrigated farmland rose during the period from a total of 174,205 acres in 1880 to 735,191 acres (U.S. Department of Commerce, *Census of agriculture*, 1880, 1920). In dryland areas, the adoption of winter wheat, planted in the fall and harvested in May and June, allowed High Plains farmers to better capture winter snowmelt and spring rains, and to expand wheat acreage. New hard winter wheat varieties, bred in semiarid conditions in southern Russia, arrived with immigrant Mennonite farmers in the 1870s. By 1919, the famed Turkey wheat variety accounted for 98% of wheat grown in Kansas (Kimbrell, 2002), helping to explain why the acres sown to wheat grew to capture 33% of cropland in the shortgrass in 1930, but corn never exceeded a maximum, reached in the same census, of 23% of total crop acreage. The growth of wheat production also helps explain the farm size adjustment that has occurred in the Shortgrass Steppe study area since the 1940s. Although the number of farms peaked in the region in 1935, and has declined in absolute terms to numbers that prevailed at the beginning of the century, farm size has more than doubled between 1930 and 1997, increasing from 605 to 1,290 acres on average (Fig. 4.5C). The ecological implications of the adjustment are not well studied. Farm size might result in a variety of cultivation differences that might affect the environment. Because they are cash constrained, small farmers generally cultivate more of their land than do larger farmers (Hansen and Libecap, 2004a,b; Libecap and Hansen, 2002). On the other hand, larger farmers may use more pesticides and fertilizers, which might have

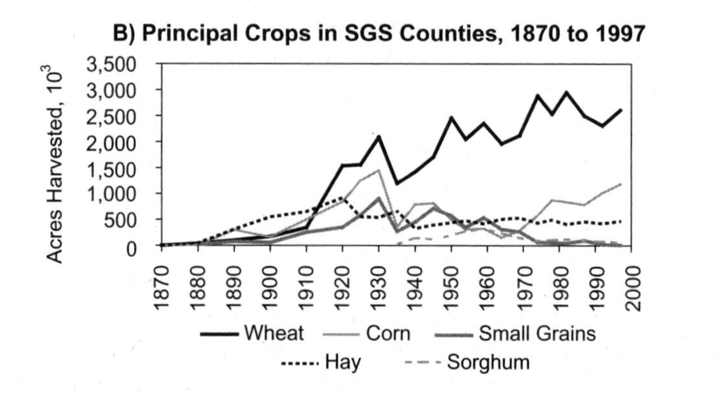

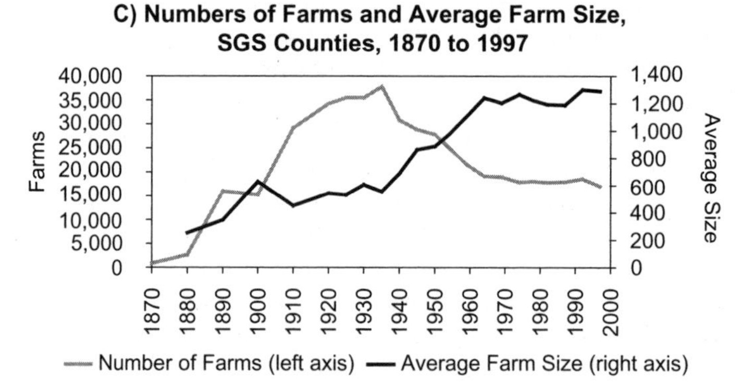

Figure 4.5 Land use, population, principal crops, number of farms, and average farm size in the Shortgrass Steppe (SGS) region. (A) Land use and population in SGS counties, 1870–1997. (B) Principal crops in SGS counties, 1870–1997. (C) Number of farms and average farm size in SGS counties, 1870–1997. Based on data from U.S. Department of Commerce, *Census of agriculture* (1870, 1880, 1890, 1900, 1910, 1920, 1930, 1940, 1950, 1954, 1959, 1964, 1969, 1974, 1978, 1982, 1987, 1992, 1997) and *Census of population* (1870, 1880, 1890, 1900, 1910, 1920, 1930, 1940, 1950, 1960, 1970, 1980, 1990, 2000).

more negative effects. These questions deserve greater attention in conservation planning in the future.

Conservation Context

To some degree, the integrity of the native shortgrass benefited from early tinkering with the public land system. Paul Gates argued that the High Plains states of New Mexico, Colorado, Wyoming, and Montana were treated differently simply because Congress intended that development should move at a slower pace (Gates, 1977). The different federal agencies responsible for land management were embroiled in extended warfare over competing visions of western development. Officials in the General Land Office were generally receptive to Horace Greeley's vision of restricting landownership to small farmers, and favored repeal of the Preemption Law. Since 1841, the Preemption Law had served as a means for settlers, ranchers, and lumber and mining interests to move in advance of the surveys, permitting claimants 33 months of free use before having to pay the standard purchase price of $1.25 an acre (Gates, 1977). High Plains farmers suggested to the Public Lands Commission in 1879 that preemption worked well because it allowed them to assemble the acres needed in the dry climate. Greeley and many Land Office officials worried that preemption could be abused by land speculators, as it undoubtedly was.

The Timber Culture Act was passed, allowing western farmers to make claims on adjacent 160-acre quarter sections in return for a promise to plant and care for trees on 40 acres within 10 years. If these duties were fulfilled, claimants could have title. However, more typically, timber culture claims were used to hold land for children or until a sale of the entry itself, known as a *relinquishment,* rose high enough in value. The receiver of the Cheyenne Land Office complained in 1888 that Timber Culture "filings are generally made to hold the land, without intention to comply with the law" (Gates, 1977, p. 113). Gates reports that 70% of all Timber Culture entries were made in Kansas, Nebraska, and the Dakotas, and that Colorado was the only High Plains state where the law was not used very much. Yet even in Colorado, these claims were used as a holding device, because only 3,789, or 27% of the state total of 27,864 claims, eventually received title.

It was equally obvious to contemporaries that the Desert Land Act of 1877 was a mechanism for avoiding the true intention of the homesteading process. The act applied to those who wished to irrigate "nontimbered" or "nonmineral" lands otherwise unfit for cultivation to file on 640 acres and pay only 25 cents an acre. The measure was certainly closer to John Wesley Powell's vision of what was required in the semiarid West, but was seen as a law bulled through the Senate by Aaron A. Sargent to help large California landowners like James B. Haggin and other corporate interests acquire alternate sections in the Imperial Valley (Gates, 1977; Rudy, 2003). Most of the original entries were made in Montana, Wyoming, California, and Colorado, and in a situation quite similar to Timber Culture lands, only a fifth were ever officially patented. In Colorado, entries were made on 3,216,311 acres and only 692,744 acres were titled (Gates, 1977). Detailed studies of whether the lands were used primarily by ranchers without much effort to irrigate the land have not been made. Interest in the institutional imprint of the public land laws on subsequent land use has waned since the 1970s.

The effect of the public land laws on subsequent land use is worth our attention in this context because the signature of the public lands is still so prominent in landscape-scale processes in northeastern Colorado. Despite numerous changes in land policy during the early settlement of eastern Colorado, each designed to induce more extensive growth, development did not proceed quickly. Why, in the absence of any real planning restrictions, did settlers resist wholesale entry on the remaining public domain? First of all, as Paul Gates reminds us, none of the western states ever made free grants of land from the portion of the public domain transferred to their control upon achieving statehood. Typically the states leased their educational lands for grazing and grain raising, and other federal grants to the states were selected by agents of state governments from what appeared to be the most promising parts of the public lands still unclaimed. In the end, roughly 18% of the public lands in Montana, Wyoming, Colorado, and New Mexico were transferred to the states or to the railroads. Congress locked the High Plains states into a pattern of slower development by insisting that if they were to sell the lands transferred to their control, higher prices for the public lands under state administration had to be charged. When Colorado was admitted to the nation in 1876, Congress required it to sell its lands at not less than $2.50 an acre. Montana and Wyoming met the test of statehood much later (in 1889 and 1890), and were required to sell at not less than $10 per acre; New Mexico gained admission in 1912, and was required to sell its eastern lands at $5 an acre and its mountain lands at $3 an acre (Gates, 1977). The states were not in a position to forgo the potential revenue, and the Congress was eager to ensure that the sale of federal public lands came first.

After 1901, several other institutional factors contributed to the persistence of this landscape-scale pattern. Historians, geoscientists, and other students of western water law and water use suggest that, increasingly, the expansion of cropland agriculture was only possible with extensive irrigation projects (Pisani, 1992, 2002; Tyler, 1992; Wohl, 2001; Worster, 1985). In the Shortgrass Steppe region, this happened much earlier in the South Platte than the Republican River basin. Several gravity flow systems were built along the South Platte River in the late 19th and early 20th centuries to expand the area of cropland with access to irrigation water. Reservoirs were dug and irrigation canals constructed, especially in the vicinity of Greeley, in southern Weld County, in a community begun as an agricultural cooperative to promote small-scale irrigation.

The success of the gravity flow systems generally resulted in more secure returns. But irrigated agriculture was ultimately constrained by the flow of water from the many headwaters of the South Platte River along the eastern slope of the Rocky Mountains. During the Depression, water commissioners began to worry about the low ebb in the water cycle. In 1931, the North and South Platte rivers measured only 55% of the mean flow measured by instrumental records between 1904 and 1940. The Colorado State Engineer's Office released figures showing that in the region between Boulder and Fort Collins, and running east to Nebraska (comprising water districts 1–6 and 64), water flow had declined 7.6% in 1930, 25% in 1931, and 5.3% in 1932. These volumes fell below estimated minimum needs of 1.25 acre-ft. of water/acre irrigated cropland, representing a steadily declining average of 1.20 acre-ft. in 1929, 1.11 acre-ft. in 1930, 0.83 acre-ft. in 1931, and 0.79 acre-ft. in 1932 (Tyler, 1992).

In the 1930s, in response to local boosters and corporate interests like Great Western Sugar (which operated 17 beet-processing plants in northern Colorado by 1933), an effort was made to divert water eastward over the mountains from the Colorado River basin. In what became an extended campaign, some 37 trans-mountain projects were eventually constructed by 1992, shifting an estimated 650,000 acre-ft. of water out of the Gunnison, San Juan, and Colorado rivers. The Colorado–Big Thompson Project was the largest diversion scheme, pulling (by various estimates) between 310,000 and 370,000 acre-ft. of water to the South Platte basin by its year of completion in 1956 (Tyler, 1992; Wohl, 2001). The diversion schemes have generally been lauded in the agricultural community, which in 1987 used 85% of Colorado's out-of-stream water for irrigation, but have drawn increasing criticism from ecologists for the alteration of stream hydrology and soil erosion caused by diversion canals and large reservoirs (Wohl, 2001). During the dry years between 1958 and 1990, an average of 34% of water used in Larimer, Weld, and Boulder counties originated in western-slope rivers. The problem remains one of maintaining an adequate volume of water flow rather than increasing flow. Water diversion, despite the scale of successive engineering projects, has only allowed farmers to keep up with water needs, not to expand irrigated cropland in the South Platte River basin. The map reproduced in Figure 4.6, for instance, reveals that the spatial imprint of irrigated cropping has changed very little during the past half century.

On the other hand, irrigated agriculture was almost unknown in the Republican River basin until the 1950s, when the invention of horizontal centrifugal pumps

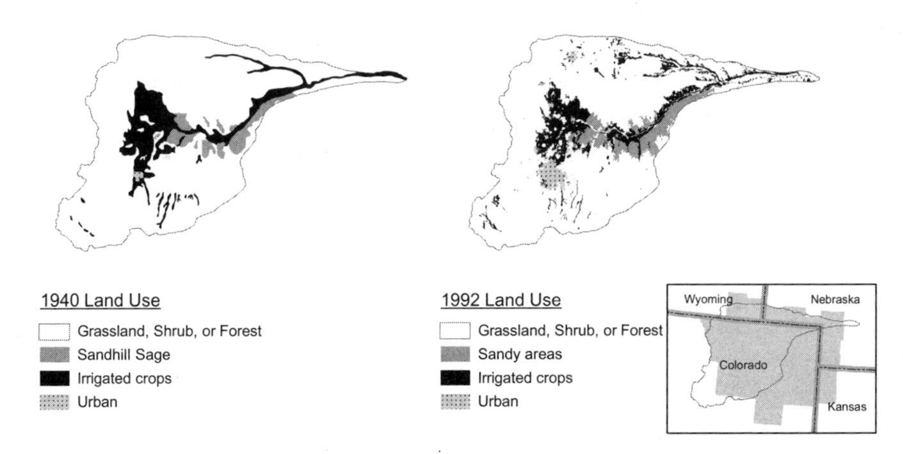

Figure 4.6 Land use in the South Platte River Basin, 1940 and 1992. Based on 1940 land-use data reported in Field Coordinating Committee 15B; land-use data derived from U.S. Geological Survey (1992).

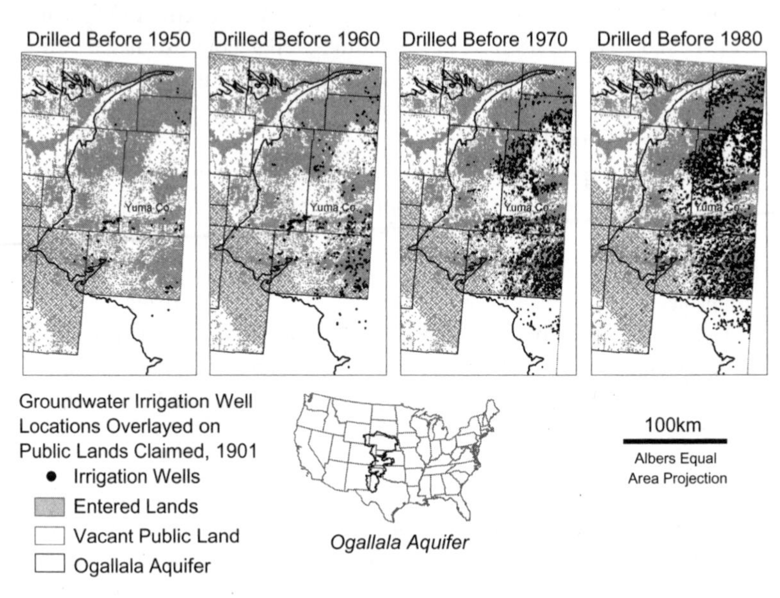

Figure 4.7 Irrigation wells in the Republican River Basin, 1950–1980. Location of irrigation wells and application rates for irrigated cropland during 1980 in the northern High Plains of Colorado. Based on map [at 1:500,000 scale] by Borman and Read (1984).

allowed wells to be dug deeper than 50 ft. The introduction of the new technology allowed cropland to expand. Remarkably, this expansion in the tablelands over the Ogallala Aquifer in eastern Colorado appears to have focused on land with higher soil quality, near the land parcels selected early during the settlement period. The location of wells dug before 1950 and the pace of wells drilled between 1950 and 1980 was documented in the early 1990s by the USGS (Borman and Reed, 1984). The study shows (Fig. 4.7) that at each time interval during the period examined by the USGS, an ever-expanding number of wells was generally situated on parcels with the best soils. Overlain on the historic public lands map shown in Figure 4.3, the well locations indicate that farmers were unwilling to risk the use of the new technology outside the proved cropping areas. The sandy plain running through the middle of Yuma County was largely avoided, as it had been before the emergence of groundwater technology.

Arguably, much of the conservation inertia can be traced to the federal government's efforts to identify land unsuitable for cropland in the 1930s through key programs, including the Soil Conservation Service (SCS) and the Resettlement Administration (Cannon, 1996; Helms, 1990, 1996). Although the land converted to grassland at the time remained relatively small, and efforts at resettlement were generally deemed modest, conservation and soil science benefited enormously from the investment in scientific enterprise.

As early as the 1920s, the undercapitalized sodbuster was increasingly a rare figure, much to the dismay of policymakers like Elwood Mead. As director of the

Bureau of Reclamation, Mead wanted to offer a new start to ordinary Americans, but eventually acknowledged that staying on the land required more local knowledge and personal wealth than most would-be irrigator–farmers possessed (Pisani, 2002). As the century proceeded, only operators inheriting substantial portions of family estates could afford to carry on in agriculture, and increasingly young farmers and ranchers were exclusively from agricultural families (Gardner, 2002; Gutmann et al., 2002). Institutionally, a steady shift away from "reclamation" of desert lands and toward conserving existing water and land resources; restoring vegetative cover; planting borders of grass, shrubs, and trees as windbreaks; adapting local practices to fit variations in topography, soil, and climate; and encouraging contour cultivation, strip cropping, and the rotation of grazing became the norm (Helms, 1990). Indeed, the last half of the 20th century witnessed a steady growth of income supports to agricultural producers, tied increasingly to participation in better management of land already in private ownership (Gardner, 2002).

Most of the work of improving conservation practice was interagency in nature, but the SCS, later renamed the Natural Resources Conservation Service, helped to negotiate the shift in attitude with farmers and ranchers. Unlike the Forest Service and Department of the Interior, the SCS was charged with working with producers and introducing ecological ideas. Arthur W. Sampson's 13 years of expertise as a range manager in National Forest lands were applied more directly to prairie ecology in privately owned range lands when the SCS began to apply his ideas of how to nurse grasslands back to their "climax" state by studying patterns of succession (Helms, 1990). The SCS used Sampson's interpretation of Frederic Clement's work to develop a classification system, and to set up floristic guides to plant populations under various range conditions. Soil Conservation Service field staff learned to inventory "decreasers, increasers, and invaders" in their effort to delineate range sites.

When drought struck again in the 1950s, the USDA's drought committee met to formulate emergency measures, and eventually recommended that any assistance be used to help farmers convert cropland back to grassland, offering farmers 50% of the cost if they agreed to keep the land in grass for at least 5 years. To discourage a return to cropland, the committee hit on the idea of long-term contracts. This was the key measure of the Great Plains Conservation Program signed into law by President Eisenhower on June 19, 1956. Under the bill, contracts of no longer than 10 years were to be entered into between the Secretary of Agriculture and producers before the end of 1971, in counties designated by the secretary as having the most serious wind erosion problems. Practices in the Great Plains, particularly the model of shared-cost contracts of 3 to 10 years' duration, influenced national conservation policies, eventually becoming the standard for other conservation programs (Helms, 1990).

President Nixon's large grain deals with the Soviet Union in the 1970s threatened to derail many of the achievements in conservation practice made since the 1930s. As the price of grain quadrupled between 1970 and 1974, Secretary of Agriculture Earl L. Butz released production controls, including annual set-asides used to lower production levels in key commodities at the outset of the Nixon administration. With the new awareness of soil conservation issues in the farm and ranch communities, many did not welcome the plow-up of grassland that followed or the dust clouds that spilled over into neighboring farms. In Colorado, 572,000 acres of grassland were planted to wheat between 1977 and 1982, prompting many grassroots campaigns against the plowing of fragile grasslands. The solution, dubbed the *sodbuster bill*, as introduced by Colorado senator William Armstrong, was to link eligibility for USDA support programs to soil conservation. Any field classified as fragile land by the SCS, and not planted to an annual crop or used as a set-aside in a USDA commodity program between December 31, 1980, and December 23, 1985, would lose its eligibility for USDA programs unless it first submitted a conservation plan. Wetlands were similarly protected from drainage under the terms of the legislation. All were rolled into the omnibus Food Security Act of 1985 (U.S. Congress, House Committee on Agriculture, 1992). Under the law, farmers had until 1990 to begin applying a conservation plan on highly erodible soils, and until 1995 to implement it fully to stay eligible for USDA programs.

None of the provisions have introduced radically different conservation measures. As this look at the structure of the landscape in the Shortgrass Steppe region throughout the course of the 20th century illustrates, choices about land use have been more incremental and local than patterns of policy formation. The biggest surprise may be that often those choices that were made earliest were the ones that have been the most sustainable. The mental model inherited from the Dust Bowl needs to make room for how bottom-up decision making shapes conservation practice. Even in the context of the 1970s export boom, local understanding of the long-term uses of the land became the basis for grassroots opposition to renewed agricultural expansion.

Nevertheless, local decisions still must interact with the policy frameworks like the public land system. Although virtually all land with any potential for agriculture has passed into private hands since the early 20th century, the architecture of the public lands system is still visible. School lands, for instance, largely remain under the control of the state land boards. These are the 1-sq. mi. parcels of state-owned land that dot the landscape in a chessboard pattern in eastern Colorado and Wyoming, as seen in Figure 4.8. Under the terms of the Land Ordinance Act, which established the township survey system in 1785, section 16 of each township was set aside for school purposes. In recent years, state authorities and nongovernmental organizations have worked together to create grassland preserves, purchasing private land adjoining the remaining school lands. The Tamarack Ranch Natural Area along the southern bank of the South Platte River is one example.

The most important opportunities for conservation exist in the persistence of privately owned grassland corridors within the Shortgrass Steppe region. With the announcement from the Secretary of Agriculture in 2003 of another round of easements, known as the Grassland Reserve Program, the emphasis on restoration and preservation continues to grow. Much of southeastern Colorado remains in grassland cover (except for land along the Arkansas River), and recent analysis of wildlife habitat in northeastern Colorado shows that historic corridors continue

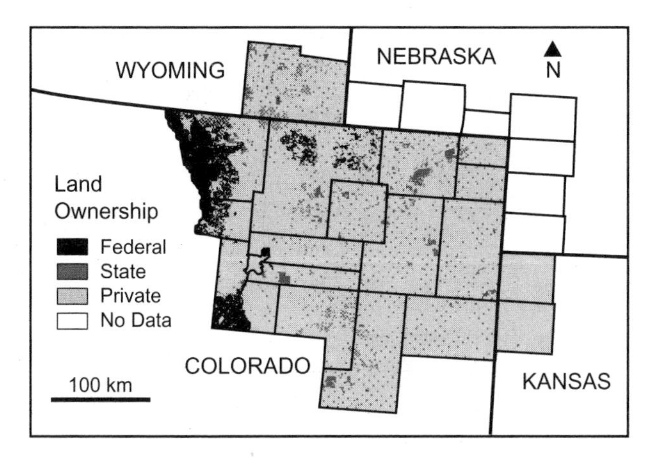

Figure 4.8 Land ownership in Shortgrass Steppe counties, 1998. Land ownership data are available in statewide land stewardship ArcInfo coverages (in compressed interchange format) from the USGS GAP analysis program, available online at http://gapanalysis.nbii.gov/portal/server.pt. Land ownership information for Colorado is from National Ecology Research Center (1993); Wyoming data are from Wyoming Gap Analysis (1996); Kansas land ownership information is from Data Access and Support Center (1994); Nebraska land ownership information is from University of Nebraska–Lincoln, School of Natural Resources, Center for Advanced Land Management Information Technologies (2005).

to exist, providing crucial routes for herbivores and bird populations (Theobald et al., 2004). The preservation of these corridors is crucial because, as ecosystem science continues to demonstrate, the mortality of the dominant C_4 perennials is driven by patch disturbances, and bunchgrass species like *Bouteloua gracilis* (blue grama) rely on the increased availability of belowground resources stimulated by gap dynamics (Coffin and Lauenroth, 1988, 1990, 1991, 1992). Plot-level experiments at CPER, in locations with sand loam soil (with 81% sand content and 1.8% organic matter), have shown bare soil microsites dug to a depth of 10 cm had more seedlings than sites where adult plants were left undisturbed (Aguilera and Lauenroth, 1995).

Similarly, the long-term survivorship of blue grama indicates that plant communities tend to be stable, to recruit seedlings, and to colonize better during drought periods than during wet years (Fair et al., 1999). Patterns of succession over several decades indicate that blue grama will eventually outcompete other species, regardless of whether they are invasive, because it is so well adapted to low precipitation and grazing (Costello, 1944; Klipple and Costello, 1960; McGinnies et al., 1991; Milchunas et al., 1998). Although fire and bison cannot function in the same way as they did in the past, studies of livestock grazing and prescribed burning have shown that they can recreate some of the patch dynamics lost historically by the bison's near demise and fire suppression (Brockway et al., 2002; Knapp et al., 1999). Free-roaming ungulates and fire tend to create a mosaic of intensely grazed land and carbon-enriched soil. That fragmentation helps to drive healthy succession dynamics of the shortgrass. An ungrazed landscape leaves the shortgrass more susceptible to plant–exotics and declining bird populations than does low to moderate grazing from domestic cattle (Milchunas et al., 1998). The main alternative use of private land, which is typically dryland wheat production, has been shown to adversely affect soil fertility (Burke et al., 1989; Ihori et al., 1995) and biodiversity (Moore 1994). However, intensified continuous cropping of winter wheat, where it has traditionally existed, has also been shown to enhance the efficiency of precipitation use and plant productivity, and hence nitrogen and carbon storage in the soil (Lauenroth et al., 2000).

Conclusion

In retrospect, the integrity of remaining corridors in the northern shortgrass are the product of a unique convergence of farm-level behavior and biophysical variation. Since the 1860s, the transformation of the Shortgrass Steppe has reduced native grassland to a fraction of its former size. It has led to the near extinction of a keystone species—the bison—and the challenges remain great. Pollution from intensified agriculture, the burdens of watershed management, and the threat of urban sprawl, all press on planning agendas. However, ecological disaster is a misnomer. The patterns of agricultural land use in the Shortgrass Steppe region have met the test of time and form the basis for a sustainable future. Institutional imprinting has largely reinforced a human signature on the landscape shaped early and often by biophysical limits. The grassland corridors that remain in the Shortgrass Steppe landscape are a unique resource for preserving biodiversity, increasing wildlife habitat, and ensuring that depleting water resources are used more wisely. Throughout many millennia, the physical extent of the shortgrass has retreated and expanded with the evolution of climate conditions. The 20th century and the Dust Bowl have taught us to respect the resilience of the shortgrass and to seek out its many dimensions.

References

Aguilera, M. O., and W. K. Lauenroth. 1995. "Influence of gap distances and type of microsites on seedling establishment in *Bouteloua gracilis*." *Journal of Ecology* 8(1): 87–97.

Bailey, R. G., and L. Ropes. 1998. *Ecoregions: The ecosystem geography of the oceans and continents.* New York: Springer.

Bennett, H. H. 1947. *Elements of soil conservation.* New York: McGraw-Hill.

Binnema, T. 2001. *Common and contested ground: A human and environmental history of the northwestern plains.* Norman, Okla.: University of Oklahoma Press.

Borman, R. G., and R. L. Reed. 1984. *Location of irrigation wells and application rates for irrigated cropland during 1980 in the northern High Plains of Colorado.* Reston, Va.: U.S. Geological Survey.

Brockway, D. G., R. G. Gatewood, and R. B. Paris. 2002. "Restoring fire as an ecological process in shortgrass prairie ecosystems: Initial effects of prescribed burning during the dormant and growing seasons." *Journal of Environmental Management* 65: 135–152.

Bureau of Reclamation, United States Reclamation Service. 1902. *Annual report: Bureau of Reclamation.* Washington, D.C.: U.S. Government Printing Office.

Burke, I. C., C. M. Yonker, W. J. Parton, C. V. Cole, K. Flach, and D. S. Schimel. 1989. "Texture, climate and cultivation effects on soil organic matter content in U.S. grassland soils." *Soil Science Society of America Journal* 53: 800–805.

Calloway, C. G. 2003. *One vast winter count: The Native American West before Lewis and Clark.* Lincoln, Neb.: University of Nebraska Press.

Cannon, B. Q. 1996. "Keeping their instructions straight: Implementing the rural resettlement program in the West." *Agricultural History* 70(2): 251–267.

Coffin, D. P., and W. K. Lauenroth. 1988. "The effects of disturbance size and frequency on a shortgrass plant community." *Ecology* 69: 1609–1617.

Coffin, D. P., and W. K. Lauenroth. 1990. "A gap dynamics simulation model of succession in a semiarid grassland." *Ecological Modeling* 49: 229–266.

Coffin, D. P., and W. K. Lauenroth. 1991. "Effects of competition on spatial distribution of roots of blue grama." *Journal of Range Management* 44: 67–70.

Coffin, D. P., and W. K. Lauenroth. 1992. "Spatial variability in seed production of the perennial bunchgrass *B. gracilis* (H.B.K.) *Lag. Ex Giffiths.*" *American Journal of Botany* 79: 347–353.

Commission for Environmental Cooperation. 1997. *Ecological regions of North America: Toward a common perspective.* Montreal: CEC Secretariat.

Costello, D. F. 1944. "Natural revegetation of abandoned plowed land in mixed prairie association of northeastern Colorado." *Ecology* 25: 312–326.

Cunfer, G. 2005. *On the Great Plains: Agriculture and environment.* College Station, Tex.: Texas A&M University Press.

Data Access and Support Center. 1994. *Kansas-GAP coverage.* Lawrence, Kans.: Data Access and Support Center, Kansas Geological Survey.

Dixon, J. 1999. *Bones, boats and bison: Archeology and the first colonization of western North America.* Albuquerque, N.M.: University of New Mexico Press.

Fair, J., W. K. Lauenroth, and D. P. Coffin. 1999. "Demography of *Bouteloua gracilis* in a mixed prairie: Analysis of genets and individuals." *Journal of Ecology* 87: 233–243.

Field Coordinating Committee 15B. Preliminary Examination Report, *Runoff and waterflow retardation and soil erosion prevention: South Platte River and its tributaries, Colorado, Wyoming, and Nebraska.* Denver, Colo.: National Archives and Records Administration, Denver Office, RG 114, Records of the Natural Resources Conservation Service, Box No. 173, River Basin Reports, Platte River.

Flores, D. 2001. *The natural west environmental history in the Great Plains and Rocky Mountains.* Norman, Okla.: University of Oklahoma Press.

Forman, S. L., R. Oglesby, V. Markgraf, and T. Stafford. 1995. "Paleoclimatic significance of Late Quaternary eolian deposition on the Piedmont and High Plains, central United States." *Global and Planetary Change* 11: 35–55.

Forman, S. L., R. Oglesby, and R. S. Webb. 2001. "Temporal and spatial patterns of Holocene dune activity on the Great Plains of North America: Megadroughts and climate links." *Global and Planetary Change* 29: 1–29.

Fredlund, G. G. 1995. "Late Quaternary pollen record from Cheyenne Bottoms, Kansas." *Quaternary Research* 43: 67–79.

Gardner, B. L. 2002. *American agriculture in the twentieth century: How it flourished and what it cost.* Cambridge, Mass.: Harvard University Press.

Gates, P. W. 1977. "Homesteading in the High Plains." *Agricultural History* 51(1): 109–133.

Gutmann, M. P., S. Pullum-Pinon, and T. W. Pullum. 2002. Three eras of young adult home leaving in twentieth-century America. *Journal of Social History* 35(3): 533–576.

Hall, S., and S. Valastro Jr. 1995. "Grassland vegetation in the southern Great Plains during the last glacial maximum." *Quaternary Research* 44: 237–245.

Hansen, Z. K., and G. D. Libecap. 2004a. "Small farms, externalities, and the Dust Bowl of the 1930s." *Journal of Political Economy* 112: 665–693.

Hansen, Z. K., and G. D. Libecap. 2004b. "The allocation of property rights to land: U.S. land policy and farm failure in the Great Plains." *Explorations in Economic History* 41: 103–129.

Hayden, F. V., and United States, Geological and Geographical Survey of the Territories. 1877. *Geological and geographical atlas of Colorado and portions of adjacent territory.* Washington, D.C.: Julius Bien.

Helms, J. D. 1990. "Conserving the Plains: The Soil Conservation Service in the Great Plains." *Agricultural History* 64(2): 58–73.

Helms, J. D. 1996. "National soil conservation policies: A historical case study of the driftless area." *Agricultural History* 70(2): 377–394.

Hunt, C. B. 1967. *Physiography of the United States.* A series of books in geology. San Francisco, Calif.: W. H. Freeman.

Ihori, T., I. C. Burke, W. K. Lauenroth, and D. P. Coffin. 1995. "Effects of cultivation and abandonment on soil organic matter in northeastern Colorado." *Soil Science Society of America Journal* 59: 1112–1119.

Isenberg, A. C. 2000. *The destruction of the bison: An environmental history, 1750–1920.* New York: Cambridge University Press.

Jordan, T. G. 1993. *North American cattle-ranching frontiers: Origins, diffusion, and differentiation.* 1st ed. Histories of the American frontier. Albuquerque, N.M.: University of New Mexico Press.

Kimbrell, A. 2002. "A blow to the breadbasket: Industrial grain production," pp. 99–110. In: Andrew Kimbrell (ed.), *Fatal harvest: The tragedy of industrial agriculture.* Washington, D.C.: Island Press.

Klipple, G. E., and D. F. Costello. 1960. *Vegetation and cattle responses to different intensities of grazing on short-grass ranges on the central Great Plains.* USDA Forest Service technical bulletin no. 1216. Washington, D.C.: U.S. Government Printing Office.

Knapp, A. K., J. M. Blair, J. M. Briggs, S. L. Collins, D. C. Hartnett, L. C. Johnson, and E. G. Towne. 1999. "The keystone role of bison in North American tallgrass prairie: Bison increase habitat heterogeneity and alter a broad array of plant, community, and ecosystem processes." *Bioscience* 49: 39–50.

Küchler, A. W. 1964. *Potential natural vegetation of the conterminous United States.* New York: American Geographical Society.

Küchler, A. W. 1993. "Potential natural vegetation of the conterminous United States. Digital vector data in an Albers Equal Area Conic polygon network and derived raster data on a 5 km by 5 km Albers Equal Area 590×940 grid." *Global Ecosystems Database Version 2.0.* Boulder, Colo.: NOAA National Geophysical Data Center.

Lauenroth, W. K., I. C. Burke, and M. P. Gutmann. 1999. "The structure and functions of ecosystems in the central North American grassland region." *Great Plains Research* 9: 223–259.

Lauenroth, W. K., I. C. Burke, and J. M. Paruelo. 2000. "Patterns of production and precipitation-use efficiency of winter wheat and native grasslands in the central Great Plains of the United States." *Ecosystems* 3: 344–351.

Lauenroth, W. K., and D. G. Milchunas. 1992. "Shortgrass steppe," pp. 183–226. In: Robert T. Coupland (ed.), *Natural grasslands: Introduction and western hemisphere.* Vol. 8A. Amsterdam: Elsevier.

Libecap, G., and Z. Hansen. 2002. "Rain follows the plow and dryfarming doctrine: The climate information problem and homestead failure in the Upper Great Plains, 1890–1925." *Journal of Economic History* 62(1): 86–120.

Madole, R. F. 1995. "Spatial and temporal patterns of Late Quaternary eolian deposition, eastern Colorado, USA." *Quaternary Science Reviews* 14(2): 155–178.

Martin, P., and C. Szuter. 1999. "War zones and game sinks in Lewis and Clark's West." *Conservation Biology* 13(winter): 36–45.

McGinnies, W. J., H. L. Shantz, and W. G. McGinnies. 1991. *Changes in vegetation and land use in eastern Colorado: A photographic study, 1904 to 1986.* Springfield, Va.: U.S. Department of Agriculture, Agricultural Research Service, ARS-85.

McHugh, T., and V. Hobson. 1972. *The time of the buffalo.* Lincoln, Nebr.: University of Nebraska Press.

Meagher, M. M. 1973. *The bison of Yellowstone National Park.* National Park Service. Scientific monograph series, no. 1. Washington, D.C.: U.S. National Park Service the Supt. of Documents, U.S. Government Printing Office.

Milchunas, D. G., W. K. Lauenroth, and I. C. Burke. 1998. "Livestock grazing: Animal and plant biodiversity of Shortgrass Steppe and the relationship to ecosystem function." *OIKOS* 83: 65–74.

Moore, J. C. 1994. "Impact of agricultural practices on soil food web structure: Theory and application." *Agriculture, Ecosystems and Environment* 51: 239–247.

Muhs, D. R., J. N. Aleinikoff, T. W. Stafford Jr., R. Kihl, J. Been, S. A. Mahan, and S. Cowherd. 1999. "Late Quaternary loess in northeastern Colorado: Part 1. Age and paleoclimatic significance." *Geological Society of America Bulletin* 111(12): 1861–1975.

National Ecology Research Center. 1993. *Colorado Land Ownership, Colorado Gap Project, 500K.* Fort Collins, Colo.: Fish and Wildlife Service, National Ecology Research Center.

Norall, F. 1988. *Bourgmont, explorer of the Missouri, 1698–1725.* Lincoln, Nebr.: University of Nebraska Press.

Opie, J. 1987. *The law of the land: Two hundred years of American farmland policy.* Lincoln, Nebr.: University of Nebraska Press.

Opie, J. 1998. *Nature's nation: An environmental history of the United States.* New York: Harcourt Brace College Publishers.

Opie, J. 2000. *Ogallala water for a dry land.* 2nd ed. Our sustainable future, vol. 13. Lincoln, Nebr.: University of Nebraska Press.

Parks, D. R. (ed.). 1989. *Ceremonies of the Pawnee by James R. Murie.* Lincoln, Nebr.: University of Nebraska Press.

Parks, D. R., and W. R. Wedel. 1985. "Pawnee geography: Historical and sacred." *Great Plains Quarterly* 5: 143–176.

Parton, W. J., M. P. Gutmann, and W. R. Travis. 2003. "Sustainability and historical land-use change in the Great Plains: The case of eastern Colorado." *Great Plains Research* 13(spring): 97–125.

Pisani, D. 1992. *To reclaim a divided West: Water, law, and public policy, 1848–1902.* Albuquerque, N.M.: University of New Mexico Press.

Pisani, D. 2002. *Water and American government: The Reclamation Bureau, national water policy, and the West, 1902–1935.* Berkeley, Calif.: University of California Press.

Powell, J. W., and Geographical and Geological Survey of the Rocky Mountain Region (U.S.). 1879. *Report on the lands of the arid region of the United States with a more detailed account of the lands of Utah.* Washington, D.C.: General Printing Office.

PRISM. 2008. PRISM (Parameter-elevation Regressions on Independent Slopes Model) Climate Data for the United States. Oregon State University. Online. Available at http://www.prism.oregonstate.edu.

Quandt, L. A., S. W. Waltman, and C. L. Roll. 1998. *Dominant soil orders and suborders: Soil taxonomy, United States.* Maps and photographs, NSSC—5502-0398-30 ed. Lincoln, Nebr.: USDA Natural Resources Conservation Service.

Rudy, A. 2003. "The social economy of development: The state of and the Imperial Valley," pp. 25–46. In: Jane Adams (ed.), *Fighting for the farm: Rural America transformed.* Philadelphia, Pa.: University of Pennsylvania Press.

Samson, F., and F. Knopf. 1994. "Prairie conservation in North America." *Bioscience* 44: 418–421.

Savage, D. A., and D. F. Costello. 1948. "The Southern Great Plains," pp. 503–506. In: U.S. Department of Agriculture. *Yearbook of Agriculture.* Washington, D.C.: U.S. Department of Agriculture.

Smith, G. D. 1986. *The Guy Smith interviews: Rationale for concepts in soil taxonomy.* U.S. Department of Agriculture, Soil Conservation Service, Soil Management Support Services technical monograph no. 11. Ithaca, N.Y.: Cornell University Department of Agronomy.

Smith, L. M. 2003. *Playas of the Great Plains.* Austin, Tex.: University of Texas Press.

Stafford, S. G., N. E. Kaplan, and C. Bennett. 2002. *Looking through the looking glass: What do we see, what have we learned, what can we share? Information management at the Shortgrass Steppe Long Term Ecological Research Site.* Presented at the 6th World Multiconference on Systemics, Cybernetics and Informatics, Orlando, Fla.

Steinberg, T. 2002. *Down to earth: Nature's role in American history.* New York: Oxford University Press.

Theobald, D. M., N. Peterson, J. Terlaak, and G. Wilcox. 2004. *Mapping and assessment of protected areas and broad-scale threats in the shortgrass prairie of Colorado.* Unpublished Report for the Colorado Division of Wildlife. Fort Collins, Colo.: Natural Resources Ecology Lab, Colorado State University.

Thomas, A. B. 1935. *After Coronado: Spanish exploration northeast of New Mexico, 1696–1727. Documents from the archives of Spain, Mexico, and New Mexico.* Civilization of the American Indian series. Norman, Okla.: University of Oklahoma Press.

Tyler, D. 1992. *The last water hole in the West: The Colorado–Big Thompson Project and the Northern Colorado Water Conservancy District.* Niwot, Colo.: University Press of Colorado.

University of Nebraska-Lincoln, School of Natural Resources, Center for Advanced Land Management Information Technologies. 2005. *Nebraska Land Cover Classification.* Lincoln, Nebr.: University of Nebraska-Lincoln, School of Natural Resources, Center for Advanced Land Management Information Technologies.

U.S. Congress, House Committee on Agriculture, and Subcommittee on Conservation, Credit, and Rural Development. 1992. *Implementation of the conservation compliance provisions of the Food Security Act of 1985: Hearing before the Subcommittee on Conservation, Credit, and Rural Development of the Committee on Agriculture, House of Representatives.* 102nd Cong., 2nd session. May 7, 1992. Washington, D.C.: U.S. Government Printing Office.

U.S. Department of Commerce. 1870, 1880, 1890, 1900, 1910, 1920, 1930, 1940, 1950, 1954, 1959, 1964, 1969, 1974, 1978, 1982, 1987, 1992, 1997. *Census of agriculture*. Washington, D.C., Bureau of the Census.

U.S. Department of Commerce. 1870, 1880, 1890, 1900, 1910, 1920, 1930, 1940. *Census of manufactures*. Washington, D.C.: Bureau of the Census.

U.S. Department of Commerce. 1870, 1880, 1890, 1900, 1910, 1920, 1930, 1940, 1950, 1960, 1970, 1980, 1990, 2000. *Census of population*. Washington, D.C., Bureau of the Census.

U.S. Department of Commerce. 1947, 1949, 1950, 1952, 1956, 1962, 1967, 1970, 1972, 1977, 1983, 1988, 1994, 2000. *County and city data book*. Washington, D.C.: Bureau of the Census.

U.S. Department of Interior, Fish and Wildlife Service, National Ecology Research Center. 1993. *Colorado Land Ownership, Colorado Gap Project, 500K*. Fort Collins, Colo.: Fish and Wildlife Service (FWS), National Ecology Research Center (NERC).

U.S. Geological Survey. 1992. *National land cover data*. Online. Available at http://landcover.usgs.gov/natllandcover.php.

Vogelmann, J. E., S. M. Howard, L. Yang, C. R. Larson, B. K. Wylie, and N. Van Driel. 2001. "Completion of the 1990s National Land Cover data set for the conterminous United States from Landsat Thematic Mapper data and ancillary data sources." *Photogrammetric Engineering and Remote Sensing* 67: 650–652.

Watenpaugh, H. N., L. A. Brown, L. A. Moorhouse, and R. T. Burdick. 1941. *Kit Carson County: A procedure for land-use classification in the central High Plains based on studies in Kit Carson County, Colo. 29 August*. Denver, Colo.: National Archives and Records Administration Archives, Denver Office, RG 83, Records of the Bureau of Agricultural economics, Colorado State Office, Box No. 2, Entry 261, 8NN-083–92.

Wedel, W. R. 1986. *Central Plains prehistory: Holocene environments and culture change in the Republican River basin*. Lincoln, Nebr.: University of Nebraska Press.

West, E. 1998. *The contested plains Indians, goldseekers, and the rush to Colorado*. Lawrence, Kans.: University Press of Kansas.

White, R. 1983. *The roots of dependency subsistence, environment, and social change among the Choctaws, Pawnees, and Navajos*. Lincoln, Nebr.: University of Nebraska Press.

Wohl, E. 2001. *Virtual rivers: Lessons from the mountain rivers of the Colorado front range*. New Haven, Conn.: Yale University Press.

Worster, D. 1979. *Dust Bowl: The southern plains in the 1930s*. New York: Oxford University Press.

Worster, D. 1985. *Rivers of empire: Water, aridity, and the growth of the American West*. New York: Pantheon Books.

Worster, D. 1992. *Under western skies: Nature and history in the American West*. New York: Oxford University Press.

Worster, D. 2001. *A river running west: The life of John Wesley Powell*. New York: Oxford University Press.

Wyckoff, W. 1999. *Creating Colorado: The making of a western American landscape, 1860–1940*. New Haven, Conn.: Yale University Press.

Wyoming Gap Analysis. 1996. *Land Cover for Wyoming*. Laramie, Wy.: Spatial Data and Visualization Center, University of Wyoming.

5

The Political Ecology of Southwest Michigan Agriculture, 1837–2000

Alan P. Rudy
Craig K. Harris
Brian J. Thomas
Michelle R. Worosz
Siena S. K. Kaplan
Evann C. O'Donnell

Flying over southwestern Michigan, the Kellogg LTER site (Fig. 5.1), one looks down on what appears to be a glorious example of the agricultural landscape of classical American populism. The farm units are discernible by their woodlots, windbreaks, and rural roads, and are fairly small, or at least approximate what some have called a human scale. There is a mixture of pasturage, grain production, fruit trees, animal enterprises, and other forms of agriculture from tree farms to glass-house nurseries. Each farm seems to have a residential homestead and many have small gardens, red barns, and white or blue silos, interspersed with more recent structures for heavy machinery. In this populist image, the farms are prosperous, independent, and sustainable, and the farm families are harmonious and fulfilled.

Three things are noteworthy about this landscape. First, appearances are deceiving. Although today's fields remain more or less continuous with historical agricultural unit sizes, much of agricultural production is now generated by farm operators who rent lands across extensive distances to maintain economically viable levels of gross production and net profitability (Sublett, 1975). At the same time, the majority of farm owner–operators are pluriactive—they and members of their households pursue multiple modes of income generation—given the difficulty of making a viable living from agriculture. Most farmers have debt loads greater than 50%, and more than half of net farm income comes from government payments. In addition, there is tremendous concern about the negative impacts of farming on the environment.

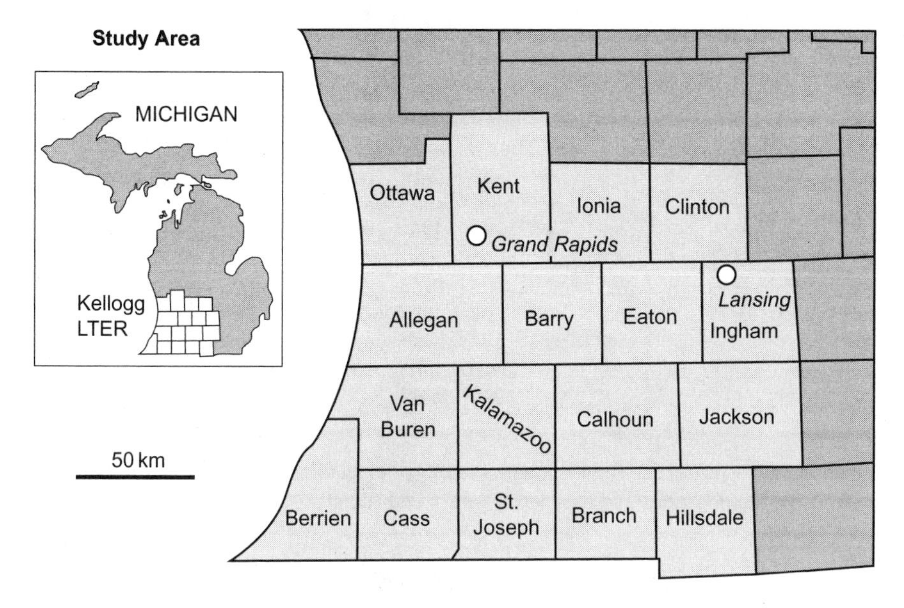

Figure 5.1 The southwest Michigan study site.

Second, new forms of exurban rural development (e.g., small aggregations of megahomes; dispersed modest rural residences on 5–10-acre parcels; clusters of inexpensive, prefabricated housing on recently converted, and still treeless, agricultural fields) are encroaching on what have historically been agricultural spaces. Within these fairly new and increasingly prevalent patterns of exurban development, woodlots hide newly constructed country homes, the residents of which are very selective in their appreciation of the sounds, sights, smells, and slow-moving machinery of agriculture.

Third, it obviously has not always looked this way. Michigan's rural landscape is the contemporary product of historical agricultural activity, including its diversity of enterprises, its golden ages and financial crises, its ecological harmonies and discordances, and its social affinities and contradictions. In fact, the state's agricultural landscape is the accreted and residual product of a series of agricultural periods. It is the dynamics of, and transitions between, these periods that have produced the social and ecological characteristics of Michigan's diverse rural areas and a good bit of the history of the state's urban centers.

Most important for our narrative, each of the historical periods reconstructed the landscape differently. Each of the periods of Michigan agriculture is an outcome of the interaction among the previous period of agriculture, the historical and current social institutions, and the ecological context. The influences among these three realms are reciprocal and relational (Fig. 5.2), influencing and stimulating, enabling and constraining each other. These transformative processes can be seen at local, regional, state, national, and international scales in forms that shifted as the national agricultural economy developed, as technoscientific advances emerged, and as the international export of commodity surpluses

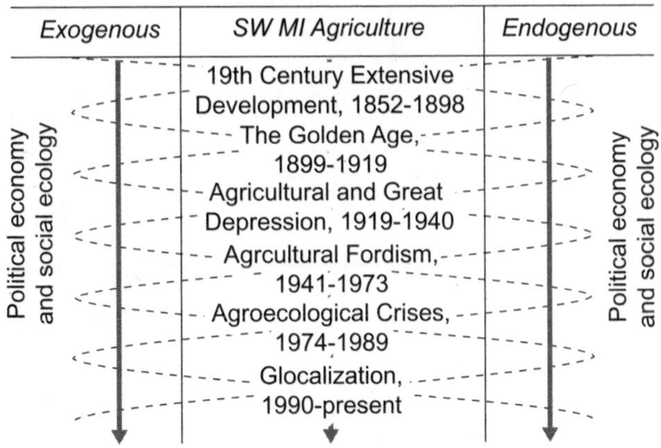

Exogenous	SW MI Agriculture	Endogenous
Political economy and social ecology	19th Century Extensive Development, 1852-1898	Political economy and social ecology
	The Golden Age, 1899-1919	
	Agricultural and Great Depression, 1919-1940	
	Agrcultural Fordism, 1941-1973	
	Agroecological Crises, 1974-1989	
	Glocalization, 1990-present	

Figure 5.2 Regional agriculture as the relational mediation of endogenous and exogenous nature and society.

became the global trade of inputs and services, bulk and specialty commodities, and fresh and processed foods. Each of the historical periods is marked by a restructuring of the agricultural landscape associated with emerging political economic processes. The region many people now call southwest Michigan is a result of these transitions in the agricultural landscape.

Pre-European Settlement History and the Geophysical

The first four billion years of the history of the land that comes to be called Michigan is a series of glacial flows and tropical ebbs. Fifteen thousand to 20,000 years ago, the last glaciers retreated across eastern and central North America. Geomorphologically, the glaciers left behind a terrain that slopes upward gradually from the lakeshore to a ridge about 50 km inland. From the ridge, the terrain falls gently to the northeast, and, after descending into a valley, rises somewhat farther to the southeast, reaching in the Muskegon highlands a height of approximately 175 m above the lake level.

The interaction of all the geology that had gone before with the retreat of the last glacier means that rich soils, wetlands, and prairies are predominant in the southwestern part of the state, and sandy soils are predominant in the west-central and northwestern part of the state. More specifically, geophysically, the southwestern region is composed of predominantly Southern Michigan and Northern Indiana drift plain, peppered with small pockets of Indiana and Ohio till plain and the Southwestern Michigan Fruit and Truck Belt soils (Fig. 5.3).

The region averages 889 mm of rain annually (Fig. 5.4), with the greatest rain- and snowfall in the west as a result of the lake effect precipitation. The southwestern corner of Michigan has an average growing season that is 20 days longer than the rest of the region (which averages 130–160 growing days), with a fairly smooth and moderate gradient from southwest to northeast across the region.

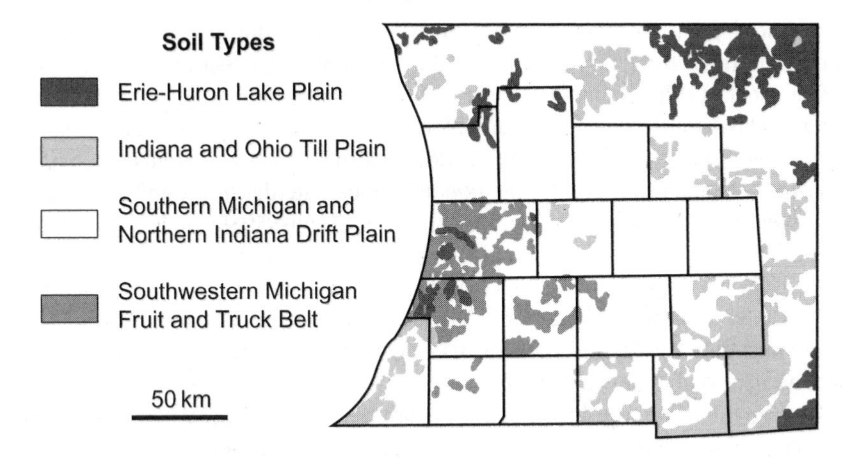

Figure 5.3 Southwestern Michigan soil types.

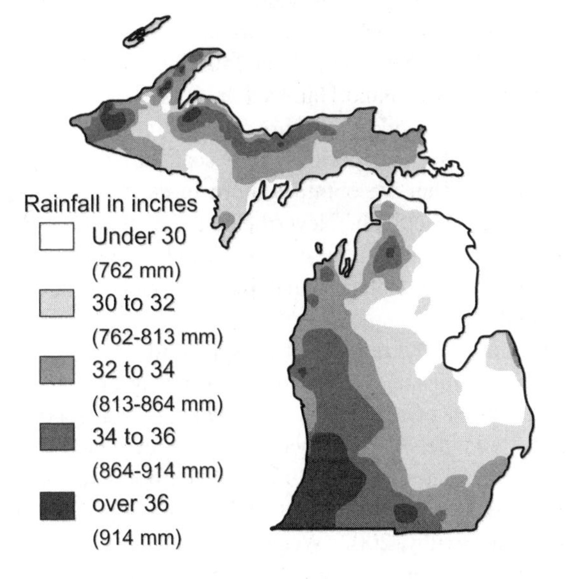

Figure 5.4 Average annual precipitation in Michigan. Data derived from Michigan's averaged annual rainfall between 1961 and 1990. The data were collected from stations that were observed by the National Oceanic and Atmospheric Administration (NOAA) cooperative and USDA-National Research Conservation Service (NRCS) SnoTel Networks, and from other state and local networks.

As a result of the combination of soil and climate, southwest Michigan was, prior to European settlement, predominantly hardwood forest and swamp, whereas the white pine forests to the north grew on fairly porous sandy loams.

The Precontact Archaeology of Southwestern Michigan

Southwestern Michigan changed dramatically at the end of the Pleistocene and the beginning of the Holocene with the deglaciation of the western Great Lakes. Diminishing ice masses, changing drainage and impoundment, variable lake levels, and shifting outlets created a dynamic environment as new biomes opened to human settlement. Southwestern Michigan was one of the first areas of the state to become free of glacial ice and available for human occupation ca. 14,000 BP.

The earliest inhabitants of southern Michigan are commonly known as paleo-Indians. The archaeological record of this time period is scant as a result of taphonomic issues associated with the downcutting of fluvial systems (Monaghan, Lovis, and Hambacher, 2004), low population and site densities, and low site visibility resulting from mobile foraging strategies (Lovis, 1988; Meltzer and Smith, 1986). The majority of the archaeological record consists of individual finds of distinctive spear points with so-called *flutes*. In addition, large hide scrapers, gravers, and an array of flake tools have been recovered. These distinctive spear points and their associated tool kit disappear from the archaeological record of southwestern Michigan ca. 9500 BP.

During this time, the local environment changed markedly, particularly with respect to water levels in the Lake Michigan basin (Larson, 2001). Vegetation changed from tundra to spruce parkland to deciduous forest; and the resident fauna (e.g., mammoth, mastodon, Scott's moose, giant beaver) became extinct ca. 10,000 BP (Holman, 1995; Lovis, 1999). The subsequent Archaic period in southwestern Michigan is largely associated with the exploitation of deciduous forest and prairie grassland from ca. 9000 BP through about 2500 BP (Kapp, 1999; Larson and Schaetzl, 2001; Monaghan, Lovis, and Hambacher, 2004).

The Archaic period is traditionally subdivided into three parts: Early, Middle, and Late. Both the Early and Middle Archaic periods have scant archaeological records. It is with the Late Archaic that we obtain a more in-depth glimpse of human adjustment to southwestern Michigan. Nevertheless, the majority of this information on the Late Archaic occupations is mixed with chronologically later Woodland occupations, making their isolation and interpretation difficult (Robertson, Lovis, and Halsey, 1999).

Faunal and floral remains reveal use of the region during spring for the exploitation of fish such as lake sturgeon (*Acipenser fulvescens*), and during late summer and autumn for intensive extraction of a range of mammals and nut foods, including walnut, hickory, acorn, and beech. Moreover, there is evidence for an indigenous cultigen, the sunflower, dating to ca. 2960 radiocarbon years BP (Parker, 1990, pp. 406–410), whereas Mesoamerican cultigens such as squash are notably absent from this region (Egan-Bruhy, 2002; Wright, 1964).

Most of the southwestern Michigan region as defined in this study falls within the Carolinian zone (Cleland, 1966). The Carolinian biotic community included oak–hickory forests along with wetter and dryer forest associations, making this an excellent zone for procurement of a variety of plants (aquatic roots, nut crops, fruits, starchy and oily seed plants) and animals (fish, various mammals, and avian species). Prairies also occurred sporadically throughout southwestern Michigan, but to date no archeological research has focused on the use of the ecozone by prehistoric peoples.

At approximately 2600 BP, the appearance of ceramic pots in southwestern Michigan signals the beginning of the Early Woodland period. These thick-walled, cord-impressed vessels were used to render and store edible nut oils and also (like proto-crockpots) to slow simmer other foodstuffs (Ozker, 1982). Most important in this period, Early Woodland groups began to aggregate seasonally at earthworks and utilize small, seasonal base camps along the St. Joseph and Kalamazoo river valleys (cf. Garland, 1986; Garland and Beld, 1999).

A synthesis of archaeological surveys on the Grand River suggests two clusters of base camps that may reflect different cultural groups. One group was focused on resource extraction on the lower portion of the Grand River and the other was located further upstream at the confluence of the Maple and Grand rivers (Brashler and Mead, 1996). A number of earthwork sites may have served to more fully integrate groups living in southwestern Michigan, and also groups from the Saginaw basin on the eastern side of the state (Beld, 1993; Beld, 1994; Brashler and Mead, 1996; Garland and Beld, 1999; Holman and Kingsley, 1996). At habitation sites in particularly advantageous spots for fishing and procurement of other aquatic resources, there may be recurrence in site location between Early Woodland and subsequent Middle Woodland mortuary mounds. This influence is most notably seen in the burial goods and interments at the mound sites on the lower Grand River, but more mundane evidence is also present in the non-mortuary pottery throughout the region.

Unfortunately, few Middle Woodland sites in southwestern Michigan have been excavated, and even fewer analyzed in a manner that would produce good botanical data. One site appears to have been intensively occupied during the warm months of the year. Intensive fishing (sturgeon) and deer hunting were the primary subsistence activities (Brashler and Holman, 2004; Brashler, Laidler, and Martin, 1998). The botanical assemblage does not indicate either intensive plant collecting or horticulture, and no agricultural tools have been recovered from the site. In fact, nowhere in Michigan during this time period is there evidence for intensive harvesting of indigenous cultigens (Monaghan, Lovis, and Hambacher, 2004).

Recent evaluation of traditional models of Middle Woodland subsistence and settlement in southwestern Michigan finds support for seasonal northward forays for hunting and gathering (Brashler and Holman, 2004). Middle Woodland groups from the Grand River appear to have made seasonal hunting and gathering forays to the Traverse Corridor, where cultural interaction with more northerly groups was possible. These groups used the transitional zone for spring fishing and foraging for fruits (grape, cranberry, nannyberry), aquatic animals (fish, clams, turtles) and plants (wild rice), and terrestrial game (elk, deer, porcupine). Within the rich resource base of southwestern Michigan, this economic strategy appears to continue well into the early Late Woodland period (ca. AD 500 to 1200).

Two cultural traditions based on the distribution of different ceramic pot styles are found within southwestern Michigan (Brashler, 1978). The Allegan tradition encompasses the St. Joseph and Kalamazoo rivers from Lake Michigan to the center of the state; the Grand River forms the southern edge of the Spring Creek Tradition (Brashler, Garland, Holman, Lovis, and Martin, 2000). Archeological data from the Kalamazoo River indicate movements of people up and down the river valleys with a "river-oriented economy [that] may have included hunting in winter, maple sap collecting in spring, fishing in both spring and summer, collecting riverine resources, and harvesting nuts in the fall" (Holman and Brashler,

1999, p. 215). The shared area included a complex mosaic of deciduous and wet forest habitats that provided hunting opportunities, but was not abundant in other resources. During early Late Woodland times we see small temporary campsites in this shared area and continued interaction with people from Saginaw Bay (Brashler and Mead, 1996).

About AD 1200, the conservative Late Woodland lifeway of southwestern Michigan was affected by interaction with Upper Mississippian groups. In some cases it is clear that these new people moved into the region; in others it seems that the local Late Woodland peoples took on some Upper Mississippian lifeways, but remained essentially Late Woodland. Changes in material culture include the presence of stylistically similar shell-tempered pottery and the widespread incorporation of the growing of corn, beans, and squash (Smith, 1992). Corn became important here, not as a replacement of indigenous horticulture, but as a complementary supplement to a diversified economy. No matter how many large, semi-permanent, relatively densely populated villages were formed, corn remained in this supplementary role.

Recent limited excavations at Moccasin Bluff (O'Gorman, 2004) document the presence of a previously unrecognized sheet midden (accumulation of organic habitation debris on an old surface of the site) and two small pit features dating to ca. AD 1440 to 1500. Recovery of dietary remains yielded no corn, but indigenous starchy and oily seeds (including goosefoot, maygrass, knotweed, amaranth, sumpweed, and sunflower) along with a variety of fruits (pin cherry, pokeberry, grape, and blackberry or raspberry) and a high density of wood charcoal are present (Adkins, 2003). Further work is needed to determine whether the seed plants are domesticated and cultivated or wild. An abundance of deer and fish, and lesser amounts of beaver, porcupine, elk, wild turkey, turtle, and mussel bone and shell remains were also recovered (Martin, 2003).

There is no clear picture of later Late Woodland in the Grand River valley after AD 1200, and Brashler and Mead (1996) suggest that there may be an actual hiatus or reduction in its use perhaps as a hunting ground with a few scattered base camps. The archaeological record is currently ambiguous in many ways about the period just prior to European contact. Critiques of old models and the addition of new data require new approaches for understanding not only late prehistory, but also the beginnings of history in this region. It seems certain that the Potawatomi were living in southwestern Michigan prior to the early 1600s, when they were encountered by the French in Green Bay after taking refuge there from Iroquoian attacks (Clifton, 1998). The historical picture is one of large, permanent villages and winter hunting forays, corn agriculture including other cultigens and wild plant materials, and selective hunting of large mammals and fishing. But the question of how to reconcile the prehistoric record of southwestern Michigan with a vastly different historical picture requires new ways of thinking about the past, as well as directed acquisition of better data (O'Gorman, 2004).

The two primary groups of interest in southwest Michigan are the Potawatomi and the Ottawa. Between 1600 and 1650, the Ottawa were located around Georgian Bay in what is now Ontario, Canada (Cleland, 1992). They were relatively sedentary, relying on whitefish (*Coregonis clupeaformis*) and longstanding, well-developed farms that produced enough yield to be stored for their winter consumption, and to supply French trappers and traders throughout the region (McClurken, 1988). The Ottawa were an integral part of the French fur industry, regularly sending hunting parties into the interior of Michigan, primarily for beaver.

The Potawatomi were part of a larger group of Woodland Indians, or *Anishnabeg*, in lower Michigan. They were "Algonquian-speaking swidden agriculturalists" (Cleland, 1992, p. 87). The Woodland Indians tended to be more mobile than the Ottawa. They lived in villages, "gardening" in the summer, and separating into smaller hunting groups in the winter. Their "hilled" intercropped gardens of corn, pumpkins, squash, and beans were developed in fields that had been cleared via tree girdling and burning (Cleland, 1992). Like the Ottawa, Potawatomi villages were located near or adjacent to waterways, but the Potawatomi fished for lake sturgeon rather than whitefish (Cleland, 1992).

Trapper, Trader, Missionary, and Early Settlement Activities

Four major rivers traverse southwest Michigan flowing from east to west—the St. Joseph, the Grand, the Kalamazoo, and the Muskegon. Two bear the names by which the Native American inhabitants of the region knew them, and two carry the names the European invaders gave them. When the first European missionaries and explorers arrived in the 1600s, they came via Lake Michigan. Because the region was largely covered by dense forest, they used the large rivers to penetrate inland. They established a few small settlements, which provided bases from which traders and trappers could operate. The earliest nonnative populations—French missionaries, fur trappers, and traders—found indigenous fruits, including crabapples, strawberries, and raspberries (Kessler, 1971).

The Potawatomi became part of the fur industry in the late 1600s and were later positioned at important waterways, giving them a geographical advantage in seeking new sources of beaver farther west. Although the French fur industry abruptly ended in 1696, the British continued to trade in fur until the industry crashed in the early 1800s (Cleland, 1992).

With the material aid of the British, the Iroquois displaced all native groups in western Ontario and Michigan's Lower Peninsula. Cleland (1992, p. 92) defines this period as the "Great Diaspora"; it is believed that there were virtually no aboriginal groups in Michigan's lower peninsula in 1670. The Ottawa went to the most northern regions of Lake Michigan and the southern coast of Lake Superior. The Woodland Indians went to the western coast of Lake Michigan, predominantly in the Green Bay region of what is now Wisconsin. After reconstituting with other displaced peoples, the group became known as the Potawatomi (Cleland, 1992).

By the early 1700s, both the Ottawa and the Potawatomi began repopulating Michigan, in part because they both sought regions with biophysical features that permitted access to fish and the ability to raise corn (McClurken, 1988). Within 60 years, both groups became reestablished in the southern Lower Peninsula. The Ottawa settled in the north and the Potawatomi in the south, with the Kalamazoo River valley as the major dividing line between them (Tanner,

1987). However, it was not uncommon for these groups to have villages near each other (Cleland, 1992).

The most prominent Ottawa settlement was a cluster of eight villages along the western portion of the Grand River that was established by 1755 and remained prominent from 1763 to 1812. The Ottawa eventually attained a population of approximately 1,200 people and remained in this region until the end of the removal period in the mid 1800s, but ventured south of the Grand River only for hunting. Four main villages scattered across the southwest Michigan region were the most prominent Potawatomi settlements (Tanner, 1987). They primarily "settled in the prairie openings and bottomlands of the Kalamazoo and St. Joseph rivers" (Cleland, 1992, p. 148).

According to Cleland (1992), a systematic census of Michigan Indians took place in the late 1830s, but it was conservative, purposely excluding some groups. He believes there may have been as many as 30,000 Native Americans across the southern third of the Lower Peninsula that were broken into smaller groups averaging around 150 people each. During this time, women became prominent because they produced marketable goods—wild rice, maple syrup, and corn—for an increasing European American population that numbered approximately 31,640 people (Cleland, 1992), of which 69% were in the southeast (McClurken, 1988). By statehood in 1837, the white population had risen to 174,543 and only 45% were in the southeast. Those migrating into the southwest Michigan region were now interested in the Grand River valley because of its rich agricultural lands. By 1840 the white population "outnumbered the Ottawa by nearly 200 to 1" (McClurken, 1988, p. 77).

As Europeans settled Michigan (as part of the overall push westward), the Great Lakes Indians fared somewhat, if only marginally, better than most Native Americans. The Ottawa were able to switch from provisioning French traders to supplying European American settlers coming from New England and New York. They were a significant part of the early settlement process, supplying pioneers with a wide range of sustenance goods including leather, moccasins, canoes, baskets, fish, deer, pigeon, and turkey. Despite uncertainty about the original population levels, it is believed that the deer and a significant portion of the edible bird species were decimated during this time.

Both for their own subsistence and for trade with the European Americans, the Ottawa gathered honey and cranberries, and cultivated 3,000 apple trees and 2,500 acres of corn, as well as a variety of vegetables (Tanner 1987). Maple syrup was one of their most important commodities. In fact, production of the latter was of such significance that it was shipped to New York, Boston (McClurken, 1988), and England (Cleland, 1992).

During the late 1830s, some Ottawa began to take on "civilized" characteristics (e.g., dress, agricultural practices, language, religion) (Cleland, 1992). Of particular importance was the purchase of lands using their annuities from the 1836 Treaty of Washington (Cleland, 1992). Land ownership was the basis on which some Ottawa were declared citizens of Michigan (not the United States) by the governor. Their farms were interspersed among the white landholdings (Cleland, 1992). Despite their importance to the settlement of the Grand River valley, 900

Ottawa were removed in 1857 and sent north. Many others lost their land as a result of a series of financial maneuvers, some legal and others not. For instance, the agent in the General Land Office swindled the Ottawa out of their legally designated homesteads (i.e., the Indian Homestead Act of 1872) across an entire township in Ingham County (Mason Township) via loan sharking and foreclosing on $20 loans (Cleland, 1992).

In 1830, the Potawatomi population was estimated at 2,500 across nine small villages in present Kalamazoo County; by comparison, Detroit had about 2,200 whites. Most of the Potawatomi were moved out of the Great Lakes region during the removal period, 1834 to 1842, but approximately 300 people who claimed to be Catholic were allowed to stay and were legally recognized as citizens of Michigan (Cleland, 1992). After statehood, many left the area and moved to northwest Michigan. The Potawatomi entered into hundreds of treaties with the U.S. government. The Treaty of Chicago (1833, ratified in 1835) was particularly important in the opening of southwest Michigan to white settlement. It essentially consolidated the remaining Potawatomi in regions away from where the new road connecting Detroit and Chicago was to be built. As these groups were pushed out of the area, pioneers complained because they, like those who moved into the Grand River valley, relied on the native population for provisioning (Cleland, 1992).

European American Settlement and Beyond

Several factors drew the first settlers to southwest Michigan in the decades before 1850. One was grain production. From the colonial period up to the early part of the 19th century, wheat production in the mid-Atlantic region dominated national production (Brigham, 1910). In the first decades of the 1800s, however, eastern market saturation and increasing land prices prompted many settlers to take advantage of low land prices in the newly opened Northwest Territory (Freedman, 1992; Gray, 1996). A second factor was transportation. The completion of the Erie Canal in 1825 facilitated the westward movement of new settlers and supplies, and led to the establishment of a land office in Kalamazoo in 1834. Settlement in the region immediately accelerated (Fig. 5.5). Transportation was also important for the movement of commodities back to eastern markets. The third factor was fruit production. The introduction of an exotic, the peach (*Prunus persica*), early during the 19th century is credited with the start of commercial fruit cultivation in southwest Michigan (Armstrong, 1993). Railroads and land speculators hyped land in southwest Michigan as the next Garden of Eden. The combination of the three factors meant that extensive farming development preceded Michigan's statehood in 1837 (Gray, 1996). On these farms, animal enterprises were important, both for self-provisioning and for petty commodity exchange, as well as for draft power.

During the 1820s, white settlers began moving into the prairies that the Potawatomi farmed (Cleland, 1992). The earliest settlers (i.e., after 1815) were keenly aware that the attractive landscape had been produced by the Native Americans. They saw fields that were either in use or abandoned. The settlers

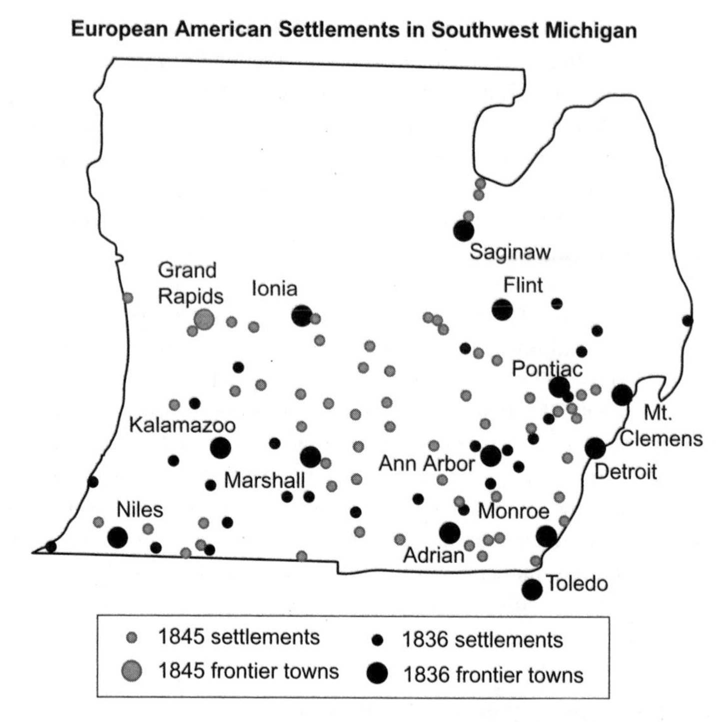

Figure 5.5 The number of European American settlements in southwest Michigan increased substantially between 1836 and 1845.

specifically sought "openings" and "prairies" because they perceived these areas to be both easier to cultivate and protected by the surrounding woods. They thought that these abandoned areas must be fertile and viable using "modern" farming practices.

"Oak openings" were said to have a park-like appearance with trees of uniform size—4–5 ft. high—with an orchard appearance, whereas the prairies were described as vast and "picturesque" (Lewis 2002). Early settlers believed the "wet" prairies were the result of lake evaporation and "dry" prairies the result of periodic ground fires from aboriginal peoples. Lewis (2002, p. 68) states that "burning remained so prevalent in the late 1820s that the territorial government passed legislation to protect settlers' property." The most desirable lands were oak openings, prairies, and timbered lands. Swamplands had a significant influence on settlement patterns in Michigan because they were entirely undesirable. Pineland also was avoided. The lands identified as Class 1 and 2 were the most desirable for mixed farming and, presumably, farmed first, as were areas believed to be free of malaria (Lewis 2002). Later settlers had to drain swamps and clear dense forests. In 1833 it was reported that wheat, Indian corn, oats, barley, buckwheat, potatoes, turnips, peas, apples, pears, plums, cherries, and peaches were easily grown in abundance in the southern Lower Peninsula.

It is important to note that early accounts of the biophysical environment are difficult to follow and use because writers had various agendas and incentives to portray Michigan in a positive light. Furthermore, many of the writers lived through the first half of the 19th century, the tail end of the Little Ice Age (1350–1850), in which there were extreme weather changes, especially with regard to winter temperatures.

Fast Forward

Almost 200 years later, agriculture is still a dominant aspect of southwest Michigan; 55% of the land in the region is in farms. The early emphasis on wheat production has shifted to crops that support industrial agriculture—especially corn and soybeans. Animal enterprises now are concentrated on a smaller percentage of the farms in fewer locales across the region. Vegetables and fruit are very important in parts of the region. Currently, almost half the fruit acreage in the state is located in the southwest Michigan region; within the region, 5 of the 17 counties account for 92% of the fruit land. These specialty crops engage growers in a unique set of agroecological, socioeconomic, and geopolitical relationships that influence spatial–temporal changes across the landscape. Indeed, it is the diversity of agriculture in southwest Michigan that is its outstanding characteristic. After describing how we delineated the region we call southwest Michigan, we turn to an analysis of the social and ecological interrelationships that produced the several transitions of southwest Michigan agriculture from its beginnings in the early 1800s to its situation at the present time.

Regionalization Methods

Since the glaciers receded, the shoreline of Lake Michigan—the western boundary of southwest Michigan—has been fairly constant. With the establishment of the Michigan Territory in 1805, and the move to statehood in 1837, the southern boundary of the region was politically and institutionally established. With the depletion of central lower Michigan's white pine forests—for the construction of Chicago and the fencing of the Great Plains (Cronon, 1991)—the southern edge of the cutover area emerged in the late 1870s as the northern boundary of the region. Lastly, it was not until the completion of post–World War II highways and the development of suburban Detroit that the eastern boundary of this region emerged, separating counties that were part of metropolitan Detroit to the east from the counties of southwest Michigan to the west. Our point is that, although an identifiable southwest Michigan region has existed for at least 150 years, the boundaries of the region have evolved and continue to do so with changing events and circumstances.

We initially constructed the region on three bases. The first was the geographical and political boundaries of the lake, the state, and the counties just noted. The second was historical geographical market orientation; before the most recent period of globalization, agricultural commodities either stayed within the region

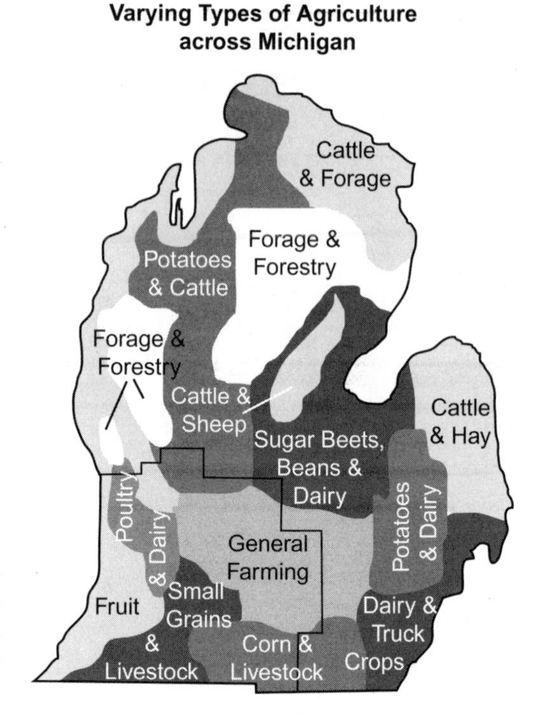

Figure 5.6 Lower Michigan's farming areas in the 1930s. Data based on a map from *Michigan History Magazine* (1938) and the Michigan Historical Center.

or flowed westward toward Chicago. Counties with similar cropping patterns to the east of this region directed their agricultural goods eastward to Detroit or Toledo. The third basis was structured by the dominant crops shared, historically and in the present, across these 17 southwestern counties. A notably different mix of crops predominates to the northeast and the north (Fig. 5.6). In the following narrative we show how the social and ecological characteristics and political economic and environmental relationships most important to southwestern Michigan's agricultural landscape have been transformed repeatedly since statehood in 1837.

Regional Agriculture

Our study of southwest Michigan focuses on the means by which agriculture draws upon, reproduces, and transforms particular ecological and social landscapes that then, in turn, affect subsequent iterations of the agroecological landscape's development. Central to our analysis is the idea that the history of agriculture lies at the heart of southwest Michigan's contemporary ecological and social conditions. In contradistinction to conventional agroecological studies (e.g., Hecht, 1985; Pfeffer, 1983), which generally start with biogeophysical characteristics and explore regional conditions of comparative advantage and agricultural opportunity, we start with agriculture and attend to its modification of, and strategies for dealing with, biogeophysical conditions. Similarly, in contradistinction to conventional sociological studies (e.g., Albrecht and Murdock, 1990; Mann and Dickinson, 1978; Murdock and Albrecht, 1998) that start with social and technical divisions of labor in agriculture and assess the regional ecological consequences and natural obstacles to agricultural development, we follow an agroecological turn (e.g., Altieri, 1995; Allen, 1993) and explore the ways in which the social and the biogeophysical interact to produce particular forms of agriculture.

We view agriculture as an absolutely pivotal arena in which the states of agro-ecological nature and rural society are produced. We argue that agriculture in southwestern Michigan is both the prime historical cause and the most significant consequence of social ecological development. It is the foundation upon which settlement first occurred and the economic base for the regional urban economy prior to the advent of intensive industrial development in the early 1900s in the Grand Rapids area and after World War II in the rest of the region.

Pivotal to our account, however, is that the social and ecological characteristics of the region have combined to produce an impressively diversified agricultural landscape. In contrast to both agroecologists and sociologists who defined and compared regions as homogeneous entities, we seek to understand a continuous diversity. We have an agricultural region, but not one that has specialized in any one commodity to any marked extent despite the region's not particularly diverse ecological and social landscape. Often in the analysis of agroecosystems, climatic, topographic and pedological diversity are understood to determine, or are opposed to, historical, technological, and political economic conditions. In rural sociology or the sociology of agriculture, the opposite is usually the case; social relations trump ecological conditions. Our approach seeks to avoid both extremes. In southwest Michigan, overemphasizing the social and underplaying nature would represent a misunderstanding, on the one hand, of the importance of small differences in microclimatic conditions and, on the other hand, of the accommodations producers and crops have historically been forced to make in and to the landscape.

The traditional North American agricultural narrative (e.g., Pfeffer, 1983; Swanson, 1988) is one that suggests region-specific processes of agricultural specialization and homogenization (e.g., the Cotton Belt, the Corn Belt), offset in some areas by the ecological capacity to foster enterprise diversity. In this traditional narrative, agriculture is generally presented as initially founded by family farms that grow crops—or at least a garden—for their own consumption, produce bulk commodities for the market, and rotate pasturage in support of draft, consumption, and market animals. Over time, each farm generates sufficient surplus that children can be established on new farms of their own in a process of social reproduction defined in Marxist terms as petty or simple commodity production (Chevalier, 1983; Friedmann, 1978) and in Jeffersonian terms as agrarian populism (Sanders, 1999; Williams, 1969). An additional component of this model, the agricultural ladder, suggests that even hired hands can accumulate enough wealth to purchase farmland when conditions are propitious (Heller and Houdek, 1996; Winters, 1978). These petty commodity/populist conditions are then seen in the traditional narrative to be whittled away during the 20th century as the technical demands of mechanical, chemical, and biological intensification, and the market demands of commodity specialization, force fencerow-to-fencerow monocropping. This process at the level of the farm is usually combined with ecological and economic conceptions of regional comparative advantage to generate processes that some describe as regional simplification and others describe as agricultural progress or economic rationalization or ecological modernization. Although we accept the basic elements of the traditional narrative, we suggest that major modifications must be made in the narrative to make it applicable to the southwest Michigan region.

Regional Periodization

We have divided the region's history into six periods. Beyond our earlier discussion of Native American inhabitants and early European American settlers, we provide few data on agricultural practices during the first period, prior to Michigan's statehood in 1837. Nineteenth-century European American surveyors and settlers observed ridged fields or garden beds, ranging in size from about 5 cm to more than 40 cm in height and from about 1 m to 2 m in breadth, that are indicative of more agricultural investment than simple horticultural pursuits and are usually associated with growing corn, beans, and squash (Schoolcraft, 1860; Hinsdale, 1931). Although these structures and the Native American transhumant agriculture of that period are of cultural value and anthropological interest, they are essentially erased by the agriculture implemented by the European Americans in the 19th century.

From 1837 to 1898, two primary processes occurred. On the one hand, the forests were gradually and partially cleared and removed, and the land was drained in the pursuit of extensive self-provisioning and partially market-oriented agriculture. With the advent of more new industrial inputs, farmers would at times extend agricultural production into lands they had previously set aside as woodlots or wasteland. On the other hand, with increasingly efficient transportation infrastructures related to the advent of Chicago as a metropolis and a rail hub with trunk lines to the east (Cronon, 1991), more intensive production developed. Most southwestern Michigan agricultural products not intended for the rural markets and emerging urban centers in the region have historically been oriented westward toward Chicago.

Grains, fruit, and livestock have historically dominated agricultural production, social practices, and emerging technoscientific institutions, although the relations between these commodities—as well as their downstream processing, distribution, and consumption—have shifted in the process of midwestern agroindustrialization (Page and Walker, 1991). Midwest agroindustrialization represents a coevolutionary process by which (1) rural agricultural goods fed urban processing and distribution industries; (2) farm implement industries, initially developed in rural areas and towns, fed both agricultural intensification and urban industrialization; and (3) these two processes were mutually reinforcing. Midwest agroindustrialization includes a necessary and parallel intensification in financial and technoscientific services. Whereas Cronon (1991) sees Chicago as dominating its hinterland, our focus on regional agroindustrialization stresses the way southwest Michigan is composed of its rural areas and its urban centers and their interactions, and deemphasizes the role of Chicago.

By stressing the temporality of regional development, rather than the dominance of one spatial actor, we are able to explore the differing historical dynamics that have contributed to the social and ecological processes and legacies of agriculture in southwest Michigan. Our periods start shortly after statehood and build to the present in five segments. The first encompasses the largely extensive forms of agricultural development during the second half of the 19th century. The second is associated with the technological, productive, and institutional intensification tied to the Golden Age of American Agriculture from the turn of the century to 1919. The third period combines the Agricultural Depression of the 1920s, and its deepening by the financial and industrial depressions of the 1930s.

The fourth period is that of postwar agricultural Fordism—when mechanical and chemical intensification, the Cold War, and the increasing integration of rural social life into urban consumer society radically alter agricultural social and ecological relationships. The last period, which can be broken into two parts, is that of agricultural and rural restructuring, starting in the early 1970s. Initially this process was driven by the new forms of government regulation, social movements, and cultural priorities associated with the environmental movement (and its coincident development alongside increasing production costs related to oil shocks and increasing debt associated with intensification). Agricultural and rural restructuring subsequently accelerated in southwest Michigan after the rise of neoconservative fiscal policies and neoliberal global economics during the early 1980s.

Before we turn to the chapters of our narrative, a couple of caveats are in order. First, agricultural labor is not a major theme in our story; although the supply of agricultural labor and the challenges of the agricultural labor movement were important at various times in other parts of the country, they were not major influences in the southwest Michigan region. Second, race and ethnicity do not appear in our narrative. We noted earlier the obliteration of Native American agriculture; although a couple of areas in the region are settled by former slaves from southern states, they are never very prominent. Certainly one could see in the efforts of the settlers an attempt to impose on the Michigan landscape the agriculture they brought with them from their European origins; but it was never possible to see a distinctively German or Norwegian or Italian farming pattern in the resulting arrangements (Thaden, 1959), perhaps because the ethnic patterns had already been modified by one or two generations of farming in the eastern United States.

Political Economy

Nineteenth-Century Extensive Development, 1837–1898

Settlement, clearing of trees, draining of land, crop experimentation, infrastructural development, the identification and utilization of microclimatic niches, and commodity development all characterize the patterns of extensive development in southwest Michigan during the boom–bust market cycles of the 19th century. Although southwestern Michigan was climatically and pedologically well suited for wheat production, settlers were confronted with a densely forested landscape and many areas of swampland; both of these forms of land cover required complex social institutions to clear and drain before they could be successfully farmed. At the same time, the lakeshore, wetlands, and mixed deciduous forest of the region provided a complex mix of microenvironments and a wide diversity of flora and fauna for hunting, crafts, and harvesting.

Small prairies sprinkled across the landscape provided the initial land for wheat production (Dunbar and May, 1995), and it was in these areas of southwest Michigan that small settlements first developed (Gray, 1996). Although the majority of the logging industry extracted white pine—the state tree—from the area to the north of this region, the hardwood forests and swamps of the southwest were predominantly cleared and drained as part of the process of the establishment of agriculture. After the Chicago fire in 1871, hardwood extraction from southwest Michigan intensified, although this boom had ended by the 1880s and forestry in western Michigan as a whole was all but over by the 1920s (Sparhawk and Brush, 1929).

Prior to the development of railroads during the latter half of the century, lakes, rivers, and canals were the primary means of transporting agricultural crops long distances. In many cases in southwest Michigan, crops were shipped first along rivers to Lake Michigan, where they were taken by steamer to Chicago or, in some cases, to eastern markets (Hartshorne, 1926). The extensive development of agricultural production was further stimulated during the latter half of the century by the development of canals (O'Kelly, 2007) and railroads, which reduced transportation costs to regional markets, although monopoly pricing power caused notable protest (Dunbar, 1969).

Early developments in the mechanization of agriculture (steam tractors, cultivators, reapers, threshers, balers) had several impacts: an increasing dependence on off-farm services, a shift from collective community labor to private labor, and the extension of production into areas of the farm that had not previously been farmed. Also during this period, farming became inextricably joined to a system of social institutions (economic, governmental, and academic), ranging from public institutions like the USDA and land grant universities, to private ones like the Chicago Board of Trade and the Grange (Stoll, 1998).

During this period, the diversified agriculture that would be characteristic of southwest Michigan began to develop. Although wheat predominated in terms of acreage, hay for animals was the second largest acreage, and fruit was not insignificant.

The Golden Age, 1899–1919

With the opening of central American and southeast Asian markets after the U.S. victory over the Spanish in 1899, agricultural transformations in southwest Michigan reflect broadly robust markets. And, with the institutionalization of federal funding for Progressive agricultural science and cooperative extension programs before and after the turn of the century, agricultural transformations in southwest Michigan also reflect the national tendency toward intensification both in mechanization and fertilization, and in plant and animal breeding. On the technical and mechanical side, what was being supplanted and displaced was the need for human labor to accomplish certain tasks (e.g., harvesting); draft animals were still needed to pull the machinery that accomplished those tasks. In particular, the establishment of statewide, county-level Cooperative Extension Service offices as part of the land grant university at Michigan Agricultural College began a century of what was intended to be close, two-way communication among Progressive farmers, extension agents, and university scientists. Production problems, technical needs, and development desires moved from the field to the laboratory, and science in the form of solutions and recommendations moved from the university to the farm.

This is the Golden Age of American Agriculture and it runs up through the end of World War I. Although the diversity of 19th-century agriculture was largely a result of the patterns of extensive development in the region, during the 20th century increasing market orientation, technological sophistication, and institutional complexity gradually reduced the on-farm diversity without homogenizing agriculture across southwest Michigan. Significant volumes of grain produced in Michigan were exported, but the majority of its agricultural commodities were consumed domestically.

The Agricultural and Great Depressions, 1920–1940

As with the rest of the nation, in the 1920s southwestern Michigan entered into an agricultural depression despite the industrial and commercial boom of the Roaring '20s. The return of European agriculture to viability after World War I depressed important markets to which U.S. agriculture had grown accustomed, and caused national overproduction crises in grain and animal sectors. Michigan grain producers were affected directly as the export segment of the total grain market decreased. When they turned to feeding livestock as a way to realize value from their grain crops, they found that livestock markets were also depressed by the loss of European outlets. In contrast, fruit growers were actively planting during this time, which indicates a sense of financial security, because it takes many years for most fruits to reach full production.

Most important, however, and partially as a response to the crisis, during this period "Progressive farmers" moved aggressively into mechanized production (tractors for plowing and cultivation, combines for harvesting, sprayers for pest control). The greater productive capacity of tractors was a boon to those farmers who could afford tractors; less time was required per acre for tillage and planting, cultivation, harvesting, and pest management, which meant that it required fewer persons to provide the labor needed for a given farm. On the opposite side of the coin, however, because intensification meant that the same number of workers could operate more acres and that land no longer had to be used to produce feed for draft animals, tractor-driven increases in productivity across the nation exacerbated overproduction (Berlan, 1991). Intensification also fostered social ecological transformations grounded in agriculture in two ways—as noted, mechanization meant that farmers no longer needed to keep draft animals, and it also meant soil compaction and erosion. Not keeping draft animals meant not needing to plant pasturage, and not receiving the soil nutritional benefits of recycled animal waste as fertilizer. Fields historically rotated with pasturage thus became available for continuous commercial crop production, increasing gross marketable production and net physical productivity of land and labor in the hopes of maintaining (or increasing) net profits, despite exacerbating already saturated market conditions.

The Agricultural Depression from 1919 to 1941 effectively ended—as did the Great Depression—with U.S. entry into World War II. With the national demands for food and fiber for the military, and later for global reconstruction in the context of the Cold War, robust agricultural markets returned. At the same time, institutional, mechanical, and chemical developments associated with Depression-era political restructuring, scientific and technological research within the land grant university complex, and the war effort laid the groundwork for the Fordist revolution in southwestern Michigan's agriculture.

Agricultural Fordism, 1941–1973

Agricultural Fordism represents two basic processes: one in production, the other in consumption. In terms of production, Fordism represents what Goodman, Sorj, and Wilkinson term *appropriationism* (Goodman et al., 1987)—the increased penetration of agriculture by capital goods (e.g., tractors, hybrid seeds, pesticides) that have the effect of increasing the commoditization of agricultural labor processes (i.e., the farm operator increasingly becomes a labor unit in the mass production of food and fiber). Processes that had been endogenous to agriculture (e.g., supplying the energy for draught power with crops raised on the farm, selecting and saving seed to maximize desirable attributes, endemic predators of pests) were moved off the farm and into the realms of industrial, commercial, and financial capital (e.g., machines that relied on petrochemical fuels, commercially bred seeds, synthetic pesticides distributed by national and transnational firms). In terms of consumption, Fordism intensified the capacity, desire, and need of farm families to increase their consumption of market goods—particularly mass-produced consumer goods and "labor-saving devices." Thus, for farm people, Fordist consumption is associated with (1) a deepening orientation to consumerism and away from even partial self-provisioning of household food and fuel, and, as a consequence of this, (2) a deepening of farmer commitments to either monocropping or very simple rotations for the purpose of productivity. The relationship between Fordist consumption and Fordist production is mutually reinforcing such that both encourage the intensification of agriculture including monocropping, very simple rotations, and cultural commitment to little else than increasing productivity. The production part of the appropriation process goes faster sooner; the consumption part of Fordist appropriationism initially proceeds slowly and then goes faster later.

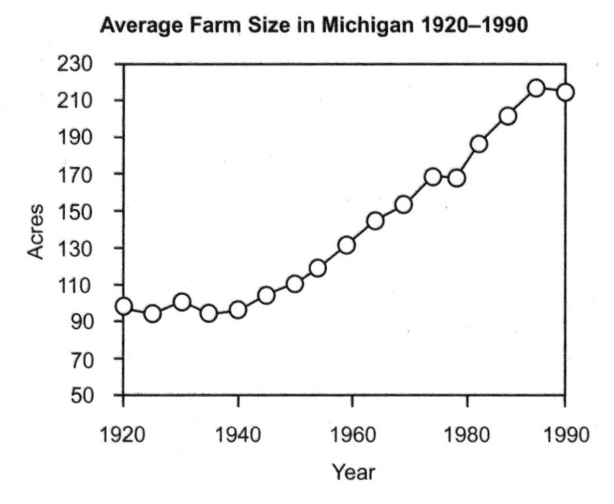

Figure 5.7 Average farm size in Michigan, 1920–1990.

Simultaneously with the promotion of mass-market commodities, however, and in a pattern more like its distant cousin, California, than its near relatives Ohio, Indiana, and Illinois, Michigan agriculture maintained and increased its agricultural diversity after World War II. Although states more central to the Corn Belt broadly reduced their crop diversity to a limited range of grains (corn, wheat, and soy predominant among them), increased the presence of feedlot livestock, and in both respects increased the size of agricultural units, southwestern Michigan's units of production remained fairly small and its agricultural diversity fairly complex, including both crops and animals. Land concentration was limited during this period, although average unit size did grow slightly and middle-range farmers declined over time in the face of larger numbers of large and small farms (Fig. 5.7).

To understand why farmland concentration in southwest Michigan was limited during this period, it helps to view the social process related to agricultural land concentration as consisting generally of two subprocesses. One subprocess makes farmland available; this may happen as current farmers and/or their heirs sell their land, as large operations downsize, or as new land is brought into farming (draining, clearing). This first process is fostered by the aging of the farm population, decreasing commodity prices, increasing tax burdens, and the current and speculative demand for land for exurban residences, as noted earlier. The second subprocess takes this available farmland into farming operations via purchase, rental, or partnership. This second subprocess is fostered by government commodity programs based on acreage, and by the increasing scale of farm machinery. Among the constraints on land concentration in southwest Michigan are (1) the different ethnic/cultural commitments to farming that keep some farmers from exiting farming and keep other farmers from increasing the scale of their operations (Salamon, 1980; Salamon and Davis-Brown, 1986; Salamon and Keim, 1979); (2) opportunities for full-time industrial work off the farm that generate income to sustain the household and to subsidize the farm operations, thus diminishing movement out of agriculture; (3) opportunities for part-time industrial, retail, and tourism work off the farm that siphon off the supply of labor that would be needed for farm expansion (Cantrell and Lively, 2002); and (4) increased productivity and/or profitability associated with new inputs, new cultural practices, new high-value crops, and on-farm value-added crop processing.

How these two subprocesses operate determines the outcome for concentration. To return to the four constraints just mentioned, the increased productivity made possible by new inputs and practices diminishes the need to get larger. Relatively small farms can be highly productive, economies of scale in production may level off at a relatively small size, and opportunities for work off the farm decrease the incentive to get larger. As important, however, is the preexisting agricultural diversity. Southwestern Michigan has rarely experienced saturated markets in grains, livestock, fruits, vegetables, and hay simultaneously, and the region has a fairly long history of moving land in and out of production, and/or from one form of cropping to another—most often, particularly in the western counties, moving between grain and fruit production.

Agroecological and Profitability Crisis, 1974–1989

The combination of the OPEC oil embargo and the associated period of economic stagflation during the 1970s provides the beginning conditions for the end of agricultural Fordism in southwestern Michigan. In addition to increasing costs of direct (e.g., fuel) and indirect (e.g., agrichemicals) petrochemical inputs to agriculture, and increasing costs associated with state and federal environmental regulatory regimes, the 1970s also seriously damaged the Michigan automobile industry and Michigan's industrial sector generally, diminishing opportunities for off-farm employment and supplemental income. This state-level economic downturn was then exacerbated by the decision of USDA officials in the Reagan administration to call in the many agricultural loans extended during the "fencerow-to-fencerow" planting strategies promoted by the Carter administration in the face of the Soviet grain deals.

These changes to the industrial and agricultural foundations of southwestern Michigan were happening at the same time that new, more ecologically oriented and input cost-reducing production practices were introduced: integrated pest management (IPM), corn–soy–wheat rotations, minimum tillage, and intensive rotational grazing. Although the costs of some inputs (e.g., lime, nitrogen fertilizer, sulfur pesticide) decreased, at least in real terms, the costs of other input factors (e.g., tractor fuel, propane for grain drying) increased, more than offsetting the potential cost savings. At the same time, real prices for agricultural commodities continued their long-term downward trend, and farm operators were squeezed between the rising costs and the falling prices. This cost–price squeeze was relieved to some extent by federal government payments. By the 1990s, federal payments constituted more than half of net farm income.

At the same time that more and more farm households were developing pluriactive strategies for staying on the land (and choosing production patterns, cropping systems, and niche crop varieties that fit with the demands of these multiple-income strategies), science, technology, and markets generated opportunities for alternative forms of value-added production and/or regional crop diversification. During the 1980s and 1990s, food consumption in the United States became more diverse, and demand grew for a wider variety of fruits, vegetables, and meats at the same time that consumers sought "healthy" options associated with polyunsaturated oils, organic production, and vegetarian diets.

Southwest Michigan's historical agricultural diversity has allowed the industry to foster some of these new consumer demands and respond to some of the market signals generated by these new consumer desires. In particular, alternative marketing systems, from farmers' markets (Bingen, 2005) to Community Supported Agriculture (DeLind, 1999), have emerged within the region's still-dominant production of corn, soy, and wheat. This has been possible because apples, grapes, and blueberries provide both industrial commodities for mass marketing and processing, and specialty crops for niche marketing and value-added processing. The intricate and uneven dynamics of the relationships between these new and changing land management styles, changing consumer dietary demand, changing agricultural land-use patterns, and changing rural development trajectories are intimately bound up with one another and engender complex and uneven consequences for social relations, ecological conditions, and rural conservation practices all around.

Globalization, 1990 to Present

Most archaeologists agree that the agricultural revolution occurred in one or more locations during a period of a few thousand years starting roughly 11,000 years ago. To self-provisioning patterns that relied on gathering, hunting, and fishing, agriculture added the cultivation of crops and the domestication of animals. These agricultural adaptations remained relatively local in scope until technological and social developments supported trading and raiding at an increasing geographical scale during a period about 5,000 years ago (Hall, 2000). These patterns of exchange remained relatively regional in scope until technological and social developments supported a shift to a global scale during a period beginning about 600 years ago (Wallerstein, 1974). What began as a contest between monarchies for global power by the middle of the 1800s had become a system of nation states, and large, private capitalist corporations operating internationally but usually based in dominant nation states (Friedmann and McMichael, 1989). During the next 100-plus years, the organization of power gradually shifted in favor of the transnational corporation as opposed to the nation state, with only rudimentary development of a transnational state that might exert countervailing power against the transnational corporation (Friedland, 2001; Friedland and Boden, 1994). By the last two decades of the 20th century, transnational corporations were able to access and move raw materials, finished goods, and financial and human capital anywhere on the surface of the globe with almost no regard for the well-being of the communities or nation states involved (Gereffi and Korzeniewycz, 1994).

Thus, by the late 1900s, agrifood systems were increasingly global in scope and local in consequences (Hecht, 1985)—hence Swyngedouw's (1997) felicitous term *glocalization*. Working at the interface of Smith's arguments about the inherent and simultaneous homogenization and differentiation of social ecological landscapes under capitalist development (Smith, 1984, 1989; Smith and Dennis, 1987), and the geographical literature on the de-/reterritorialization processes of globalization (whether global warming or free trade), Swyngedouw (1997) argues that the scalar dynamics of social and ecological processes are shifting. Glocalization represents the simultaneous homogenization and differentiation of social ecological landscapes, particularly with respect to the transformation of local processes in the context of globalization. In terms of agrifood systems, this means that the relatively stable national dynamics of Fordist agricultural markets, policy, and research have been destabilized—with direct consequences for local and regional agroecological conditions. It also means that relations between agricultural production, distribution, markets, and consumption are being restructured as the relations between these entities and the patterns and governance of rural/urban development, environmental standards, and agrifood regulation are transformed by globalization.

For southwest Michigan, this means that historical rural/agricultural and urban/industrial dynamics are shifting rapidly. This raises questions about the extent to which and the ways in which agroecological land use and conservation measures can adapt at a similar pace. The promotion of and struggles over the social and ecological consequences of seven major developments—(1) alternative agriculture; (2) genetically modified crops; (3) niche or specialty crop production; (4) value-added agritainment, craft production, and rural vacationing; (5) exurban sprawl; (6) the environmental and human health implications of agrichemical use generally and pest management particularly; and (7) industrial animal farming—are all developing along international and regional lines, with widely diverse consequences for different localities in the glocalization of southwest Michigan.

We noted at the outset of this chapter that the distinguishing characteristic of the agriculture in the southwest Michigan region is its diversity. Our job in the wider study for which this chapter is a start is to explore the historical (amount, composition) and spatial (location, patchiness) patterns of this agricultural diversity as well as its sociocultural and ecological influences and consequences. These days in southwest Michigan, farms are increasingly growing houses where they used to grow grains, fruits, and vegetables. Increasingly, the use of the land is changing from farming to low-density residences, strip malls, light industry, government facilities, and highways. Whereas during the Fordist epoch suburbanization was broadly associated with expansion at the boundary between the urban core and the suburban ring, these days urban expansion is associated with spotty and noncontiguous exurban development, whether as medium-density prefabricated housing developments or low-density megahome construction. How this set of social transitions of the agricultural landscape is playing itself out in relation to new cropping systems and forms of animal agriculture—and how each is affecting the environment—will be central to our developing understanding of southwest Michigan's contemporary agricultural transformations.

Grain Commodities

Nineteenth-Century Extensive Development, 1837–1898

Grain production in the region expanded rapidly as more and more land was cleared and drained. Between 1854 and 1904, the acreage planted to wheat and corn increased more than 150% from 452,595 to 1,153,512 acres (U.S. Department of Commerce, *Census of agriculture*). Wheat, planted on 22% of improved acreage in 1884, and corn, planted on 12% of improved acreage, were major parts of the agroecosystem during this period. Minor grains such as oats, barley, buckwheat, and rye were planted on approximately 7% of improved acreage in 1884, meaning that these six grains covered 40% of the improved landscape. Forty percent is more significant than it might seem, because all farms had to maintain extensive acreages rotating as pasturage.

Despite the development of rail transport and the concentration of midwestern grain marketing and processing in Chicago, practically all the wheat that was produced in Michigan during the 19th century was milled within the state. Milling centers of southwest Michigan were Grand Rapids, Alma, Coldwater, Lansing, and Jackson (Wood, 1914). Presumably this flour was consumed by growing urban populations in Michigan. The Civil War also had a significant impact on grain production as serendipitously timed bumper crops were sent to feed Union armies (Dunbar and May, 1995).

The Golden Age, 1899–1919

By the end of the 19th century, wheat production had already spread into the Great Plains (Brigham, 1910), and Michigan farmers now had to compete with other regions in a context of declining prices. For this reason, the expansion of grain production in the region slowed decidedly compared with the previous 50 years. In fact, although acreage in grain grew slightly from 1,702,481 to 1,758,372 acres between 1884 and 1904, acreage in grain as a percentage of improved land actually declined slightly from 41% to 39%.

From 1898 to 1919, agricultural expansion was starting to level off. Although the change in grain acreage was relatively small, the composition of the grain that was being produced continued to change. Most notably, there was a movement away from wheat production, and toward corn and other grains. Corn was often preferred by farmers because they could sell it when corn prices were high or use it as feed for hogs or cattle grown on the farm (Hart, 1986).

Many farmers, in fact, preferred to feed their corn to livestock because livestock production allowed for productive use of labor during winter months (Hart, 1986). Approximately two thirds of corn was used as livestock feed (Dunbar and

May, 1995). Equally important, alcohol consumption, and corn whiskey in particular, was also starting to increase during this period, providing new outlets for excess production of corn (Pollan, 2003).

The Agricultural Depression and the Great Depression, 1920–1940

The decrease in grain prices made cash grain farming extremely unprofitable, and many areas of grain farming were simply taken out of production as farmers across southwest Michigan were unable to pay taxes (Barnes, 1929). Farmers were also faced with declining relative income compared with manufacturing jobs (Alstou and Hatton, 1991). These conditions were exacerbated by the arrival of the industrial and financial depression in 1929.

The mechanical intensification of agriculture during the Depression era affected the environment by increasing soil erosion in many areas. Furthermore, the intensification of modern, monocultured grain crops—corn in particular—created new opportunities for agricultural insect pests. The European corn borer, introduced into North America around 1910, spread quickly across Canada and entered Michigan in the early 1920s (Larrimer, 1928). In response to the threat of the corn borer, in the spring of 1927, Congress passed a $10,000,000 corn borer cleanup campaign. Within Michigan, this campaign was largely an educational movement stemming from the agricultural section of the Cooperative Extension Service (Dibble, 1936; Musselman, 1928) that sought to reduce the amount of crop residue left on farm fields. The campaign largely failed to control the corn borer. However, the corn borer did represent, at least to some entomologists at the time, a larger structural problem. Dr. Charles Brues, a Harvard professor, suggested that increasing farm size and concentration of farmland was the source of increasing insect pest problems because large, reliable sources of food were allowing pest populations to grow (Anonymous, 1929). He also argued that a reduction in farm size would be the only way to control pest populations.

The combination of local environmental degradation, increased pest infestation, and national and global depression had a significant impact on both the amount and composition of grain production. As noted earlier, declining prices and decreased grain productivity drove many farmers out of business as their main cash crop became increasingly unprofitable. To maintain their farms, many farmers shifted out of grain production, and, from 1904 to 1940, the amount of farmland used for grain production declined from 1,314,259 acres to 948,614 acres, and grain's share of cropland declined from around 40% to less than 25%. This drastic decline was the result both of taking land out of production and of diversifying production into crops other than grains.

By the end of the 1930s, new opportunities for grain production began to emerge with the development of soybeans as a potential crop that could be rotated with corn and wheat as a means to reduce soil exhaustion, control pest infestations, and take advantage of new market demand for soy oil and soy meal. Although even as late as 1940 the fraction of farm acreage used for soybean production in the region was less than 1%, there were indications that soybeans would become critical for southwest Michigan agriculture. Significant changes took place first technologically in 1934, when the hydrogenation of soybean oil became possible on an industrial scale; and then politically in 1935, when margarine manufacturers decided to use only domestically produced oils and fats (Berlan, 1991). These key decisions, buoyed by the discovery that soy meal could be used as a high-protein animal feed, opened the door for the expansion of soybean production in the coming decades.

Agricultural Fordism, 1941–1973

In the 1940s, the face of grain production in southwest Michigan changed drastically. First, grain production, which had declined significantly during the previous era, found new opportunities in markets in post–World War II Europe. Stimulated by new demand both internationally and nationally, grain farmers in southwest Michigan responded by increasing the amount of farmland used for grain production. In this era, however, corn and not wheat was the dominant grain produced. In part because of the Agricultural Depression, farm ownership became markedly more concentrated and farm size increased, whereas the diversity of grains produced declined. In 1974, the amount of soybeans planted was only 5% of total farm acreage; however, it increased rapidly through the next time period. Although there continued to be some farming of field crops other than corn, wheat, and soybeans during the early part of this era, by the 1970s it was clear that other grains were of relatively little importance. By 1974, less than 3% of farmland in the southwest Michigan region that was planted in grain or legume crops, was planted in crops other than corn, soy, or wheat.

Equally important, grain production in this period was defined by new mechanical and chemical developments that increased productivity and provided new positive economies of scale. Although tractors were available prior to World War II, they were not produced in sufficient quantities to meet demand, and many farmers in the region continued to rely on animal traction throughout the war years (see Animal Commodities, below). After World War II, as industrial production reoriented to include nonmilitary as well as military uses, fossil-fuel-driven farm machinery quickly became the norm for grain production. Within the entire state, the number of corn pickers alone more than doubled—from 10,681 to 23,514 from 1950 to 1954. By 1969, there were 13,053 self-propelled grain and bean combines and more than 70,000 tractors. The replacement of horses and the few other draft animals by self-propelled farm machines did not represent simply a labor-saving development. It altered the landscape by facilitating the further specialization of farms and by increasing the separation of animal systems from grain production. Farmers no longer had to feed and house animals, but at the same time the agricultural fields of farmers who had no other animal enterprises no longer benefited from the animal manure.

When livestock was removed from the grain production system, it was essential that new sources of fertilizer be found. Around World War I, two German scientists developed a way of synthesizing ammonia through the Haber-Bosch process. This meant that nitrogen fertilizer could be produced industrially, rather then relying on Chilean guano sources. The industrial production of ammonia was further expanded within the United States during World War II because it was an essential part of munitions production. In 1930, 140,082 tons of commercial fertilizer was applied in Michigan. By 1964, this quantity had reached 679,519 tons.

The use of industrially produced agricultural chemicals was not without a cost, but neither did it lead to a wholesale domination of the landscape. This was particularly evident with regard to pesticides. In the case of the corn borer, the use of pesticide as a management technique proved to be extremely limited, because pesticide had to be applied when corn was nearly full grown, which made scouting and spraying for pests time-consuming and physically inconvenient. On the other hand, management of pests such as the western corn rootworm, which entered Michigan during this period already resistant to organochlorine pesticides, was initially successful through second-generation, organophosphate-based soil insecticides. Inevitably, however, the rampant use of pesticides developed during this era began to affect nontarget species and degrade agricultural and nonagricultural environments alike. Insecticides such as dichlorodiphenyltrichloroethane (DDT) had serious and lasting impacts on avian, aquatic, and terrestrial wildlife populations (Frank at al., 1981; Heinz et al., 1985, 1994).

The new tractor-driven, monocropped, chemically intensive, densely planted system could not have succeeded without structurally homogenous crops that were able to utilize high nutrient levels. Hybrid corn fit the bill. Although research into hybrid corn had begun in the 1930s, it was not until this period that its commercial use expanded. By 1940, more than 10% of corn acreage in southwest Michigan had been planted in hybrid corn (Griliches, 1960). By 1959, 90% to 100% of all corn varieties planted were hybrids. However, the spread of hybrid corn did not have an equitable effect on all agricultural operations. Rather, the yield improvements disproportionately benefited those who had large, relatively successful farms (Griliches, 1960). Larger farms not only had greater capital to buy the new hybrid seeds, but they could also purchase the host of chemicals and machines that worked in tandem with the hybrid crops to make production so high.

Agroecological and Profitability Crisis, 1974–1989

Through the post-Fordist era, from 1974 to 1989, the composition of grain production continued along similar trajectories as the previous era. Grain as a percentage of farmland continued to increase. Wheat acreage declined, whereas land planted to corn and soybeans increased. The factors driving these trends, however, were beginning to change. As a result of the homogenization of grain landscapes in previous eras, new environmental problems were being encountered. The homogenization of landscapes and the development of pesticide resistance led to increased pest infestations. Application of highly concentrated synthetic fertilizer began to pollute water sources; mechanization made possible increased tillage, which fostered soil erosion. Confronted with heightened public awareness of environmental issues and new federal regulations, farmers had to seek new ways to manage their fields.

Although serious grain pests, such as the western corn rootworm, were initially managed with soil insecticides, the rapid spread of resistance was making chemical management increasingly ineffective. At the same time, the homogenized landscape was providing pest populations with a large area of highly concentrated, high-quality food to support their populations. The solution to managing the corn rootworm in southwest Michigan was to rotate field crops to disrupt the lifecycle of the pest. This strategy not only was an effective means of pest management, but also allowed farmers to take advantage of the growing market for soybeans, and also had a further benefit in that it fixed nitrogen in the soil. The continued expansion of soybeans throughout this era is a testament to their use by farmers as a strategy for pest management, economic diversification, and nutrient management.

The problem of soil erosion was addressed through the expansion of no- and low-till farming. As with the expansion of soybeans, the benefits to grain farmers were multifaceted. Not only did the reduced disruption of the soil reduce the amount of wind and water erosion, but it also reduced the amount of labor time and fuel that were necessary to work the soil. Unfortunately, the concomitant effect of no-till farming was that it required more herbicide use to control weeds and maintain the clean, untainted fields that had become the cultural norm for southwest Michigan. Especially on the sandier soils and during the spring planting season, the herbicides leached into the groundwater. Ironically, no-till farming necessitated that farmers unlearn the lesson of eliminating crop residue that had been taught during the corn borer cleanup campaign in the 1920s (above).

Farmland concentration accelerated throughout this era, so fewer farmers were working larger tracts of land. Farm machinery itself became larger so that corn pickers that once picked a single row could now pick five or more rows simultaneously. Economic pressure resulting from low grain prices forced many farmers to "get big or get out." Farmland that had been in families for generations became available as a result of death or retirement, and was sold to settle the estate or rented out to provide a pension for the surviving spouse. In most cases, land that was rented was used for grain crops because of the quick return and low labor requirement compared with fruit or vegetable production.

Globalization, 1990 to Present

If grain production from 1940 to 1973 was defined by technology, and grain production from 1974 to 1989 was defined by environmental degradation, the final era of grain production, from 1990 to the present, is defined by globalization and increasing conflicts with nonagricultural populations. During this era, the forces driving grain production have shifted away from local or regional conditions and toward global-level influences. Certainly, extraregional factors have always had an impact on southwest Michigan grain production, from competition with wheat-producing states in the Great Plains and the West to the decline in prices of the Depression to new markets opened by World War II. However, in many ways, grain and legume production in this era is driven by the global supply of (e.g., from Brazil), and global demand for (e.g., by Mexico), cash grains and legumes in ways that had not previously been the case. The market is not only affected directly by global supply and demand; in addition, grain and legume production is indirectly affected by markets for nontraditional uses (e.g., sweeteners, fuel alcohols, starch utensils, biodiesel fuels), and thus by the global supply of and demand for crops and other commodities that historically had not been seen as fungible and thus competitive with grains and legumes (e.g., cane and beet sugar, fossil fuels, wood pulp).

By 1997, the grain–legume landscape in southwest Michigan was dominated by corn and soybeans. Soybeans represented almost 20% of all farm acres planted, and corn was planted on 28% of all farmland. Wheat had decreased to only 5%. Despite the continued homogenization in the types of grains produced, there has been, at least to a limited extent, a diversification in the varieties of those types of grains that are produced (e.g., high-lysine corn, high-oil soybeans), and in the ways in which these grains are produced (e.g., genetically modified crops, organic crops).

Biotechnology has had a significant impact on grain production during this period. By 2003, an estimated 73% of soybean acreage and 35% of corn acreage was planted in genetically modified crops. For corn, the dominant genetically modified variety incorporated a plant-expressed insecticide (the *Bt* toxin) that controlled the European corn borer. Since the entry of the corn borer to the region nearly 100 years earlier, *Bt* corn represented the first highly effective control. Genetically modified soybeans consisted of a variety resistant to the Roundup herbicide. This allowed for easy weed control of soybean fields. Interestingly, however, despite the claims of the biotechnology industry, neither of these varieties contributed to a reduction in chemical application in the region.

Although transgenically modified crops (GMOs) were beneficial to farmers in that they reduced crop loss and made weed management easier, they were not without problems. With the expansion of international trade, international markets have become increasingly important to grain production in the region. Resistance to transgenically modified crops in countries in Europe and elsewhere has barred some southwest Michigan grain from some markets. In addition, some authors (e.g., Rissler and Mellon, 1996) have expressed concern that the constant and universal expression of the *Bt* toxin in corn will rapidly lead to resistance (loss of susceptibility) in the pest species. Other authors have suggested that the increased use of glyphosate (Roundup) will lead to resistance in the weed species, and/or that the herbicide tolerance genetic construct may be transmitted to weed species and may become incorporated in one or more weed genomes, creating so-called superweeds—weeds that are resistant to one of the major herbicides. Although there is no indication that any of these outcomes has occurred yet in southwest Michigan, the potential exists for these major environmental impacts.

In contrast to transgenically modified crops, organic production represents a relatively small part of southwest Michigan agriculture. Statewide slightly more than 40,000 acres, less than 1% of the total farmland in the state, were in organic production in 2001. However, organic production has grown rapidly in Michigan during the past 10 years, increasing more than 280% between 1997 and 2001. Furthermore, the 2002 adoption of national organic standards and the growth of organic production nationally speak to its future potential. The exact amount of organic production in southwest Michigan is unknown; however, one extension agent stated that the majority of organic production in the region is in soybeans. The production of organic soybeans has been stimulated in part by the demand for edible organic soybeans in international markets. Organic production techniques emphasize the use of naturally occurring fertilizers that release nutrients more slowly than synthetic fertilizers; thus, less nutrient is available for leaching and erosion. Organic farming emphasizes the use of "naturally occurring" substances for pest management, such as sulfur and toxins produced by the *Bt* bacterium; thus, there is expected to be less impact on nontarget species. It seems likely that as organic production continues to expand, organic production of soybeans, as well as other crops, will likely increase in southwest Michigan in the future.

In addition to changes in technology and international trade, the expansion of residential developments into rural areas in the region is having an impact on how grain is produced. During the past two decades, movement into rural southwest Michigan has increased, as urbanites from as far away as Chicago have taken advantage of low land prices to build vacation homes or year-round houses in the country. Small "hobby farms," bed and breakfasts, and other tourist ventures now dot the landscape, seeking to take advantage of the pastoral ideal of country living. The implications of these residential developments for grain farming have been significant. First, to reduce complaints, grain farmers must now take into account adjacent nonfarm residences when timing the application of manure or pesticides. Furthermore, surveillance of the environmental impacts of agricultural practices has increased as the new exurban residents bring with them nonfarm aesthetics, environmentalist perspectives, and affiliation with environmental organizations. One of the reasons the production of wheat has decreased in the southwest region is the increased difficulty of the aerial application of pesticides and fertilizers that had been common.

Despite the increased social contention surrounding grain production in southwest Michigan, in some areas grain production has increased to provide a sink for manure from ever-larger and more concentrated animal operations. In general, grain land remains an important sink for the manure that livestock operations produce, at the same time that manuring remains an important mechanism for recycling many of the nutrients back to the land in a form that releases them relatively slowly. Of course, because of the high water content of manure, transportation costs for distant disposal are very high, and manure must be disposed of locally. As farmland has been taken over by residential areas, as animal production has become more concentrated spatially and organizationally, and as legislation has been put in place regulating manure application, there has developed a shortage of land on which to apply manure. It is this dynamic that has led to the bringing of new land into grain production to serve as a sink for manure. At the same time, grain production in the region continues to be an important source for animal feed in the region. Although livestock and grain production have become more separated and specialized at the level of the farm, they remain closely integrated at the regional level.

Fruit Commodities

Nineteenth-Century Extensive Development, 1837–1898

Eastern North American and European varieties of apple, peach, and pear seedlings were brought to southwest Michigan by early settlers and planted in fencerow corners. Some growers grafted these varieties onto wild plum stocks to speed fruiting. They also gathered indigenous small fruits from wild species such as the pawpaw (Armstrong, 1993). Blueberries were one of the few indigenous fruit species that became a commercial fruit crop (Kessler, 1971). Early economic successes encouraged growers widely to begin construction of formal peach orchards typically intercropped with staple crops like potatoes. A series of severe frosts killed large numbers of young trees in the 1840s and set into motion both on-farm research and demands by local horticultural societies that the state help to determine how best to cultivate fruit in the region. In this context, fruit growers were involved in the initial negotiations with the state legislature to open an agricultural college, culminating in the establishment of the Michigan Agricultural College in 1855, and in the negotiations to establish a department of horticulture at the college, finally realized in 1883.

By the mid 1800s, several of the biogeophysical conditions (e.g., sandy loam soils, seasonal variation in temperatures) that make the region suitable for fruit were formally identified and widely promoted (e.g., Winchell, 1865). One key feature is Lake Michigan. It was found to extend the growing season along the coast by cooling the air in the spring and summer, which delays budding and moderates the hottest days, and by warming the air in the fall and winter, which delays autumn frosts and creates heavy, insulating snows (Hill, 1939). Those counties that benefit most from the lake's moderating effects—Berrien, Van Buren, Allegan, Ottawa, and Kent—were identified as composing the Fruit Belt (Winchell, 1866). In addition, Lake Michigan provided the moisture necessary for production. Unlike some of the major fruit-growing regions of California (Stoll, 1998), Michigan growers did not require the development of extensive irrigation systems. In fact, research on fruit irrigation did not begin for another 100 years.

Fruit was initially grown for friends and family. Early commercial growers typically had small orchards on mixed farms (Kessler, 1971). In 1853, the Detroit and Milwaukee Railroad linked Grand Rapids and Detroit, which was significant for the development of the northern portion of the southwest Michigan fruit-growing region. Equally important events for the fruit industry were the opening of the Benton Harbor Fruit Market in 1860 and the subsequent completion of the Chicago–Lake Michigan Shore rail line in 1871 (Kessler, 1971). By the end of the 19th century, southwest Michigan was an established fruit-growing region, serving several significant markets (primarily Chicago, and secondarily Detroit, Milwaukee, and Grand Rapids), and famous for its peaches but anchored by its other crops, especially apples. In 1884, there were 173,251 acres of apples, peaches, and grapes being grown, which represents 4.12% of the improved farmland in southwest Michigan.

As increasing numbers of formal peach, apple, and pear orchards came into production, and the production of fruit expanded, it caused significant changes in the biophysical environment of the region. The varieties of apples and peaches and grapes that were planted were European and eastern North American varieties, exotic to the region. Although none escaped from cultivation to become an invasive nuisance, they did replace a significant percentage of the native vegetation. The sizeable and concentrated planting of these varieties made it possible for native and introduced insect and microbial pest species to increase to the point of being significant problems.

As the fruit landscape became more homogeneous, widespread damage became apparent as the transmission of diseases and the rapid spread of damaging insect pests increased. For instance, the Peach Yellows virus, first found in Michigan in 1866 in Berrien County, in combination with the severe winters of 1873 and 1879 severely damaged the peach industry. Yellows causes premature ripening and red spots from the skin through the flesh. Although the cause of Peach Yellows was not understood during this period, it was believed that it was contagious and that the only preventative measure was to remove and burn affected trees immediately (Wilcox and Smith, 1911). Control of Yellows was so critical that it prompted the passing of the Insect and Plant Disease Act of 1875, the first state legislation in the nation with the objective of controlling a plant disease (Kessler, 1971).

A central consequence of the problems fruit growers had with frosts and freezes, pests and diseases was a concentration of fruit acreage in the southwest Fruit Belt counties along Lake Michigan. Seventy-five percent of the 94,000-acre increase in fruit acreage between 1874 and 1904 was in these five counties. The other 13 counties in the region together lost more than 7% of their fruit acreage.

The Golden Age, 1899–1919

Despite the fact that the total acreage of fruit crops in southwest Michigan reached its all-time peak, 213,993 acres, in 1904, the Golden Age of Agriculture was not the golden age of fruit in southwest Michigan. Nevertheless, fruit plantings, especially along Lake Michigan, were used as an economic development tool to promote regional tourism during the early years of the 20th century (State of Michigan, Public Domain Commission and Immigration Commission, 1914).

Enormous economic successes in commercial production had encouraged a "peach-planting frenzy between 1884 and 1906" (Armstrong, 1993, p. 16). Yet, a particularly devastating storm on October 10, 1906, killed 73% of the peach trees across the region. This freeze devastated the industry, which never fully recovered. Although peaches were still planted across southwest Michigan, they became spatially concentrated in the southernmost county (Berrien) where the lake effect is strongest and the growing season is 20 days longer than in the rest of the region (Schaetzl, n.d.). Apples, a comparatively hardy crop, continue to be planted across the region to this day, whereas grapes and cherries are far more concentrated because of their greater need for specific ecological conditions. By 1904, more than 9% of the improved land in the southwest Fruit Belt was planted in apples, peaches, and grapes; and secondarily, strawberries, pears, plums, and cherries.

Additionally, during this period Michigan fruit producers had to respond to increasingly tough competition from producers in the Pacific Coast states. Extending the Progressive concern with productive efficiency, the development of the Farm Bureau and Cooperative Extension offices intensified the already tight relationship between fruit producers and land grant university scientists at Michigan State College, as it was then called. Thus, attention was turned toward alternative strategies, including new markets such as processed foods and beverages (Kessler, 1971), increasing yield and efficiency, decreasing production costs, and improving cosmetic appearance.

In response to quality concerns, Michigan State College focused its efforts on developing scientific, rational, and profitable agricultural techniques, using the newly established Cooperative Extension Service to reach the farmers whom the university agents believed most likely to adopt new technologies and techniques (Rosenberg, 1997). By this time, entomologists had already introduced chemistry to the orchard (Houck, 1954). Agrichemicals (e.g., lead, arsenic, sulfur) were in full use by 1900. When, early during the 20th century, the plum industry nearly succumbed to the plum curculio, one of the major insect pests of tree fruits, the industry was rescued by the introduction of lead arsenate paste (Kessler, 1971). In addition, variety testing and long-range breeding programs began to be developed and supported at several new regional experiment stations (Kessler, 1971).

During this period, the fruit industry began to learn how to control and adapt to nature—the nature of the biophysical environment as well as the nature of sociopolitical and economic conditions—each of which influenced the distribution of fruits across the landscape.

The Agricultural and Great Depressions, 1920–1940

Between the height of the Golden Age (1904) and the end of the Agricultural Depression era (1940), southwest Michigan lost nearly 63,000 acres of fruit, approximately 29.2% of its acreage. This overall decline was largely the result of two countervailing trends. First, fruit production shifted out of the areas that were not prime fruit land. Although fruit acreage in the Fruit Belt declined only 6.2%, fruit acreage inland declined by 67.1%. Second, there was an increase in berry acreage between 1929 and 1940 across southwest Michigan of almost 5,000 acres; 72% of this change was in the southwest Fruit Belt. During the Depression era, overproduction, increased competition, and decreased market prices challenged the southwest Michigan fruit industry. Members of the industry—growers, government agencies, land grant university, processors—collaborated on several on-farm strategies to mitigate these challenges, and to keep both the growers farming and the land in production.

1. *Increase quality.* Continuing from the previous era, the fruit industry concentrated on increasing quality as a way to preserve market share, which often meant increasing applications of heavy metal pesticides. State and federal grades and standards were enacted to force growers to "protect" crops in particular ways. Cherry growers, for instance, were required to

spray their orchards, on particular dates, with lead arsenate to control cherry fruit fly. In 1921, efforts were also put into enforcing the Insect and Plant Disease Act of 1875 by expanding the role of the Plant Industry Division in the Michigan Department of Agriculture, which was charged with inspecting nurseries, and later with removal of "nuisance" fruit plants (Kessler, 1971). Nevertheless, some crops such as pears rapidly declined during this era because of the difficulty controlling fire blight (a bacterial disease) and psylla (an insect pest) (Kessler, 1971).

2. *Grow for the market.* As a result of high market prices, grape planting between 1918 and 1920 markedly increased. Kessler (1971, p. 145) states that "the sale of grapes for home wine making, because of Prohibition, caused additional planting." During the same time, a major juice processor moved into the area. At the market level, at least two approaches were used to increase quality. First, there were organized efforts, beginning in 1923, to keep immature fruit out of the market (Kessler, 1971). Second, a shift in varieties grown was encouraged. For instance, the primary plum grown in the 1920s (the Damson) was so tart that it was suitable only for processing into jams and jellies. In 1926, the Stanley prune plum was introduced and soon became the dominant variety. At the same time, consumer interest in maraschino cherries during the 1930s made it profitable to try growing sweet cherries, even though sweets are susceptible to frost damage and to cracking from excessive moisture.
3. *Increase use of technoscience.* Although some commodities (e.g., grains) became highly mechanized during earlier eras, many fruits were found to be too fragile; hand harvesting was required to maintain quality. Furthermore, growers had a readily available supply of cheap labor. Thus, technological developments consisted primarily of tractors, sprayers, and new pesticides. In addition, existing breeding programs were expanded and many new ones developed. The Bluehaven blueberry, for instance, was a mid-size bush developed specifically for southwest Michigan that was easier to pick. The first commercial planting of blueberries was established in 1928, "demonstrating that blueberries are well adapted to thousands of acres of sandy, acid soil, which had until then been considered wasteland" (Kessler, 1971, p. 146). Today, competition from residential uses increasingly compromises access to this previously undesirable land.
4. *Elevate efficiency in the orchard.* At the beginning of the 20th century it was common practice for orchards to have multiple fruit varieties, but it was argued that efficient orchards require monoculture plantings (e.g., so they ripen at the same time). By the early 1930s, the Montmorency variety of tart cherries (i.e., one variety of one cultivar) was grown almost exclusively. Historically, peaches were typically interplanted with apples, but during this period, as the peach trees aged and became unproductive, they were pushed out and replaced by apples. In addition, the number of apple varieties grown was reduced from nearly two dozen to only a few. Thus, harvest became easier to time and to manage.

In addition to the production strategies just described, industry efforts were focused on marketing moreso than in any previous period. Marketing had several effects on the landscape, including the planting of new fruit cultivars, changing cultivar varieties, and the homogenization and standardization of fruit land. At the same time, the pressures of competition from other regions pushed fruit out of less productive areas and into the Fruit Belt counties. In addition, pressures of insect and disease pests forced greater concentration on those fruits for which the region had a comparative advantage—apples, blueberries, grapes, and cherries.

Agricultural Fordism, 1941–1973

Up to the 1940s, increasing fruit production was primarily based on extensive cultivation. After World War II, the application of capital and technology allowed the industry to use the natural environment maximally (Stoll, 1998), with little concern about the impacts of agroindustrialization on the rural countryside. These new practices had several effects on the land. On the one hand, they created a visually stunning, uniform landscape, especially around bloom and at harvest. On the other hand, industrialization of the landscape also meant the loss of biodiversity, increased pest problems, and a continuous need for new pesticides and pest control technologies. These conditions developed as a result of design strategies, pest management practices, and new forms of mechanization.

1. *Design.* Growers changed the landscape with changes in cultivars and cultivation techniques. For instance, commercial blueberry cultivation became important during this period. They were initially gathered from the wild; now, most Michigan blueberry acres are cultivated in the southwest Fruit Belt counties. During the Fordist period, orchard plantings also came to be "designed," their space calculated, and trees planted for efficiency, technological compatibility, and maximum yield. Research on size controlling rootstocks began in 1937; by the mid to late 1960s they were used on approximately 60% of new trees (Kessler, 1971). By 1974, dwarfing and semidwarfing apple trees allowed growers to increase density from 33 trees/acre during the Golden Age to 109 trees/acre during the Fordist era—an increase of 232%.
2. *Pest management.* After World War II, DDT, as well as other broad-spectrum pesticides (e.g., organophosphates), became widely available to fruit growers. They were immediately accepted and their value and effectiveness were not questioned (Russell, 1996). With growers following the spray calendar (e.g., Mitchell et al., 1953) and directions from other "experts" (e.g., chemical company representatives), pest control became a routinized process; pesticides were applied on a regular schedule, regardless of whether pests were present (Perkins, 1982). At the same time, both state and federal agencies codified grades and standards for quality. For example, formal rules about tart cherries established what were the minimum grades of fruit that processors (i.e., pitters) could receive, purchase, sell (Michigan Department of Agriculture, 1953), can, and/or preserve

(Michigan Department of Agriculture, 1962), as well as the specifications for the products that have been pitted (U.S. Department of Agriculture, 1946 [1941]), canned (U.S. Department of Agriculture, 1949a), and/or frozen (U.S. Department of Agriculture, 1949b). Over time, wholesalers and retailers reinforced these standards through their preferences for cosmetic appearance and by promoting the idea that consumers will tolerate nothing less than perfection (Pimentel et al., 1993). The point is that these grades and standards left growers with little choice but to attempt to grow perfect fruit with the only means available—agrichemicals (Worosz, 2006).

3. *Mechanization.* For several fruit crops, one of the most significant changes was the development and introduction of harvesting equipment. The mechanical cherry harvester (shaker), for instance, was promoted as the "solution" to the labor "problem." However, its use required extensive orchard modification—wider tree rows, special pruning, and leveling of the ground (Childers, 1975). It also became necessary to alter the character of the fruit with growth regulators so that they would be ready for simultaneous harvesting. Additional agrichemical applications were necessary to care for the cherries that were inevitably left on the trees, as well as to care for the tree itself.

Whereas trends in fruit land use during previous time periods focused on industry development (1852–1898), extensive cultivation (1899–1919), and marketing concerns (1920–1940), what stands out in the Fordist period (1941–1973) is the increasing intensity with which the land was used. Industrial, commercial fruit production is synonymous with increasing calculation and control. In fact, contemporary southwest Michigan Fruit Belt growers recount a time in which pests were controlled to the extent that nothing was alive in their orchards—no bugs, no birds, and no grass—just rows and rows of trees on barren soil. By the 1950s, the wide-scale use of synthetic, broad-spectrum agrichemicals began to raise concerns on numerous bases, including pest resistance and secondary outbreaks (Pickett, 1949).

Agroecological and Profitability Crisis, 1974–1989

Although the fruit growers in the southwest region had their crisis of profitability at the end of the Fordist era, from the 1970s to the present the cost/price squeeze and increasing land values throughout the Fruit Belt counties in southwest Michigan have continued to contribute to the overall loss of fruit farm acreage. High capital investment, and delayed return on investment, meant that orchard lands could not be rapidly shifted either between different enterprises or in and out of production. Fruit growers who continued to farm used several strategies directly related to land use, including enterprise selection, pest management, and risk reduction.

1. *Changing crops and/or varieties.* Some fruit commodities (e.g., pears) that once were strong faded almost entirely from the landscape during this period. In the case of grapes, however, some growers chose to develop

new vineyards and/or to change grape varieties. Although they were well adapted to the biophysical environment, native grapes (e.g., concords) are valued "about one-fifth that of Vinifera and one-third of French-American varieties" (Baxevanis, 1992, p. 203). During the 1980s and early 1990s, Vinifera acreage increased more than 26% statewide (Michigan Department of Agriculture, 1995).

2. *Reducing pest management costs and/or impacts.* During the early 1970s, many southwest Michigan apple growers were introduced to IPM by the State Cooperative Extension Service and by private crop consultants. The goal was to promote biologically based timing of agrichemical applications that would minimize unnecessary use and maximize effectiveness, thus minimizing environmental burden and maximizing efficiency. Other fruit crops were perceived to be too difficult to manage with alternative techniques; for example, federal regulations imposed zero tolerance for fruit fly damage in cherries (see earlier mention). Thus, most growers continued to rely on a "conventional," routinized program of pest control. Although the official IPM program succumbed to a lack of institutional support in the mid 1980s, most growers continued to attempt to reduce numbers and rates of pesticide applications, and some growers continued to practice scouting for pests (Harris and Worosz, 1998).

   Pesticides were once applied by hand, a tree/bush/vine at a time. After World War II, sprayers were automated, but their application was less exact, emitting the same quantity regardless of plant size or presence. During the 1970s and 1980s, multiple developments in sprayer technology (e.g., electronic eye) greatly increased precision. This precision meant that less product was lost to the air and/or soil. In addition, broad-spectrum pesticides were increasingly replaced with pest-specific substances. Growers interviewed during the 1990s remarked that there were more living organisms on their farms at that time than they had ever seen in their lifetime.

3. *Reducing risks from "natural" hazards.* Prior to the post-Fordist era, the biophysical environment was highly managed and manipulated. Practices such as windrows, tile drainage, smudge pots, and wind machines were already in use by the 1970s. However, cultural practices such as fertilization, tree bark painting (to reflect winter sun away from tree trunks), and slope alteration (e.g., to increase growing season sun exposure and/or air drainage) continued to increase the intensity with which the land was used. For example, up until this period, all plant material was removed from the orchard/vineyard/field floor to reduce competition for water and/or nutrients. For most crops, this practice ended when the use of herbicides and irrigation became feasible. Today, various grasses are typically used to cover the soil between the rows, both to conserve moisture and to provide habitat for pest predators. From 1974 to 1997, the acreage of fruit land in the southwest Michigan region under irrigation increased 44.2%, from 6,462 to 9,318 acres.

Globalization, 1990 to Present

During the current epoch, land in fruit production has continued to decline across the region; 17% of the land in fruit production in southwest Michigan was converted to nonfruit agriculture or to nonagricultural uses between 1974 and 1997. The spatial distribution of this decrease in fruit acreage further concentrated fruit production in the Fruit Belt counties. Since 1940, the non–Fruit Belt counties in southwest Michigan have lost more than 91% of their fruit farm land, whereas the Fruit Belt counties in the southwest region have lost only 43% of their fruit acreage. These two components imply that fruit acreage in the region as a whole declined 61% from 1974 to 1997. Today, the region has approximately 94,000 fruit acres, which is most highly concentrated in Berrien (19,768 acres) and Van Buren (22,259 acres) counties (Michigan Agricultural Statistics Service, 2001). To continue production, growers are, again, faced with increasing size and/or efficiency, changing crops, and/or finding niches. The competition for land in fruit production is particularly fierce because the same attributes that make land good for fruit production (i.e., slope, proximity to Lake Michigan) also mean that the land offers scenic vistas that are highly desired for exurban residences. Anecdotal evidence suggests that in response to these pressures, some growers have reoriented their production away from commodities for processing and toward the higher value fresh market. Therefore, requirements for cosmetic appearance, and hence for pest management, are intensified.

1. *Pest management.* Nearly every grower in the region is believed to use at least some aspect of IPM currently, but the extent to which growers have adopted IM techniques is unclear. Furthermore, the adoption of IPM has in some cases meant the application of additional pesticides. Several growers stated that monitoring and scouting found previously unobserved pests that required treatment. Furthermore, alternative methods are incompatible with some practices and their pest issues. For instance, as noted during the Fordist period mentioned earlier, apple density increased from approximately 32 trees/acre in the 1880s to 140 trees/acre in the 1990s, with the highest density orchards having more than 500 trees/acre. As density increases, pests spread more rapidly and are more difficult to treat. In 2000, the fireblight bacterium (*Erwinia amylovora*) killed nearly 400,000 trees in southwest Michigan (Longstroth, 2002). Control was compromised both by planting dwarfing varieties that are more susceptible and by planting them closer together, which decreased spray penetration and facilitated the rapid spread of the disease (Fig. 5.8).
2. *Population and sprawl.* Regional newcomers are especially concerned about pesticide residues and environmental contamination; their concerns are heightened by the federally mandated roadside warning signs that indicate the location of recent agrichemical applications. At the same time, it should be noted that these concerns are not unwarranted. Contaminated soils from heavy metal use (e.g., lead, arsenic) have been found in similar regions across the country (Jones and Patterson, 2003). The rapid

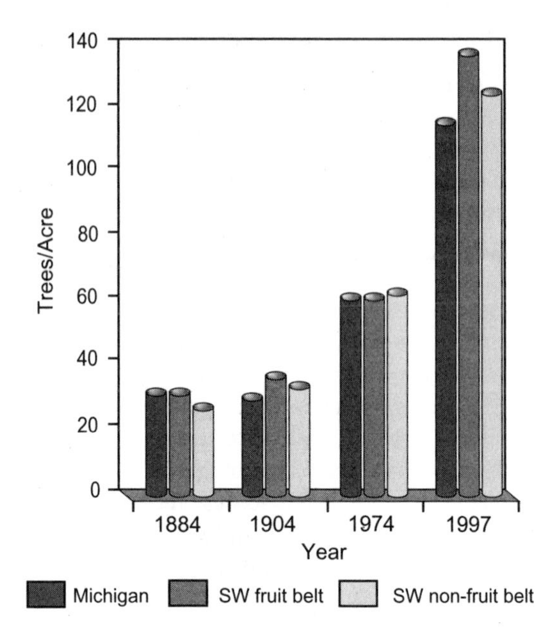

Figure 5.8 Apple orchard density in Michigan, 1884–1997. SW, southwest.

expansion of the urban, suburban, and exurban populations during this period exacerbated conflicts. Because good fruit land is attractive for rural residences for the reasons noted earlier, the population in southwest Michigan Fruit Belt counties increased 27% between 1980 and 2000, in contrast to an 8% increase in the non–Fruit Belt counties in the region. The proximity of these new rural residents to the agricultural operations, the traffic congestion, and new community mandates have altered the rural social landscape. Long-term residents, both farm and nonfarm families, perceive increasing tourism and urban sprawl as the most problematic land-use changes. They feel that these newcomers, who are unfamiliar with typical farming practices, change the tone of local politics and increase conflict (e.g., over zoning rules). Nevertheless, the visual imagery of the landscape has had and does have an impact on tourism and responses to sprawl (i.e., farmland conservation and preservation). At the state level, Michigan allows farmland owners to place their farmland in a program that indemnifies them against property tax increases for as many as 20 years. Although it has not yet happened in the southwest region, several counties near the region have implemented programs to purchase development rights from farmland owners. Both of these efforts have the effect of freezing in place a particular relationship between land in farms and the off-farm biophysical environment.

The state has also attempted to interpose its authority between farm operators and nonfarm rural residents in conflicts over the off-farm impacts of farming practices. For each major agricultural enterprise in Michigan, the state government has established Generally Acceptable

Agricultural Management Practices (GAAMPS), which establish minimal standards that farming practices must meet or exceed in order not to be considered an actionable nuisance under law. The establishment of GAAMPS has made it difficult for rural residents to seek redress for pesticide spray drift or odors from animal operations (the two most frequent complaints) (Gunter et al., 1999). Again, the GAAMPS program has had the effect of cementing in place a particular relationship between agriculture and the surrounding biophysical environment.

3. *"Nature" and "ruralness."* The viewscape is how the new residents come to know the agroecological environment, and it is reflected in their values and attitudes (Redclift and Woodgate, 1994). Their notions of agriculture paradoxically tend to center around an image of "nature" that is constructed as rolling hills with blooming fruit trees. The fruit itself has been iconized into a symbol of agriculture (Worosz, 2006), and fruit production is the definition of "ruralness." Thus, land use is built on a socially constructed definition of "nature" that does not include the messiness of industrialized agriculture. On the one hand, new rural residents construct nature as "agriculture"; on the other hand, they construct agriculture as "nature"—thus the paradox. In essence, both the "natural" environment and farming become little more than illusions, and the legacy of what was agriculture is preserved in the names of roads, subdivisions, and shopping centers (Thompson, 2000), while seasonal events become nothing more than an economic development tool for the nonagricultural community (Aronoff, 1993). For example, blossom festivals were initially used to request divine intervention (e.g., prevent pest damage, ensure yield), but are now agritainment (e.g., parades, carnival rides) complete with the naming of queens, courts, and king's men (Worosz, 2006).

   At the same time, these notions are reinforced by the fruit industry itself and are codified by government agencies. The biophysical characteristics of the landscape, for instance, are used to promote local products to tourists, such as wine labeled "Lake Michigan Shore." The Michigan wine industry has come to rely heavily on tourists; they account for up to 95% of the overall business, and much of the tourist trade is repeat customers.

4. *Niche markets.* In addition to reorienting production, some growers have also sought to capture niche markets—agritainment (e.g., u-pick, corn mazes, hayrides), value-added (e.g., branded pies and jams), and direct sales (e.g., farmers' markets, roadside stands) (Cantrell and Lively, 2002). The presence of consumers in and around their orchards/vineyards/fields, again, increases concentration on the appearance not only of the fruit, but also of the farm (Busch and Tanaka, 1996). For example, the planting of seasonal grasses in orchards that was recommended during the 1980s and 1990s as a technique to manage pests is discontinued so that consumers can see a pristine orchard. Thus, orchard-level pest/predator manipulation is sacrificed for "beauty."

Animal Commodities

Nineteenth-Century Extensive Development, 1837–1898

Draft animals provided the principal source of traction on the farms in the region during this period, and a significant portion of transportation of commodities and goods, both in rural and in urban areas. During this period, the number of horses on farms in the region increased from 23,000 to 221,000. The breeding of horses was important, both to replenish the farm stock and to supply nonfarm uses. The increasing production of small grains and hay, noted earlier, fed both the farm animals and the nonfarm animals.

During this period the number of hogs quintupled, as the farms of the region supplied fresh pork, cured ham and bacon, and lard for baking to a growing urban population. Initially, hogs were allowed to graze freely in the woodlands, and later they were allowed to root freely in fields of root crops planted for that purpose. Slaughter and curing were done either on the farm or by commercial butchers.

The number of milk cows quadrupled during this period, as the farms of the region supplied dairy products to a growing urban population. Milk was sold raw, and butter and cheese were produced on the farms. The growth in other cattle tracked the number of milk cows almost perfectly as male calves were raised for beef. Slaughtering was done locally, either on the farm or by commercial butchers, and hides were processed into leather. Cattle were fed largely on pasture and with hay during the winter.

During this period, the number of sheep on farms in the region almost quadrupled, but the number of farms on which sheep were raised increased only slightly. Sheep were raised both for wool, which was spun in the household, and for meat, which was slaughtered on the farm. Sheep grazed either on pasture or on unimproved land, and were fed hay during the winter.

Poultry (chickens, ducks, geese, and turkeys) provided both eggs and meat for farm consumption, and for sale in the rural towns and the urban centers. Although most poultry ranged freely in the barnyard and farm pond, the food they could obtain on their own was supplemented with various grain mixtures.

The animal agriculture of this period in southwest Michigan could be characterized as petty commodity production—a balanced mixture of self-provisioning and production for largely local markets, relying primarily on family/household labor with perhaps some small amount of full-time, long-term hired labor. The animals and animal products that were sold supplied the nonfarm residents of towns and cities in the region.

As noted earlier, horses, milk and beef cows, and sheep all relied on extensive areas of grass pasture for grazing, and extensive areas of hay for additional feeding. This area supported populations of arthropods, birds, and small and medium mammals. At the same time, the grazing animals deposited manure in the fields very slowly and gradually, plop by plop, in a highly dispersed pattern. In addition, horses and poultry consumed corn and small grains. These amber waves of grain supported bird populations much larger than had existed before the clearing and planting of the land (Neumann, 1985). The manure from poultry, from swine when they were penned, and from horses and cattle when they were kept near the barn in the winter was spread on the croplands to replenish the fertility of the soil. Because all the animals' feed had come from the cropland, this did not cause problems of excessive phosphorus or nitrogen.

The Golden Age, 1899–1919

Although draft animals continued to be the principal source of traction on farms during this period, the number of horses on farms reached its peak in 1890, and declined slowly during the next 30 years as steam tractors become more prevalent. Off the farm, horses remained important for the transportation of people and goods. Thus, farms in southwest Michigan continued to produce hay and grain to feed draft animals both on and off the farm. As a result of hunting and other efforts, the large mobile flocks of birds that had threatened grain yield in the past were beginning to decline during this period.

As the urban population in Michigan and the three nearby states (Illinois, Indiana, and Ohio) continued to increase, the market for eggs and poultry meat grew apace. In this period the number of poultry on southwest Michigan farms increased by 48.7%, with chickens constituting 95.3% of those flocks. Ration for poultry continued to be largely produced on the farm. Poultry was mostly sent to market live. Like the birds themselves, eggs produced increased 47.4%. Anecdotal evidence suggests that eggs provided "pin money" for farm wives and perhaps younger farm children.

It was in 1916 that Carl Sandburg (Sandburg, 1916, p. 1) described Chicago as the "hog butcher for the world," and we would suggest that many of those hogs came from southwest Michigan, transported by rail to the stockyards in less than 24 hours. Hog production was a petty commodity enterprise for many farms. During this period the number of swine on southwest Michigan farms increased by only 13.6%, roughly the same relative increase as the number of farms in the region. Swine continued to receive a variety of roots and greens produced on the farms as their feed.

The production of milk remained important during this period. The number of milk cows in the region increased by 53.1%. In 1910, farms still sold milk and cream separately, and also sold butter and cheese. During this period, the number of other cattle continued to be about the same as the number of milk cows. This would suggest that these were largely integrated farm operations, breeding the mature cows, and raising the calves either for beef or for milk production. The production of beef cattle in the region supplied both local slaughterhouses and the Chicago meatpackers. Feed for cattle was still mostly pasture and hay, with some movement toward more specialized forages and hays during this period.

Because of the reliance on grazing for a significant portion of the year, much of the cattle manure was deposited directly on the land in the pastures. Manure that was collected from the barnyards and animal buildings was simply piled up to await spreading in late winter or fall, after which it would be tilled into the soil. The relatively low density of animals per unit of land in farms in the region diminished the potential negative environmental impacts of this practice, and recycled nutrients from the animal waste back into the soil in a form that could be gradually converted into nutrients that the next year's crop could take up.

The Agricultural and Great Depressions, 1920–1940

The number of horses on farms in the region decreased by 48.2% during this period. Many farms gave up at least some of their horses and purchased internal combustion engine tractors. During the first two decades of this period, farm people left farming and shifted to industrial employment. This decreased the amount of internal and local labor available for farms, and contributed to the shift to more powerful self-propelled farm machinery. At the same time, the reliance on horses for transportation and draft power in the nonagricultural sectors of society greatly decreased, so raising draft horses was no longer a profitable enterprise. Because the average farm had about three horses, and each horse ate the grain from about 10 acres, giving up the horses meant that 30 acres of small grains and pasture could be converted to other uses. Although the demand for barley for beer was increasing, the demand for small grains (especially oats and rye) for food manufacturing was decreasing, so generally the cropland was allocated to some other enterprise; in some cases, the number of other animals was increased. The decrease in small-grain acreage doubtless further decreased the graminivorous bird populations in the region.

During this period, the production of chickens increased by more than one third (from 1920 to 1940, the increase was 33.0%), and became more concentrated. From 1925 to 1940, the number of chickens produced on farms producing chickens increased by 48.1%. In contrast, the number of eggs produced increased 28.9% from 1910 to 1940, while the number of eggs produced on farms producing eggs increased by 26.7% from 1925 to 1940, so concentration was not happening as much in egg production as it was in poultry meat production.

The years around 1910 represent an initial peak in hog production in the southwest region. During this period, the number of swine in the region decreased by 56.0%, and did not regain a comparable level until the late 1970s. It would appear that this decrease was only partly the result of farmers dropping swine from their mix of enterprises; from 1925 to 1940, the number of farms with swine decreased by only 9.7%. However this appearance may be misleading. In fact, probably three trends were occurring. First, farms were tending to drop swine from their mix of commercial enterprises, and some of these farms gave up swine altogether. Second, some of the farms that dropped commercial swine continued to raise a few hogs for self-provisioning and extended family support. Thus, the number of swine on farms with any swine reaches its lowest value (6.7) in 1935. Although some of the selling down may have been a response to the lack of markets during the depths of the Depression, this would be all the more reason for a farm to continue to feed the number of hogs that the resident and extended family members could consume. Third, some of the farms that stayed in commercial swine expanded their hog operations. The number of hogs on farms with hogs almost doubled every decade after 1935.

Although the size of the dairy herd continued to grow during this period, the latter part of the 1930s and the first part of the 1940s were the peak of dairy farming in southwest Michigan. Since the mid 1940s, the number of dairy cows in the region has declined continuously up to the present. In fact, the decline had its origins during this period. From 1925 to 1940, the number of farms with dairy cattle decreased by 12.0%. Although the number of dairy cows per farm with dairy cows increased by 25.6% from 1925 to 1940, the mean number of dairy cattle per farm was still less than six. Because one or two cows would have supplied the dairy needs of the farm family, it would appear that a significant amount of petty commodity production was still occurring. Butter was still being produced on the farm for sale, and cream was still being separated from the milk and sold separately.

Although 1910 continued the previous pattern of equal numbers of dairy and nondairy cattle, from 1920 on, the number of nondairy cattle in the region was consistently less than the number of dairy cattle by about a third. This would suggest that female cattle were being kept because dairy production was profitable, whereas young male cattle were being sold for slaughter or for finishing operations elsewhere.

Agricultural Fordism, 1941–1973

The number of horses on farms continued to decline during the first part of this period, reaching a low point in the late 1950s or early 1960s. In 1954, most of the farms that had horses had only one horse. Although some of these animals would still have been workhorses, presumably most of them were being kept for pleasure riding. At the same time that horses were being eliminated from crop farms, the late 1950s also saw the return and increasing development of horses as a farm enterprise. By 1969, the number of horses in the region had increased 66.6% from the low point in 1959, and the average number of horses on farms with horses had increased from slightly more than one in 1954 to almost four in 1969. These horse farms would have maintained some acreage for pasture, some acreage for hay, and presumably some acreage for riding trails.

It was during this period that the poultry sector became highly concentrated. On the one hand, the number of chickens raised increased by about 100%. At the same time, the number of farms raising chickens decreased by approximately 90%. Although the average number of birds raised per chicken farm increased about 20-fold to 3,000, the average broiler operation was raising 7,000 birds per year. Similarly, the number of eggs produced increased by 60.3% from 1940 to 1964, while the number of farms producing eggs in southwest Michigan declined by 89.9%. Not only was egg production concentrated on fewer farms, it was also spatially concentrated. One county, Allegan, produced 12% of all the eggs produced in Michigan. Because the population of Michigan increased 69.0% from 1940 to 1970, the increase in egg production barely kept pace with the increased population. This reflects the long-term decline in egg consumption that began during this period. From its peak in 1945, per-capita egg consumption in the United States had decreased 26.7% by 1970 as consumers shifted away from cholesterol, and food manufacturers found substitutes for eggs.

It was also during this period that the raising of turkeys shifted from small-scale petty commodity production to industrial-scale operations. Although the number of farms raising turkeys decreased by 97%, the number of turkeys produced increased by more than ninefold (938%). Half the farms and four fifths of the production were in one county (Ottawa). The increase in turkey production reflected an increase in turkey consumption in the United States. During this 30-year period per capita consumption increased from its long-term average of less than 2 lb. to 6.6 lb., as consumers sought to avoid beef and pork because of their image of high fat and cholesterol. For these large, concentrated poultry operations, the disposal of manure began to be a problem.

The number of hogs and pigs on southwest Michigan farms continued to increase slowly during this period, growing 79.9% from 1940 to 1969. The number of farms with swine continued to decrease during these years, from 30,848 to 5,661 (81.6%). Thus the average number of pigs and hogs on a swine farm increased by 878% during this period. Because the land area of these hog farms did not increase to that extent, the disposal of hog manure was an increasing problem.

The number of dairy cattle in the region started its long-term decline during this period, decreasing by 46.0% from 1940 to 1969. Because the number of farms with dairy cattle declined even more precipitously during these years, by 88.1%, the average number of dairy cattle on dairy farms increased from 5.79 to 26.2. In 1970 the conventional wisdom was that one adult farm operator could manage about 80 cows, so these farms were probably a mixture of small-scale petty commodity production and commercial dairy farms. Concomitant with the decline in dairy cattle and dairy farms, the number of nondairy cattle increased by 89.8%, and the number of farms with nondairy cattle was at least double the number of farms with dairy cattle. So while the beef sector was beginning to emerge in 1969, the average head per farm was still less than 50, and manure management was not yet a problem.

Agroecological and Profitability Crisis, 1974–1989

Although the number of horses in southwest Michigan increased slightly during this period, the number of farms with horses declined by about a third. The average number of 6.1 horses on a horse farm reflects the increasing specialization of horse farming in breeding and raising horses for racing and recreation.

The shift to horse farming for racing and recreation is symptomatic of the general withdrawal of southwest Michigan from traditional agriculture during this period. From a high point of slightly over a million birds in 1969, production fell 35% by 1987. This occurred despite the increase in per capita consumption of turkey meat in the United States during that period from 6.6 lb. to 11.6 lb. Similarly, chicken sales from southwest Michigan decreased by 2.5% during this period, despite the 50% increase in per capita consumption of chicken meat in the United States from 26.3 lb. to 39.4 lb. At the same time, the number of poultry farms in the region decreased by about 75%, so the remaining production was concentrated on a much smaller number of farms. Although the Census of Agriculture stopped collecting data on egg production and sales after 1969, information from extension agents and specialists suggests that the region also gradually withdrew from egg production. This pattern of withdrawal is probably the result, at least in part, of the difficulty of disposing of the manure being generated by large-scale, concentrated poultry operations.

This same pattern of withdrawal occurred in dairy and beef cattle during this period. Reduction in the dairy herd in the region was fostered both by the federal dairy buyout program in 1980, and the introduction of synthetic bovine growth hormone that increased milk production per cow by about 10%. In contrast, swine production in the region increased by almost 100%, stimulated by a state policy favoring concentrated hog-feeding operations ("hog hotels"). Although these operations were being discouraged in other states where environmental problems had occurred (e.g., North Carolina), Michigan was welcoming them with financial and legal (permitting) assistance.

Globalization, 1990 to Present

During the last decade of the 20th century and the first decade of the 21st century, the previous trends in animal agriculture in southwest Michigan continued. The number of horses in the region increased by 73.9%, and the number of farms with horses increased by 55.9% to 5,488. Thus, more than a quarter (28.4%) of southwest Michigan's farms had horses, and the average horse farm had almost seven horses.

The number of milk cows in the region continued to decline slowly (about 1% per year), but the number of dairy farms declined by more than 50% as the consequences of the dairy buyout, the use of bovine growth hormone, and economies of scale were felt. Thus, the average dairy herd more than doubled, from 58 in 1987 to 126 in 2002. Because the land operated by dairy farms did not increase that much, disposal of local concentrations of manure became more of a problem.

In contrast to the slow decline in dairy cattle, the numbers of nondairy cattle in the region plummeted during this period, decreasing by 70.5% between 1987 and 2002. Since the number of nondairy cattle farms only decreased by about 25% during this period, manure management is not the increasing problem that it is for dairy operations.

Whereas hogs show the general pattern of withdrawal from traditional animal agriculture, they also show the increasing intensity of dairy. While the number of hogs in the southwest region declined by 18.7% from 1987 to 2002, the number of hog farms declined by 68.4% to 879. Thus the average number of hogs per farm increased by 157% to more than 800, and disposal of local concentrations of manure was an increasing environmental and social problem.

Commodity Summary

The Agricultural and Great Depressions, 1920–1940

During this period, the steep decline in the use of animals for draught power led to declines in the use of farmland for pasture and hay and small grains. These land uses were replaced by annual crops, especially corn and incipiently soybeans. These changes in land use implied more tillage (for annual crops) and more cultivation (for row crops), and thus more compaction of the soil and soil erosion. The shift to cultivated annual crops meant that less carbon was being sequestered by agriculture, and that greenhouse gases were increasing. The greater use of tractors and combines meant that more greenhouse gases were being produced because internal combustion engines have greater flatulence than horses and other beasts of burden. The shift to combines for harvesting meant that less grain was being scattered in the fields for birds and other animals. The increasing spatial concentration of monocultures of grain and fruit led to more problems with insect and disease pests. The increased importance of cosmetic standards and legal regulations led to more spraying of heavy metal pesticides. At the same time, mechanical sprayers caused greater off-target deposition, and thus greater impacts on off-target species.

Agricultural Fordism, 1941–1973

More than any other, the post–World War II period marks the industrialization of southwest Michigan agriculture. By the end of this period, animal draught power was almost entirely gone, replaced by internal combustion engines. Tillage and cultivation were intensively practiced, leading to soil compaction and soil erosion. Fewer than half the farms in the region included animal enterprises, so fertilization was increasingly in the form of synthetic nitrogen and mineral phosphates and potash, which resulted in increased runoff to surface waters and leaching to groundwater. These, in turn, increased the eutrophication of lakes and rivers in the region. Pest management relied on persistent, broad-spectrum insecticides that generated pest resistance and secondary outbreaks, and that negatively affected nontarget species of birds and fish. These various negative impacts set the stage for the environmental backlash against agriculture in the following period.

Conclusion

Since settlement, grain production has been one of the most significant parts of agriculture in southwest Michigan. During the first era, the ecological suitability of the region resulting from the ease of transportation and the availability of small, open prairies helped to draw settlers and to orient them to grain production. As agriculture expanded, settlers altered the landscape by cutting down trees and draining wetlands. Expanding national (new uses) and international (new users) demand led to increasing intensification of production until finally soil erosion, pest infestations, aquifer reduction, and other environmental limits were felt. Technological and institutional developments overcame some of these environmental limits as grain production became defined by machines, chemicals, and hybrids. Again, however, environmental degradation resulted from these new technologies and farmers were forced to alter production practices to incorporate practices such as crop rotation and reduced tillage. Finally, encroachment at a global scale through technologies such as biotechnology and locally through exurbanization has once again altered grain production. This tension has both created new opportunities for grain production and placed new constraints on decision making by farmers. Inevitably, these new opportunities and constraints have once again altered and recreated the ecological landscape on which the corn, wheat, soybeans, and other crops are grown.

References

Adkins, C. L. 2003. "Evidence for corn agriculture in southwestern Michigan? New botanical evidence from Moccasin Bluff." *Michigan Archaeologist* 49: 17–32.

Albrecht, D. E., and S. H. Murdock. 1990. *The sociology of U.S. agriculture: An ecological perspective.* Ames, Iowa: Iowa State University Press.

Allen, P. (ed.). 1993. *Food for the future: Conditions and contradictions of sustainability.* New York: Wiley.

Alstou, L. J., and T. J. Hatton. 1991. "The earnings gap between agricultural and manufacturing laborers, 1925–1941." *Journal of Economic History* 51(1): 83–99.

Altieri, M. A. 1995. *Agroecology: The science of sustainable agriculture.* Boulder, Colo.: Westview Press.

Anonymous. 1929. "Says pests imperil big-scale farming." *New York Times,* July 9, section 1: 26.

Armstrong, W. J. 1993. "Berrien County's great peach boom!" *Michigan History Magazine* May/June: 10–17.

Aronoff, M. 1993. "Collective celebration as a vehicle for local economic-development: A Michigan case." *Human Organization* 52(4): 368–379.

Barnes, C. P. 1929. "Land resources inventory in Michigan." *Economic Geography* 5(1): 22–35.

Baxevanis, J. J. 1992. *The wine regions of America: Geographical reflections and appraisals.* Stroudsberg, Pa.: Vinifera Wine Growers Journal.

Beld, S. G. 1993. "Site 20IA37 (Arthursburg Hill Earthworks), Lyons Township, Ionia County, Michigan," pp. 3–82. In: *Lyons Township archaeological survey, S-92-313. Report on file in the Office of the State Archaeologist.* Lansing, Mich.: Bureau of Michigan History, Michigan Department of State.

Beld, S. G. 1994. "Site 20IA37, Lyons Township, Ionia County, Michigan," pp. 2–39. In: *Ionia County archaeology, phase II - S9-319. Report on file in the Office of the State Archaeologist.* Lansing, Mich.: Bureau of Michigan History, Michigan Department of State.

Berlan, J.- P. 1991. "The historical roots of the present agricultural crisis," pp. 115–136. In: W. H. Friedland, L. Busch, F. H. Buttel, and A. P. Rudy (eds.), *Towards a new political economy of agriculture.* Boulder, Colo.: Westview Press.

Bingen, R. J. 2005. *Farmer's markets in Michigan: Preliminary results from a survey of market managers.* East Lansing, Mich.: Michigan State University.

Brashler, J. G. 1978. *Boundaries and interaction in the Early Late Woodland of southern lower Michigan.* PhD diss., Michigan State University, East Lansing, Mich.

Brashler, J. G., and M. B. Holman. 2004. "Middle Woodland adaptation in the Carolinian/Canadian transition zone of western lower Michigan," pp. 14–29. In: W. A. Lovis (ed.), *An Upper Great Lakes archaeological odyssey: Essays in honor of Charles E. Cleland.* Detroit, Mich.: Wayne State University.

Brashler, J. G., and B. Mead. 1996. "Woodland settlement in the Grand River basin." In: M. B. Holman, J. G. Brashler, and K. E. Parker (eds.), *Investigating the archaeological record of the Great Lakes state: Essays in honor of Elizabeth Baldwin Garland*, vol. 181–249. Kalamazoo, Mich.: Western Michigan University.

Brashler, J. G., E. B. Garland, M. B. Holman, W. A. Lovis, and S. R. Martin. 2000. "Adaptive strategies and socioeconomic systems in northern Great Lakes riverine environments: The Late Woodland of Michigan," pp. 543–579. In: T. E. Emerson, D. L. McElrath, and A. C. Fortier (eds.), *Late Woodland societies: Tradition and transformation across the midcontinent.* Lincoln, Nebr.: University of Nebraska.

Brashler, J. G., M. R. Laidler, and T. J. Martin. 1998. "The prison farm site (20IA58): A woodland occupation in the Grand River basin of Michigan." *Midcontinental Journal of Archaeology* 23: 143–198.

Brigham, A. P. 1910. "The development of wheat culture in North America." *Geographical Journal* 35(1): 42–56.

Busch, L., and K. Tanaka. 1996. "Rights of passage: Constructing quality in a commodity subsector." *Science, Technology and Human Values* 21: 3–27.

Cantrell, P., and J. Lively. 2002. *The new entrepreneurial agriculture: A key piece of the farmland protection puzzle.* Traverse City, Mich.: Michigan Land Use Institute.

Chevalier, J. M. 1983. "There is nothing simple about simple commodity production." *Journal of Peasant Studies* 10(4): 153–186.

Childers, N. F. 1975. *Modern fruit science: Orchard and small fruit culture.* New Brunswick, N.J.: Horticultural Publications, Rutgers University.

Cleland, C. E. 1966. "The prehistoric animal ecology and ethnozoology of the upper Great Lakes region." *Museum of Anthropology Anthropological Papers No. 29.* Ann Arbor, Mich.: University of Michigan.

Cleland, C. E. 1992. *Rites of conquest: The history and culture of Michigan's Native Americans.* Ann Arbor, Mich.: University of Michigan Press.

Clifton, J. A. 1998. *The prairie people: Continuity and change in Potawatomi Indian culture, 1665–1965.* Iowa City, Iowa: University of Iowa.

Cronon, W. 1991. *Nature's metropolis: Chicago and the Great West.* 1st ed. New York: W. W. Norton.

DeLind, L. B. 1999. "Close encounters with a CSA." *Agriculture and Human Values* 16(1): 3–9.

Dibble, C. B. 1936. *Corn borer control by good farming.* Michigan State College Extension Bulletin no. 59. East Lansing, Mich.: Michigan State College.

Dunbar, W. F. 1969. *All aboard! A history of railroads in Michigan.* Grand Rapids, Mich.: W. B. Eerdmans.

Dunbar, W. F., and G. S. May. 1995. *Michigan: A history of the Wolverine State.* 3rd rev. ed. Grand Rapids, Mich.: W. B. Eerdmans.

Egan-Bruhy, K. C. 2002. "Floral analysis of sites 20BY28 and 20BY387," pp. 6.1–6.31. In: W. Lovis (ed.), *A bridge to the past, the post Nipissing archaeology of the Marquette Viaduct replacement project sites 20BY 28 and 20BY387, Bay City, Michigan.* East Lansing, Mich.: Michigan State University Museum and Department of Anthropology.

Frank, R., T. T. Davies, R. L. Thomas, H. E. Braun, and D. L. Gross. 1981. "Organochlorine insecticides and PCB in surficial sediments of Lake Michigan (1975)." *Journal of Great Lakes Research* 7(1): 42–50.

Freedman, E. 1992. *Pioneering Michigan.* Franklin, Mich.: Altwerger and Mandel.

Friedland, R., and D. Boden. 1994. *NowHere: Space, time and modernity.* Berkeley, Calif.: University of California Press.

Friedland, W. H. 2001. "Reprise on commodity systems methodology." *International Journal of Sociology of Agriculture and Food* 9(1): 82–103.

Friedmann, H. 1978. "Simple commodity production and wage labor in the American plains." *Journal of Peasant Studies* 6(1): 71–100.

Friedmann, H., and P. McMichael. 1989. "Agriculture and the state system: The rise and decline of national agricultures, 1870 to the present." *Sociologia Ruralis* 29: 93–117.

Garland, E. B. 1986. "Early Woodland occupations in Michigan: A lower St. Joseph Valley perspective," pp. 47–83. In: K. B. Farnsworth and T. E. Emerson (eds.), *Early Woodland archaeology. Kampsville seminars in archeology, volume 2.* Kampsville, Ill.: Center for American Archaeology.

Garland, E. B., and S. G. Beld. 1999. "The Early Woodland: Ceramics, domesticated plants, and burial mounds foretell the shape of the future," pp. 125–146. In: J. Halsey (ed.). *Retrieving Michigan's buried past: The archaeology of the Great Lakes state. Bulletin 64.* Bloomfield Hills, Mich.: Cranbrook Institute of Science.

Gereffi, G., and M. Korzeniewicz. 1994. *Commodity chains and global capitalism.* Westport, Conn.: Greenwood Press.

Goodman, D., B. Sorj, and J. Wilkinson. 1987. *From farming to biotechnology: A theory of agro-industrial development.* New York: Basil Blackwell.

Gray, S. E. 1996. *The Yankee West: Community life on the Michigan frontier.* Chapel Hill, N.C.: University of North Carolina Press.

Griliches, Z. 1960. "Hybrid corn and the economics of innovation." *Science* 132(3422): 275–280.

Gunter, V. J., M. Aronoff, and S. Joel. 1999. "Toxic contamination and communities: Using an ecological–symbolic perspective to theorize response contingencies." *Sociological Quarterly* 40(4): 623–640.

Hall, T. D. 2000. *A world-systems reader: New perspectives on gender, urbanism, cultures, indigenous peoples, and ecology.* Lanham, Md.: Rowman and Littlefield.

Harris, C. K., and M. R. Worosz. 1998. *A fruitful experience: The practices of IPM and organic growers.* Michigan Agricultural Experiment Station Research Report 553. East Lansing, Mich.: Michigan State University.

Hart, J. F. 1986. "Change in the corn belt." *Geographical Review* 76(1): 51–72.

Hartshorne, R. 1926. "The significance of lake transportation to the grain traffic of Chicago." *Economic Geography* 2(2): 274–291.

Hecht, S. B. 1985. "Environment, development and politics: Capital accumulation and the livestock sector in eastern Amazonia." *World Development* 13(6): 663–684.

Heinz, G. H., T. C. Erdman, S. D. Haseltine, and C. Stafford. 1985. "Contaminant levels in colonial waterbirds from Green Bay and Lake Michigan 1975–1980." *Environmental Monitoring and Assessment* 5(3): 223–236.

Heinz, G. H., D. S. Miller, B. J. Ebert, and K. Strongborg. 1994. "Declines in organochlorines in eggs of red-breasted mergansers from Lake Michigan, 1977–1978 versus 1990." *Environmental Monitoring and Assessment* 33(3): 175–182.

Heller, C. F., Jr., and J. T. Houdek. 1996. "Farm tenants and landlords in nineteenth-century southern Michigan." *Agricultural History* 70(4): 598–625.

Hill, E. B. 1939. "Types of farming in Michigan." Michigan State College Agricultural Experiment Station. East Lansing, Mich.: Michigan State College.

Hinsdale, W. B. 1931. *An archaeological atlas of Michigan.* Ann Arbor, Mich.: University of Michigan Press.

Holman, J. A. 1995. *Ancient life of the Great Lakes basin.* Ann Arbor, Mich.: University of Michigan Press.

Holman, M. B., and J. G. Brashler. 1999. "Economics, material culture, and trade in the Late Woodland lower peninsula of Michigan," pp. 212–220. In: J. Halsey (ed.), *Retrieving Michigan's buried past: The archaeology of the Great Lakes state. Bulletin 64.* Bloomfield Hills, Mich.: Cranbrook Institute of Science.

Holman, M. B., and R. G. Kingsley. 1996. "Territoriality and societal interaction during the Early Late Woodland period in southern Michigan," pp. 341–382. In: M. B. Holman, J. G. Brashler, and K. E. Parker (eds.), *Investigating the archaeological record of the Great Lakes state: Essays in honor of Elizabeth Baldwin Garland.* Kalamazoo, Mich.: Western Michigan University.

Houck, W. E. 1954. *A study of some events in the development of entomology and its application in Michigan.* PhD diss., Michigan State College, East Lansing, Mich.

Jones, T., and N. Patterson. 2003. "EPA targets soil at Barber orchard." Online. Available at www.Citizen-Times.com.

Kapp, R. J. 1999. "Michigan Late Pleistocene, Holocene, and presettlement vegetation and climate," pp. 31–58. In: J. Halsey (ed.), *Retrieving Michigan's buried past: The archaeology of the Great Lakes state. Bulletin 64.* Bloomfield Hills, Mich.: Cranbrook Institute of Science.

Kessler, G. M. 1971. "A history of fruit growing in Michigan," pp. 114–147. In: *One Hundredth Annual Report of the Secretary of the State Horticultural Society of Michigan for the Year of 1970,* vol. 100. East Lansing, Mich.: State Horticultural Society of Michigan.

Larrimer, W. H. 1928. "America's corn crop and the corn borer." *Scientific Monthly* 27(5): 424–433.

Larson, R. P. 2001. "Cultural practices for cherry mechanization," pp. 687–697. In: B. F. Cargill and G. E. Rossmiller (eds.), *Fruit and vegetable harvest mechanization: Technological implications*, vol. 1. East Lansing, Mich.: Rural Manpower Center, Michigan State University.

Larson, G., and R. Schaetzl. 2001. "Origin and evolution of the Great Lakes." *Journal of Great Lakes Research* 27: 518–546.

Lewis, K. E. 2002. *West far to Michigan: Settling the lower peninsula 1815–1860.* East Lansing, Mich.: Michigan State University Press.

Longstroth, M. 2002. *The fireblight epidemic in southwest Michigan.* Paw Paw, Mich.: Michigan State University Extension, Van Buren County.

Lovis, W. A. 1988. "Human prehistory of southwestern Michigan: A paleogeographic perspective," pp. 43–50. In: G. Larson and G. W. Monaghan (eds.), *Wisconsin and Holocene stratigraphy in southwestern Michigan. 35th field conference guide. Midwest Friends of the Pleistocene.* East Lansing, Mich.: Michigan State University.

Lovis, W. A. 1999. "The Middle Archaic: Learning to live in the woodlands," pp. 83–94. In: J. Halsey (ed.), *Retrieving Michigan's buried past: The archaeology of the Great Lakes state. Bulletin 64.* Bloomfield Hills, Mich.: Cranbrook Institute of Science.

Mann, S. A., and J. M. Dickinson. 1978. "Obstacles to development of a capitalist agriculture." *Journal of Peasant Studies* 5: 466–481.

Martin, T. J. 2003. "Animal remains from the 2002 investigation of the Moccasin Bluff site, Berrien County, Michigan." *Michigan Archaeologist* 49.

McClurken, J. M. 1988. *We wish to be civilized: Ottawa-American political contests on the Michigan frontier.* PhD diss., Michigan State University, East Lansing, Mich.

Meltzer, D. J., and B. D. Smith. 1986. "Paleoindian and Early Archaic subsistence strategies in eastern North America," pp. 3–32. In S. Neusius (ed.), *Foraging, collecting and harvesting, Archaic period subsistence and settlement in the eastern woodlands.* Occasional paper no. 6. Carbondale, Ill.: Center for Archaeological Investigations, Southern Illinois University at Carbondale.

Michigan Agricultural Statistics Service. 2001. *Michigan rotational survey: Fruit, 2000–01.* Lansing, Mich.: Michigan Department of Agriculture.

Michigan Department of Agriculture. 1953. *Amended, establishing the grade of red sour cherries that may be sold, offered for sale, purchased or received for canning and/or preserving purposes, when such canned or preserved cherries are to be resold, and requiring grade certification of the same.* Lansing, Mich.: Bureau of Marketing and Enforcement.

Michigan Department of Agriculture. 1962. *Regulation no. 600 as amended.* Lansing, Mich.: Plant Industry Division.

Michigan Department of Agriculture. 1995. *Michigan rotational survey: Fruit, 1995.* Lansing, Mich.: Michigan Agricultural Statistics Service.

Mitchell, A. E., F. Sherman III, and D. Cation. 1953. *Spraying calendar.* East Lansing, Mich.: Michigan State College, Cooperative Extension Service.

Monaghan, G. W., and W. A. Lovis (with contributions by M. J. Hambacher). 2004. "Modeling archaeological site burial in southern Michigan: A geoarchaeological synthesis." In: *Environmental research series no. 1, Michigan Department of Transportation.* East Lansing, Mich.: Michigan State University Press.

Murdock, S. H. and D. E. Albrecht. 1998. "An ecological investigation of agricultural patterns in the United States," pp. 299–316. In: M. Micklin and D. L. Poston Jr. (eds.), *Continuities in sociological human ecology.* New York: Plenum Press.

Musselman, H. H. 1928. *Plowing for European corn borer control.* Michigan State College Extension Division Bulletin no. 55. East Lansing, Mich.: Michigan State College.

Neumann, T. W. 1985. "Human–wildlife competition and the passenger pigeon: Population growth from system destabilization." *Human Ecology* 13(4): 389–410.

O'Gorman, J. A. 2004. "The myth of Moccasin Bluff—Rethinking the Potawatomi pattern as a model for ancient Potawatomi history." In: *Files of the Michigan State University Consortium for Archaeological Research.* East Lansing, Mich.

O'Kelly, M. E. 2007. "The impact of accessibility change on the geography of crop production: A reexamination of the Illinois and Michigan Canal using GIS." *Annals of the Association of American Geographers* 97(1): 49–63.

Page, B., and R. Walker. 1991. "From settlement to Fordism: The agroindustrial revolution in the American Midwest." *Economic Geography* 67(4): 281–315.

Parker, K. E. 1990. "Botanical remains from the Eidson site," pp. 396–410. In: E. B. Garland (ed.), *Late Archaic and Early Woodland adaptation in the lower St. Joseph River Valley. Michigan cultural resource series volume 2.* Lansing, Mich.: Michigan Department of State, Michigan Department of Transportation, and the Federal Highway Administration.

Perkins, J. H. 1982. *Insects, experts, and the insecticide crisis: The quest for new pest management strategies.* New York: Plenum Press.

Pfeffer, M. J. 1983. "Social origins of 3 systems of farm production in the United States." *Rural Sociology* 48(4): 540–562.

Pickett, A. D. 1949. "A critique on insect chemical control methods." *Canadian Entomologist* 81(3): 67–76.

Pimentel, D., C. Kirby, and A. Shroff. 1993. "The relationship between 'cosmetic standards' for foods and pesticide use," pp. 85–105. In: D. Pimentel and H. Lehman (eds.), *The pesticide question: Environment, economics and ethics.* New York: Chapman and Hall.

Pollan, M. 2003. "The (agri)cultural contradictions of obesity." *New York Times Magazine*, Oct 12, 2003.

Redclift, M., and G. Woodgate. 1994. "Sociology and the environment: Discordant discourse?" pp. 51–66. In: Michael Redclift and Ted Benton (eds.), *Social theory and the global environment.* New York: Routledge.

Rissler, J., and M. G. Mellon. 1996. *The ecological risks of engineered crops.* Cambridge, Mass.: MIT Press.

Robertson, J. A., W. A. Lovis, and J. R. Halsey. 1999. "The Late Archaic: Hunter-gatherers in an uncertain environment," pp. 95–124. In: J. Halsey (ed.), *Retrieving Michigan's buried past: The archaeology of the Great Lakes state. Bulletin 64.* Bloomfield Hills, Mich.: Cranbrook Institute of Science.

Rosenberg, C. E. 1997. *No other gods: On science and American social thought.* 2nd ed. Baltimore, Md.: John Hopkins University Press.

Russell, E. P. 1996. "'Speaking of annihilation': Mobilizing for war against human and insect enemies, 1914–1945." *Journal of American History* 82(4): 1505–1529.

Salamon, S. 1980. "Ethnic differences in farm family land transfers." *Rural Sociology* 45(2): 290–308.

Salamon, S., and K. Davis-Brown. 1986. "Middle-range farmers persisting through the agricultural crisis." *Rural Sociology* 51(4): 503–512.

Salamon, S., and A. M. Keim. 1979. "Land ownership and women's power in a midwestern farming community." *Journal of Marriage and the Family* 41(1): 109–119.

Sandburg, Carl. 1916. *Chicago poems.* New York: H. Holt.

Sanders, E. 1999. *Roots of reform: Farmers, workers, and the American state, 1877–1917.* Chicago, Ill.: University of Chicago Press.

Schaetzl, R. J. n.d. "Contemporary land uses: Fruit production." In: *Geography of Michigan and the Great Lakes Region.* East Lansing, Mich.: Michigan State University, College of Social Science, Department of Geography.

Schoolcraft, H. R. 1860. *Archives of Aboriginal Knowledge. Containing All the Original Papers Laid before Congress Respecting the History, Antiquities, Language, Ethnology, Pictography, Rites, Superstitions, and Mythology of the Indian Tribes of the United States, Volume I.* Philadelphia, Pa.: J. B. Lippincott and Company.

Smith, N. 1984. *Uneven development: Nature, capital, and the production of space.* New York: Blackwell.

Smith, N. 1989. "Uneven development and location theory: Towards a synthesis," pp. 142–163. In: R. Peet and N. Thrift (eds.), *New models in geography: The political-economy perspective.* London: Unwin-Hyman.

Smith, B. D. 1992. "Prehistoric plant husbandry in eastern North America." In B. D. Smith (ed.), *Rivers of change: Essays on early agriculture in eastern North America.* Washington, D.C.: Smithsonian Institution.

Smith, N., and W. Dennis. 1987. "The restructuring of geographical scale: Coalescence and fragmentation of the northern core region." *Economic Geography* 63(2): 160–182.

Sparhawk, W. N., and W. D. Brush. 1929. *The economic aspects of forest destruction in northern Michigan.* Washington, D.C.: U.S. Department of Agriculture.

State of Michigan, Public Domain Commission and Immigration Commission. 1914. *Michigan: agricultural, horticultural and industrial advantages.* Lansing, Mich.: State of Michigan.

Stoll, S. 1998. *The Fruits of natural advantage: Making the industrial countryside in California.* Berkeley, Calif.: University of California Press.

Sublett, M. D. 1975. *Farmers on the road: Interfarm migration and the farming of noncontiguous land in three midwestern townships, 1939–1969.* Chicago, Ill.: University of Chicago Press.

Swanson, L. E. 1988. *Agriculture and community change in the U.S.: The congressional research reports.* Boulder, Colo.: Westview Press.

Swyngedouw, E. 1997. "Neither global nor local: 'Glocalization' and the politics of scale," pp. 137–166. In: Kevin Cox (ed.), *Spaces of globalization.* New York: Guilford Press.

Tanner, H. H. 1987. *Atlas of Great Lakes Indian history.* Norman, Okla.: University of Oklahoma Press.

Thaden, J. F. 1959. Ethnic Settlements in Rural Michigan. "1946." *Michigan Agricultural Experiment Station Quarterly Bulletin* 29: 102–111. Accompanying map, 1945.

Thompson, J. 2000. "Environment as cultural heritage." *Environmental Ethics* 22(Fall): 241–258.

U.S. Department of Agriculture. 1946 [1941]. "United States standards for grades of red sour cherries for manufacture." Washington, D.C.: Production and Marketing Administration.

U.S. Department of Agriculture. 1949a. "United States standards for grades of canned red sour (tart) pitted cherries." Washington, D.C.: Agricultural Marketing Service.

U.S. Department of Agriculture. 1949b. "United States standards for grades of frozen red sour (tart) pitted cherries." Washington, D.C.: Production and Marketing Administration.

U.S. Department of Commerce. *Census of agriculture.* Washington, D.C.: Bureau of the Census.

Wallerstein, I. M. 1974. *The modern world-system: Capitalist agriculture and the origins of the European world-economy in the sixteenth century.* New York: Academic Press.

Wilcox, E. V., and C. B. Smith. 1911. *Farmer's cyclopedia of agriculture.* New York: Orange Judd Publishing.

Williams, W. A. 1969. *The roots of the modern American empire: A study of the growth and shaping of social consciousness in a marketplace society.* New York: Random House.

Winchell, A. 1865. "Soils and subsoils of Michigan." In: *Annual Meeting of the Michigan State Agricultural Society.* Lansing, Mich.: State of Michigan Legislature.

Winchell, A. 1866. "The fruit-bearing belt of Michigan." In: *15th American Association for the Advancement of Science.* American Association of the Advancement of Science.

Winters, D. 1978. *Farmers without farms.* Westport, Conn.: Greenwood Press.

Wood, L. H. 1914. *Geography of Michigan.* Kalamazoo, Mich.: Horton-Beimer Press.

Worosz, M. R. 2006. *Pits, pests, and the industrial tart.* PhD diss. Michigan State University, East Lansing, Mich.

Wright, H. T. 1964. "A transitional Archaic campsite at Green Point (20 SA 1)." *Michigan Archaeologist* 27: 87–91.

6

Agrarian Landscape Transition in the Flint Hills of Kansas

Legacies and Resilience

Gerad Middendorf
Derrick Cline
Leonard Bloomquist

The Flint Hills of east-central Kansas contain the largest remaining area of unplowed tallgrass prairie in North America (about 1.6 million ha). What remains today is a small fraction of the estimated pre-European extent of the tallgrass prairie, which stretched over substantial portions of what is now Illinois, Missouri, Iowa, Minnesota, and the eastern edges of the Dakotas, Nebraska, and Kansas. Before European immigration into the area, the region was home to various Native American tribes, migrating buffalo, and other large ungulates that fed off the abundant grasses. The Native Americans depended on the buffalo as a means of sustenance, and recognized that management of the grasses ensured their return in the spring.

The prairie can still be found in the Flint Hills for both biophysical and socioeconomic reasons, and it has been one of the key elements in the development of the region. In addition to being steeply sloped in some places, much of the Flint Hills uplands has a layer of cherty limestone near the surface. The shallow and rocky soils precluded plowing by early European American settlers and their successors. Since the significant arrival of cattle in the 1860s and 1870s, two key range management practices (burning and grazing) have helped to maintain the structure and function of the tallgrass ecosystem. Yet the land-use regimes have changed since European American arrival, and the human signature on the land is by no means a static one. This chapter documents the salient landscape transitions in the region during the past 150 years, and how these transitions have coevolved with changes in the socioeconomic context.

We examine changes in the political economy of the region from early settlement to the present, and link these transitions with alterations in the agrarian landscape.

Those changes include (1) the transition from Native American to European American land-use patterns, (2) the expansion of the agricultural economy during the late 19th and early 20th centuries, (3) drought and depression during the inter-war years, (4) agricultural intensification of the post–World War II period, and (5) a set of current issues we discuss under the rubric of conservation, including urban edge development, fire suppression, soil erosion, invasive species, and the current institutional context within which contemporary land-use decisions are made.

Our approach begins with the recognition that ecosystems and human social systems have conventionally been conceptualized separately from each other, even though there has long been recognition of their interconnectedness. Humans are increasingly seen as being part of virtually all ecosystems—in most cases for a very long time—and thus it is now less tenable to attempt to conceptualize them separately. Indeed, there is increasing evidence that the tallgrass prairie of the Flint Hills was long coevolving with those peoples who had migrated over the Bering Strait and southward into the Great Plains (Reichman, 1987). Conventional accounts have at times either portrayed these lands as generally underutilized by humans before European arrival, or alternatively as kept in equilibrium by "ecologically minded" Native Americans. Viewing this environment as a social ecosystem allows for the recognition that Native Americans and European Americans brought different experiences, worldviews, and land-use practices, and therefore altered their environment in varying ways, creating different signatures on the land. At the same time, the biophysical elements of the environment shape society by providing opportunities, resources, and limitations, all of which are perceived differently depending on the mental models of the humans interacting with those elements.

Scholars have conceptualized the study of society and environment in various ways. Worster (2003) has emphasized the cultural element of the human–environment equation, arguing, for example, that it was not a process of humans learning to adapt to the environment but rather the "economic culture" of entrepreneurialism and opportunism that drove landowners to plow up millions of acres of grasslands in the 1910s and 1920s to sow in wheat. Norgaard (1994) framed societies and environments as coevolving systems in which the coevolutionary process involves changing relations between components of those systems (e.g., values, organization, technology, knowledge, environment), and in which the systems themselves are in a constant process of change.

Scholars who have studied various aspects of the Flint Hills and the surrounding area have of course brought their varying perspectives. For example, Malin (1942) and Wibking (1963) drew on the notion of human adaptation to the biophysical environment. For Malin (1942), the human–environment dialogue was one in which newcomers to an ecosystem (e.g., settlers) went through an early exploitative stage during which they "experimented" with the environment, caused initial ecological disturbance, but then moved into a less destructive stage as their knowledge, tools, and practices "improved" with experience. Their engagement in the economy would guide progress toward geographical adaptation. Kollmorgen (1969) argued that a variety of geographical misconceptions on the part of "American woodsmen" of the frontier era led them to attempt to impose a small-plot, grain-cropping system on western grasslands—as opposed to extensive ranching—leading to destruction of forage.

Wibking's (1963) geography of the cattle industry in the Flint Hills is manifestly a story of an industry adapting to optimize its relationship with the environment. Similarly, Wood (1980), writing from the perspective of the beef industry, described the history of cattle in this region as a linear, uninterrupted march of progress, in which shrewd, pioneering risk takers invested in pure-bred cattle, made their fortunes, and brought the environment under their financial domain. These two narratives are accounts of rational humans using reason to "adapt" to their environments while overcoming any biophysical barriers or environmental damage. Worster (2003), on the other hand, saw agricultural capitalism and the culture of entrepreneurialism as the force driving environmental maladaption in the Great Plains.

In this chapter we approach the human–environment relationship as an ecological dialogue that includes both biophysical and social elements in constant interaction with each other. They mutually shape each other—in a sense, coevolving—although as Worster (2003) suggests, coevolution is not a natural, apolitical process; rather, it is driven by human interests as much as biophysical factors. Two of the salient themes in this narrative are legacy and resilience. The role of legacy in this case study involves transition, which incorporates contradictory elements. Big bluestem and other tallgrasses and native grasses are the central legacy of the Flint Hills uplands. They are central not only to the prairie ecosystem of the region but also to the agrarian and other human systems that have developed there. The grasses were the main source of food for native ungulates (primarily *Bos bison*), which in turn were the primary protein source of the hunting and gathering societies of Native American tribes living in the region. Similarly, the tallgrass prairie proved to be a fertile ecological base for the ranching systems established by the European American settlers who displaced the Native Americans.

The social legacies of the agrarian and other human systems established in the Flint Hills have had complex relationships with the region's tallgrass prairie ecosystem. Of particular concern is the extent to which human practices have sustained or threatened the viability of the region's ecosystem. From this perspective, most human systems established in the Flint Hills have had contradictory relationships with its tallgrass ecosystem. For example, although the livelihood of Native American tribes in the region depended on sustaining the bison, Sherow (1992) contends that High Plains Indians had severely depleted the bison herds prior to the massive hunts by European Americans, thus playing a role in the extirpation of bison. On the other hand, European American ranchers, who adopted sustainable practices such as spring burnings that benefit the tallgrass ecosystem, also adopted agricultural practices such as double stocking that, if not done prudently, threatened the viability of the ecosystem.

Humans' contradictory relationships with the tallgrass ecosystem developed in the context of the region's incorporation into the world system. The introduction of horses into the High Plains by European colonists helped make Native Americans more efficient hunters. More important, Native Americans began hunting beyond their subsistence needs as they became increasingly dependent on trade of bison products to European Americans (Sherow, 1992). Similarly, the beef cattle industry developed by European Americans was dependent on urban markets in Chicago and elsewhere (Cronon, 1992). Because their income was largely dependent on the number of cattle they could bring to the market, ranchers adopted double stocking and other practices designed to maximize the number of cattle produced from the tallgrass ecosystem. Research on the long-term ecological impact of overgrazing suggests, however, that this practice reduces biological diversity in the tallgrass ecosystem (Hoch, 2000).

More recently, there has been a trend of urban residents moving into exurban and rural areas for a more "natural" living experience (Hoch, 2000). This urban–rural migration, however, ends up fragmenting prairie by suppressing the application of fire, and eventually leads to a transformation of the landscape from one defined by grass species to one dominated by woody species (Briggs et al., 2002). A common theme throughout this chapter is the idea of legacy as an enduring quality to maintain the features of the tallgrass prairie. Conflict arises when this legacy is challenged by elements that stand in contradiction to the goals of landscape conservation.

Another theme that emerges in this case study is the resiliency of the prairie. At various points during the past 150 years, this social ecosystem has exhibited remarkable resilience during episodes of both drought and overgrazing. Moreover, at times when portions of the tallgrass prairie had been assumed dead, it proved to have recuperative abilities when precipitation levels increased and/or when grazing pressure was reduced. This resilience of the bluestem pastures, as part of a social ecosystem, also has implications for stability in some social patterns. We recognize that it is also a social landscape that has long been coevolving with human systems.

Description of the Region

Study Area

The local study area is the Flint Hills region in Kansas. The Flint Hills contain the largest remaining contiguous tract of unplowed tallgrass prairie (Knapp and Seastedt, 1998). Within the study area in southern Riley County is Konza Prairie, a 3,487-ha native tallgrass prairie preserve owned by TNC and Kansas State University (KSU). Figure 6.1 shows the counties that compose this study and the major cities that surround the area.

The Flint Hills encompass more than 50,000 km^2 and the range of hills is about 32 km. wide. The upland terrain is relatively steep sloped and overlain by shallow, chert-bearing limestone soils unsuitable for cultivation. The maximum local elevation is about 457 m. The largest flow of water in the region is the Kansas River in the north, with several other rivers draining the uplands. The larger of these river systems contain relatively flat and fertile bottomland that generally lacks shallow limestone on the surface.

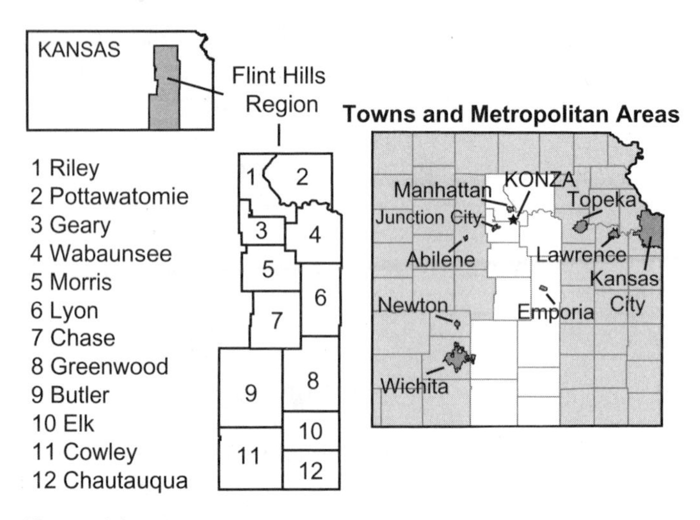

Figure 6.1 The geographical descriptions of the Flint Hills study area. Map components are based on data from the Kansas Geospatial Community Commons, Kansas Applied Remote Sensing (KARS) Program (1993).

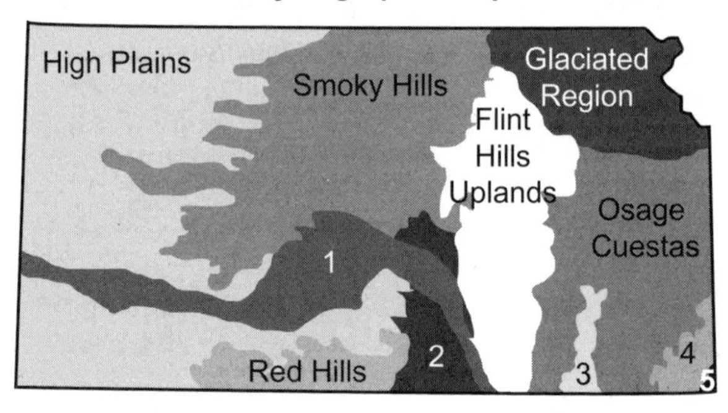

Figure 6.2 This map shows the general physiographic regions in Kansas. Based on data from www.kgs.ku.edu/Physio/physio.html.

Topographically, the Flint Hills differ from the adjacent physiographic regions (Fig. 6.2). The Flint Hills stand out in the landscape for their stream-dissected hills, sharp reliefs and escarpments, and bench-and-slope topography. Although this topography was sculpted to a substantial degree by fluvial processes over a long, geological timescale, this is not to say that the modern landscape is static. Oviatt (1998) explains that the superficial erosional processes in the Flint Hills are dynamic, with the rates and extent of soil erosion depending on climate, fire, and grazing regimes, as well as the timescale of interest. On the shortest timescales (decades to years), the main issue is soil erosion and fill in stream channels and steeply sloped hillsides.

Land Cover: Climate, Fire, and Grazing

> There were wild flowers, hundreds of kinds of wild flowers, blooming in their place and season. There were elk and shaggy bison, and prairie chickens booming out their mating call on brisk April mornings. Great trees hugged the stream channels and floated like islands on distant horizons. And there was grass in abundance, dozens of kinds of grass. Eight feet tall on favored sites, belt high in most places, it was green and bronze and wine and gold, rippling and shining in the sunlight. It's almost gone now, that shining, swirling landscape. Other prairie survives.... But the tallgrass prairie, the king of prairies, became the corn belt. Became Chicago, became Des Moines, became home for 25 million people. As the homesteaders' steel plows sliced through its matted roots, it all but vanished in a ringing, tearing sound. (Farney, 1980, p. 38)

The predominant land cover of the Flint Hills uplands is tallgrass prairie (Küchler, 1974), which once dominated the U.S. Midwest (Fig. 6.3). During the 19th century Europeans found the soil beneath the grasslands extremely fertile for agricultural production. Currently, less than 1% of native tallgrass remains—approximately 1.6 million ha—most of which is in the Flint Hills (Knapp and Seastedt, 1998). Cropland is primarily found in the bottomlands near the rivers. Where the hill slopes are dissected by streams, in upland riparian areas, woody species such as red cedar are found in some abundance. Various tree species are also common as windbreaks for agricultural crops. The flatter bottomlands of the larger streams and rivers have mostly been plowed for row crop production. The native bluestem pastures have survived in the Flint Hills because of the inability of early settlers to plow the sod of the uplands. Limestone at or near the surface precluded cultivation and played a key role in the social and ecological legacy of the region.

There are two main varieties of native bluestem: little bluestem (*Andropogon scoparius*) and big bluestem (*A. furcatus*). The latter can reach a height of more

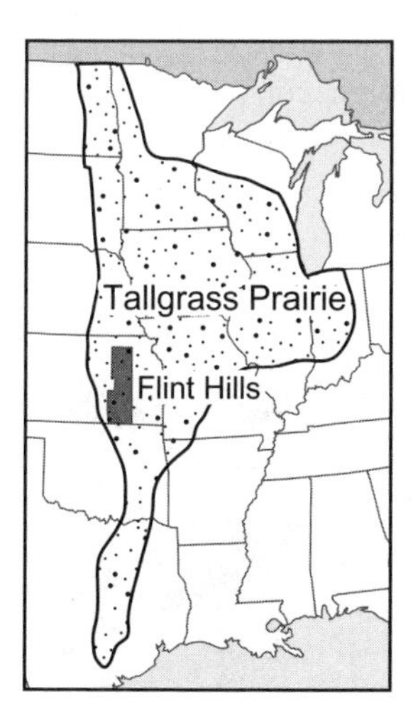

Figure 6.3 This map shows the historical extent of the tallgrass prairie across the United States. Based on data from the Konza website (www.konza.ksu.edu).

than 3 m in the most productive years, but as a result of fire and grazing regimes, this is uncommon (Reichman, 1987). The growing season for the bluestem is approximately 180 days, from late April to late October. Peak growth is during May and June, after which the grasses begin to die off in early July, when precipitation decreases and temperatures increase.

The bluestem is the predominant species in the area for a number of reasons (Reichman, 1987). First, the limestone surface of the area allows for the percolation of water, which the roots of the bluestem can reach, often more than 3.7 m below the surface. Second, the aboveground biomass from previous seasons decomposes to provide nutrients for growth during subsequent seasons. Likewise, the storage of rhizomes under the surface helps to stimulate new growth. Third, the bluestem grows in high density and towers over other varieties of grass, allowing it to crowd out competing species. Finally, there are three key processes that regulate and sustain the tallgrass prairie: variable continental climate, periodic fire, and ungulate grazing. Because these elements are fundamental to the structure and function of the tallgrass ecosystem, let us briefly consider each of them in turn.

Climate

The temperate, midcontinental Flint Hills of eastern Kansas receive greater precipitation than places farther west as a result of air masses bringing moisture from the Gulf of Mexico. Figure 6.4 provides the average amounts of precipitation for the state of Kansas [data drawn in part from Konza LTER (2004)]. At Konza, precipitation totals exhibit considerable temporal variability, with an average annual precipitation of 835 mm, of which about 75% occurs during the growing season (about 180 days). Intense thunderstorms during the spring and early summer can bring a multitude of meteorological phenomena, including lightning, hail, heavy rains, strong winds, and tornadoes. During dry months, precipitation can be less than 25 mm, whereas at the other extreme, intense rainfall of 100 mm or more per hour often floods the bottomlands.

The yearly mean temperature is 13 °C, but throughout the year temperatures can vary from –9 to 27 °C. When studying these climate patterns, it is important to understand that certain biomes persist essentially as a reflection of the general climatic conditions, or at least it can be argued that the climate is a strong determinant of biome-scale patterns of vegetation (Hayden, 1998).

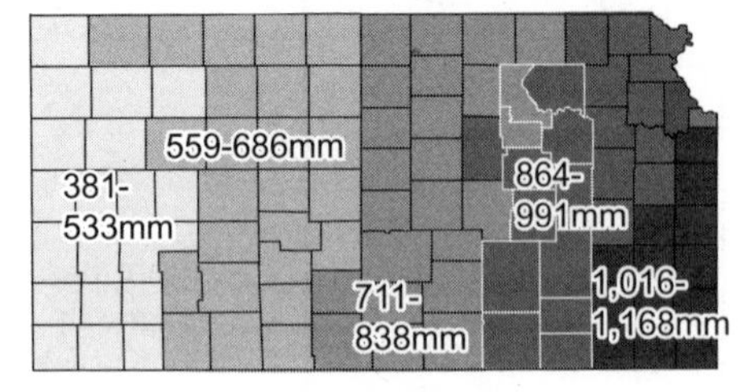

Figure 6.4 Average precipitation for Kansas in millimeters. Based on data from the High Plains Regional Climate Center (www.hprcc.unl.edu/index.html).

Fire

Fire is an essential element in the legacy of the tallgrass ecosystem, and Native Americans were the first to burn the prairie regularly for many reasons, one being that they found that burned areas grew much greener and attracted buffalo (Aldous, 1934; Pyne, 1984; Sauer, 1975; Unrau, 1971). Whether through lightning strikes or planned burning regimes, fire eliminates debris and competitor species (Pyne, 1984).

Without roads or other boundaries for containment, burning often got out of control, causing problems for nonpracticing neighbors. Over time, though, widespread regular burning was adopted by nearly all ranchers in the Flint Hills. Eventually, the practice was supported by KSU research, although it remained controversial (Aldous, 1934). A continuing legacy of prescribed burning for the management of the tallgrass prairie is essential. Among the reported benefits of prescribed burning are improvements in grazing distributions, reduction in litter, the recycling of nutrients, and the control of woody species (Ohlenbusch and Hartnett, 2001). Indeed, the elimination of woody species is a major goal of prescribed burning (Owensby et al., 1973).

Grazing

Grazing by ungulates is the third essential element in the tallgrass prairie. Before the near extirpation of the plains buffalo (*B. bison*) during the 19th century, that species played a key role in the Great Plains (Knapp et al., 1999). With a population numbering into the tens of millions, the plains buffalo were a major component of sustenance for most Plains Indians, including the Kansa tribe indigenous to the study area (Unrau, 1971). During the European American settlement period and extension of railroads, the bison population was reduced to mere thousands, and in their place domesticated cattle were brought in, which adapted well to the prairie grasses (Malin, 1942).

At Konza Prairie, watersheds are divided into experimental plots for the purpose of investigation on two levels: burn frequency and grazing. Bison were reintroduced in 1987 under low management conditions to understand better their role in the ecosystem. Since that time, researchers have found that interactions between fire and grazing by bison are important elements in the composition and spatial heterogeneity of vegetation in the tallgrass prairie (Vinton et al., 1993). Researchers discovered that compared with ungrazed areas, plots grazed by bison have a reduction in the dominance of C_4 (warm season) grasses, which allows for more subdominant C_3 (cool season) shrubs and forbs, resulting in richer diversity and community heterogeneity (Knapp et al., 1999).

Although grazing was naturally accomplished through the life cycle of bison and other animals like antelope, very few buffalo and antelope remain on the prairie today, having been replaced by domestic cattle. Since European American settlement of the Flint Hills, owners and operators of the region have replicated the natural disturbance elements of the prairie through agricultural practices (burning and grazing) geared toward cattle ranching.

The three essential natural elements of the tallgrass prairie—climate, fire, and grazing—protect the grassland by eliminating the threat of succession to a landscape dominated by wooded species. Both Native Americans and European Americans adapted practices involving fire and grazing, thus engaging in close interactions and processes that simultaneously shaped both their societies and the ecosystem.

Native American Period: Land-Use Patterns of the Kansa Indians

Historians believe the Kansa (a.k.a., the Konza or Kaw) traditionally identified themselves as "people of the Southwind." French explorers Marquette and Joliet were the first Europeans to make reference to the Kansa (in 1673), placing them in an area west of the Mississippi River (Unrau, 1991). Sharing a Dhegiha Siouan heritage with the Osage, Quapaw, Omaha, and Pawnee, the Kansa had migrated from possibly as far east as present-day Virginia and North Carolina. Their first known settlement in the Flint Hills region was in the northeastern corner of what is now the state of Kansas, near the mouth of the Kansas River.

The Kansa were a seminomadic tribe that raised crops in the bottomlands. Inhabiting the forest fringe, the Kansa subsisted on a mix of food crops in combination with animal protein obtained from hunting the prairies to the west. Among other crops, the Kansa planted corn, squash, and beans in small plots. The most important animal protein in their diet came from bison meat, which was supplemented by elk and deer. A documented expedition to the Kansa camp in 1819 found the tribe's diet consisted heavily of corn products and soup with beans seasoned with buffalo meat (Parrish, 1956).

The bison hunt was a tradition undertaken twice a year in Kansa society. During this time, the tribe would leave their permanent settlements on the river bottomlands to traverse the upland prairies to the west in search of bison herds. The Kansa hunted buffalo throughout the tallgrass of the Flint Hills for more than 200 years until their forced removal in 1873. The "Thunder People," a special group of Kansa, were granted permission to burn the prairie for ceremonial purposes (Unrau, 1971). The burns had two purposes in the context of Kansa culture: a call for rain and a way to attract bison to the succulent grasses that emerged after a burn. Thus, the Kansa's reasons for burning may have been framed in terms of culture and sustenance; in effect, the practice helped to preserve the tallgrass ecosystem by perpetuating two key elements of that system—fire and grazing.

Unfortunately, the Kansa endured a troubled existence in the region before they were eventually forced out by European Americans. Unrau (1971, p. 25) succinctly summarizes their experience:

> [T]he Kansas were forced to live and exist mainly as a survival culture. Starting in the seventeenth century and continuing for nearly two hundred years, they encountered a superior number of alien people and difficult conditions that militated against any natural increase in population. A refinement of their traditional way of life was virtually impossible. By the time they had experienced their final forced

> removal to Indian Territory in 1873, the Kansas had been reduced to less than half their original number; meanwhile, their culture had been so radically modified as to be almost unrecognizable.

Their troubles stemmed from encounters with other Native American tribes as well as Europeans. The Osage had claimed lands near the southern boundaries of the Flint Hills, thus putting them in potential competition with the Kansa to their north. The Osage were more numerous than the Kansa during the 18th century, totaling around 6,500 people in 1750 (Burns, 1989; Unrau, 1991). In contrast, the Kansa numbered fewer than 5,000 and experienced a precipitous decline throughout the century, totaling only 1,500 by 1806 (Unrau, 1971). One reason for the decline in the Kansa population was its prolonged conflicts with the Osage in the 1790s. Fear of the more numerous Osage to the south and east, and European encroachments from the north, prompted the Kansa to abandon their village at the mouth of the Kansas River and move west to where the Big Blue River joins the Kansas River—just east of present-day Manhattan, Kansas. Establishment of the Blue Earth village placed the Kansa in closer proximity to another enemy, the Pawnee, resulting in additional conflict and another move to near present-day Topeka. The intertribal conflict between the Kansa and Pawnee was fueled by their competing claims for the land and the bison that grazed on it.

The French were the first Europeans to come into contact with the Kansa (and Osage). The French set out to befriend the two tribes by opening trade with them. The Native Americans provided animal pelts and rival Native American slaves, which the French desired, whereas the Native Americans received guns, ammunition, and alcohol—not to mention cholera and smallpox, which killed hundreds of Kansa in outbreaks in 1839 and 1849 (Unrau, 1971). The ravaging effects of disease contributed greatly to the decline of the Kansa and Osage populations, as was true for most Native American tribes (Snipp, 1989). Contacts with Europeans increased exponentially after the Louisiana Purchase, whereby the French ceded control over the land occupied by the Kansa, Osage, and Pawnee to the United States. Soon, a growing number of European Americans traveled through the Flint Hills on their way to trade in Santa Fe.

In 1825, the U.S. government negotiated a treaty with the Kansa, whereby the tribe relinquished claim to all lands in Missouri and some land in Kansas. In addition, the U.S. government sought to change Kansa culture by discouraging hunting and trapping while promoting agriculture, in the hopes of establishing a more sedentary lifestyle and a European American land-use pattern among the tribe's members. Along with several hundred cattle, the tribe was given hogs, poultry, and agricultural implements (Parrish, 1956). To assist in their transformation, a blacksmith and agriculturalist were commissioned to live among them, and formal education was offered to their children. The Indian Agency was established and Daniel Morgan Boone, son of the frontiersman, became the government farmer for the Kansa tribe (Petrowsky, 1968). Religious missionaries also became involved, seeking to convert the Kansa to Christianity. These attempts to induce cultural change among the Kansa were largely unsuccessful (Unrau, 1971). The Kansa and Pawnee consequently began to hunt beyond their subsistence needs to have furs they could trade for European commodities.

During this period, the U.S. government renegotiated its treaty with the Kansa. The result was the Mission Creek Treaty of 1846, which included an agreement of the Kansa tribal leaders to move all their (52 km^2) settlements to a reservation near the newly formed village of Council Grove. They also relinquished the tribe's rights to the two million acres they had been granted in 1825, receiving a payment of approximately 10 cents an acre (Unrau, 1971). The new treaty stated the Kansa tribe would have use of the Council Grove site forever, despite its placement on a strategic juncture of the Santa Fe Trail. In this case, "forever" lasted less than 30 years, with the Kansa forcibly moved to Indian Territory in present-day Oklahoma in 1873.

A key reason for the short-lived promise of the U.S. government was the Kansas/Nebraska Act of 1854, which established Kansas as a U.S. territory, opening its land to claims by U.S. citizens. Although the land made available to them was not supposed to include Indian reservation land, squatters quickly moved in to establish residence on land purportedly reserved for Native American tribes. A census taken in 1855 found 30 European American families had established illegal claims on the Kansa reservation. The next 5 years would see the population of the Kansas Territory grow from 8,500 to almost 100,000 (Unrau, 1971). This massive in-migration was driven in part by the slavery issue, as both proponents and opponents of slavery moved to the Kansas Territory to influence its political position on the slavery issue (Miner, 2002). The state of Kansas was officially recognized as a free state on January 29, 1861—just months before the beginning of the Civil War. It also harkened the beginning of the end for the Kansa Indians in the state that bears their name.

In sum, the Kansa and other Native Americans of the region did indeed leave a signature on the landscape. Their culture was local resource dependent, and their practices were adaptive. Perhaps most important, the practice of burning the prairie was embedded in their culture. This practice meant that part of the Kansa legacy in the Flint Hills was to perpetuate the structure and function of the tallgrass prairie, in part by mitigating against succession to woody species. Moreover, their relatively small population—declining after European contact—suggests two important things: (1) that their cropping systems were not extensive, and (2) that although bison provided a major dietary source of protein, the small size of the Kansa population relative to the reported size of buffalo herds meant that their bison harvest was probably sustainable, holding other factors constant. Moreover, the fact that they were seminomadic meant that their impact on the land was spatially and temporally dispersed. This would change with European contact, which brought pressure on the Kansa to end their seminomadic lifestyle and to overharvest bison as their dependence on trade with Europeans increased.

European American Settlement to 1900

Key Legislation

Social policy was key in laying the foundation—and to some extent driving—the agricultural development of the Flint Hills during the second half of the 19th

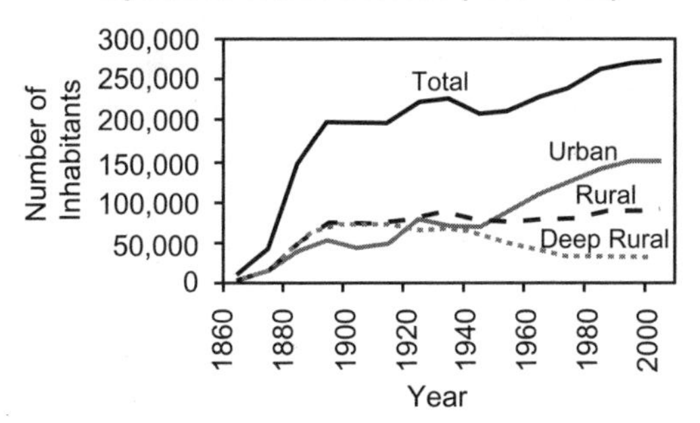

Figure 6.5 Population in the Flint Hills (1860–2000). The following counties are urban: Butler, Geary, and Riley; rural: Cowley, Lyon, and Pottawatomie; and deep rural: Chase, Chautuaqua, Elk, Greenwood, Morris, and Wabaunsee. Based on data from U.S. Census.

century. Three bills passed the U.S. Congress in 1862, each contributing to the agricultural transformation of the newly incorporated frontier. The Homestead Act provided 160 acres of land to private individuals, allowing them to secure title to the land after 5 consecutive years of improvement. The Homestead Act was intended to encourage the population of the plains, which it did. After 1862, the number of settlers in the region grew rapidly.

Also in 1862, the Pacific Railroad Act allotted large amounts of land to railroad companies for infrastructure development. The Atchison, Topeka and Santa Fe Railway (ATSF) rail lines that would eventually run throughout the Flint Hills were important developments in enabling the movement of cattle through and from the region. Greater exposure to the region through expanding rail service during the period from 1870 to 1880 precipitated a threefold increase in population (Fig. 6.5), and a more than fourfold increase in the number of farms in the region.

Finally, the legislative foundation of the land grant college complex was formed by the Morrill Acts of 1862 and 1890, the Hatch Act of 1887, and the Smith-Lever Act of 1914, which addressed teaching, research, and extension, respectively. During the mid 19th century, there was a significant demand among the public as well as educators for government-supported colleges that would serve the needs of the "common man" (Kerr, 1987). These "people's universities" would be practical and democratic, and would offer a vocational education revolving around the needs of the rural population. When settlers arrived, they would be provided institutional support to be productive on the land. Kansas State Agricultural College was established in Manhattan, Kansas, in 1863 as the state's land grant institution. Later the Smith-Lever Act would provide funding for county extension agents who were charged with transferring science-based information to agriculturalists and their families.

General Farming Period (1854–1880)

Following Wibking (1963), the historical development of the Flint Hills can be divided into the period of Native American influence (pre-1855, discussed earlier), the General Farming period (1854–1880), and a post-1880s ranching period, each of which is characterized by a different land utilization emphasis. The General Farming period began just after Kansas was organized into a territory in 1854, when settlers (from areas such as Ohio and Indiana) began to move in rapidly, typically settling in the bottomlands along rivers first, then in the uplands, which were purchased from railroads. The 1860 U.S. population census shows that 8 of the 12 counties in the study region were populated in that year. The total population at that time was 10,707, and the number of farms was 1,448 (U.S. Census, 1860). These new settlers practiced "Ohio-style" farming, reproducing agricultural techniques on lands similar to those with which they were familiar. This meant cropping corn in the bottomlands and small grains in the uplands, as well as some hay for cattle. Additionally, many had some livestock such as cattle, hogs, and sheep. According to Wibking (1963), this was a period during which settlers "experimented" with the land to ascertain what crops, tools, and practices would prove successful.

A new wave of immigrants settled the area after 1870, some establishing operations in the uplands, where land was more plentiful. However, cropping small grains in the uplands proved to be difficult. Isern (1985) argues that upland farmers succeeded, not from their attempts at cash cropping, but through their limited grazing and social ties with bottomland farmers. The passage of herd laws in some Flint Hills counties banned transitory cattle. One of the challenges of the time was keeping free-range livestock from destroying fields and gardens. Prior to the introduction of barbed wire, farmers relied on board or stone fences, or had to maintain hedges. These options were either time-consuming or costly to establish and maintain. Growing adequate hedges could take several years, and timber in the uplands for fencing was scarce.

After herd laws were enacted, there was a perception that the uplands would always be publicly available for cattle. One example is an advertisement for land in Chase County from 1874:

> For young men with small means and willing hands, Chase County is paradise. Here you can have the great unsettled commons for stock to range upon that costs you nothing except a little looking after. Why not come to it, own a few stock and help graze it down, and thus make wealth out of that which is now, every fall, committed to the destructive element known as prairie fire. (Miner, 2002, p. 139)

Two points worthy of note in this text are the notion of converting grass to wealth (indicative of worldview) and the assumption of fire as destructive to the prairie. These free-range areas would disappear within 10 years, after barbed wire made fencing pastures cheap and efficient. By 1880, general farming had declined significantly in the region and began to give way to large-scale ranching. One explanation of the decline in significance of the (bottomland and upland) small farm was that many small farmers did not want or were not able to acquire large tracts of upland area. The uplands were not viewed as important because of their lower value for cropping, and because it was often assumed they would remain open range. Those with the capital were able to accumulate large amounts of pasture grass through their ability to acquire land from the railroads. After barbed wire became available, the free range was fenced in, and small farmers had neither room to expand nor access to grazing land (Wibking, 1963).

The final blow for small farmers were droughts in the 1880s, which forced many of them to sell to expanding ranchers, who were accumulating land in bottomlands and uplands, and combining them into larger holdings (Wibking, 1963). Yet, despite the decline in small farms, during the next 10 years the region as a whole grew in population nearly fourfold. By 1870, the number of inhabitants had grown to 43,918, and the number of farms to 4,941. The 1880 census recorded 147,569 people and 19,911 farms. Thus, despite the transition away from general farming and toward ranching as the predominant land-use pattern, the region as a whole was growing rapidly (Fig. 6.5). Moreover, it is clear that the numbers of people and farms were now well beyond the old levels of the Kansa Indians. This means the landscape was undergoing change as more bottomland was broken out for crops, and upland prairie quickly moved toward private enclosure patterned according to the rectangular, checkerboard imprint of the Homestead Act and delineated by the paths of barbed wire fencing.

Loss of the Buffalo and the Proliferation of Cattle

During this same period, the slaughter of migrating buffalo herds in the plains of Kansas and other prairie states was well under way. Although the conventional history of buffalo decline tends to emphasize railroad expansion, profit taking, and sport shooting, Sherow (1992) argues that Native Americans by this time were hunting well beyond their sustenance requirements and that the plains buffalo herds were already in significant decline. Certainly, the hunting expeditions led by the railroads to make way for oncoming rail, cattle, farms, and towns accelerated the decline of the species. Killing the buffalo proved financially successful for railroads. Buffalo skeletons were converted into fertilizer (100 skeletons/ton fertilizer), which was sold to farmers (Bryant, 1974). Between 1872 and 1874, the ATSF profited from shipping "459,453 buffalo hides, 2,250,400 pounds of buffalo meat and over 10 million pounds of bones" (Bryant, 1974, p. 33). Clearly, the loss of buffalo was a blow to the Plains Indians. Tribes lost a means of sustenance and the primary natural capital that sustained their culture. The Kansa made their last buffalo hunt in 1873, the same year they accepted a 100,000-acre reservation in north-central Indian Territory, in present-day northeastern Oklahoma (Unrau, 1971).

The precipitous decline of buffalo herds and the concomitant increase of available transportation networks allowed for a rapid transformation of the region toward agriculture focused on the cattle industry. Growth of the Kansas City cattle market accelerated in 1855, when the Missouri legislature enacted a ban on Texas cattle during the summer months. Ranchers in that state were concerned about Texas fever, a disease spread by ticks carried on Texas cattle. The longhorn breed was immune, but domesticated English stock was extremely susceptible to the deadly disease. Thousands of cattle died in Missouri, and within a few years Kansas farmers began losing thousands of head as well. The first Kansas quarantine law was enacted in 1859, to enforce a ban on Texas, Arkansas, and Native American stock between June and November. The quarantine laws and onset of the Civil War brought a temporary end to the cattle drives and, by the end of the war, millions of wild longhorns were roaming the open Texas landscape (Gard, 1954). During the following years, Texas cowboys rounded up their herds and were ready to drive them to market. The question was the route by which they would drive cattle to market to avoid opposition. A revision to the Kansas quarantine laws in 1867 provided a loophole. Under certain circumstances that guaranteed reimbursement for death of local cattle, Texas drovers were allowed to enter Kansas from the southwest. The Union Pacific Railroad had built west of St. Louis and Kansas City, which made the drive north shorter and hopefully more profitable. Malin (1942) notes that although quarantine laws prevented Texas cattle from traveling through the Flint Hills, drovers continually ignored the laws.

In 1871, more than 600,000 cattle arrived in Abilene over land, and many of them were then shipped by rail to the east (Bryant, 1974). The increase in the number of cattle driven north was accelerated by market prices in the east. Texas cattle were worth three times more in Chicago and St. Louis, and 5 to 10 times more in New York, than in Texas (Gard, 1954). The northward drives were made possible by changes in the quarantine laws, westward expansion of rail service, and the realization of the local economic growth precipitated by the cattle trade.

The route commonly traversed by longhorns and cowboys north through Indian Territory and into north-central Kansas was known as the *Chisholm Trail.* Inside Kansas, the trail headed north on the west side of the quarantine line. The Flint Hills region was east of this area, but nonetheless saw a flurry of activity associated with the trail. Destinations for cattle varied, but included stock raising, finishing facilities, and slaughterhouses. Nor were all cattle that arrived immediately sold. The ticks that spread Texas fever were not a factor in the winter, and because quarantine laws did not apply between November and April, some cattle were wintered on local pastures, which included Flint Hills bluestem. This wintering process continued in the Flint Hills during the following years, despite the end of Abilene as a cattle nexus in 1872.

The Texas cattle drives established a culture wherein the Flint Hills were recognized as an important region for the grazing of transient cattle. The upland bluestem pastures were recognized for their potential as grazing lands because of their high-nutrient grasses and resiliency. Cattle were driven to the Flint Hills for finishing. The combination of rich upland grasses, fertile bottomlands for grains, sufficient rain, and railroads provided the key elements for the development of an agricultural economy in the region that was geared to cattle production.

Barbed Wire and Enclosure: The Establishment of the Predominant Land-Use Patterns

A major transition period in the agrarian landscape of the Flint Hills was the rapid enclosure of the upland tallgrass prairie. Using barbed wire, the grasslands were transformed from a common grazing area to large tracts of privately held, fenced land within a span of a few years. This late-19th-century transference of public grazing land to private owners in the Flint Hills signaled a growing distinction along the lines of geography, land use, and ownership. The availability of cheap and low-maintenance barbed wire fencing precluded the need for herd laws and common areas, and allowed for the settlement and eventual grazing of upland prairie pasture.

The growth of the cattle industry in the Flint Hills after the enclosure movement took two general directions. The first was the continuance of the transitory cattle system, which now relied on rail for cattle in-shipments, rather than overland trail routes. The Flint Hills served as the final finishing grounds for cattle, which were fattened on the protein-rich native grasses.

The second development was the breeding of fine stock year-round—purebreds, which were brought in by immigrant settlers (Wood, 1980). Cattle inventories (other than milk cows) for the Flint Hills after 1860 show a remarkable increase from just 5,781 in 1860 to more than 545,673 in the 1890 census. The number of farms in each county illustrates the heterogeneity among the 12 counties in relation to cattle inventories and geographical characteristics. For instance, Cowley County in the southern Flint Hills has the least amount of upland grass (6%) according to government surveys, and correspondingly the lowest ratio of cattle to farms in 1890 (14:1). On the other hand, Chase County leads the region in upland grass (88%) and had an average cattle-to-farm ratio of 66:1 in 1890. This example illustrates the land-use distinctions that arose from Flint Hills geography. The large tracts of unspoiled upland prairie turned into pastures for a growing cattle business.

During all this rapid development, there was little chance for the upland farmer who faced adversity because of the local climate, geography, and the increasing price of land. Many small farmers who believed grazing lands would always be available, or those unable to buy pastureland, either "made individual accommodations" (Isern, 1985, p. 264) or were "forced to sell out" (Malin, 1942, p. 12). The new settlers succeeded with cattle on the uplands, and the end result was a clear distinction: The bottomlands were extensively used for cropping, whereas the uplands were prairie pastureland. The terms of ownership also developed along geographical distinctions: The bottomlands were typically smaller and owner operated, whereas the uplands were distinguished by large urban-owned lease arrangements.

Settlement to 1900: An Overview

From the European American settler's point of view, the latter decades of the 19th century were generally a boom period for the Flint Hills. The human population increased dramatically, growing from about 11,000 in 1860 to nearly 200,000 in 1900. Throughout most of this period, both urban and rural populations were growing (Fig. 6.5). Similarly, the number of farms in the region increased dramatically, from about 1,500 farms in 1860 to more than 25,000 farms in 1900. Accordingly, the acres of land in farms expanded from about 250,000 acres in 1860 to more than 6.2 million acres in 1900. And, as would be expected from our discussion of the decline of the small farmer and the establishment of larger ranching operations, the average farm size also increased steadily, from an average of about 158 acres in 1880 to 256 acres in 1900.

The expansion of corn acreage can be attributed to the growth of the population and number of farms planting corn for human consumption, but corn was also

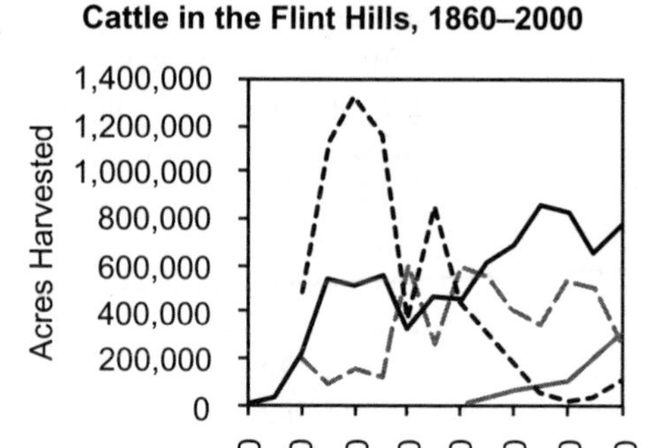

Figure 6.6 Selected crops (in acres harvested) and inventory of nontransient cattle in the Flint Hills, 1860–2000. Based on data from U.S. Census and U.S. Department of Agriculture. Note that the 1900 data is from the Kansas Board of Agriculture, and the 1960 data is from the 1959 census.

grown extensively as feed for cattle during the winter months. The number of cattle in the region was also growing rapidly, from about 6,000 in 1860 to more than half a million head in 1900 (Fig. 6.6). Thus, the area experienced a decline in general farming; an increase in farming focused on corn for cattle feed; the raising of fine, registered stock (especially among the larger ranchers); and the grazing of transient cattle during the summer months (the predominant land-use pattern in the central Flint Hills).

By 1900, the Flint Hills had become best known for its nutrient-rich grasses, and the grazing of transient cattle was established as a major business of the region. The transient cattle system was deceptively simple. In April and May, as the bluestem pastures begin to green up, cattle were shipped in and set out on the pastures to graze. The length of the grazing season is determined by the protein levels of the grass, which begin to drop off by July. During the grazing season—in the herds of the late 19th century—cattle would put on 90 to 136 kg on grass alone. The original longhorns from the southwest were less efficient at converting grass to body weight. In the 1890s, ranchers began the process of herd improvement by bringing in purebreds from England and Scotland—shorthorns, Angus, and Herefords. Over time, the herds were bred for more efficient and rapid weight gain, laying on meat faster than unimproved herds, thus saving feeding costs. Purebreds would mature in 3 years at 680 to 770 kg, which was 227 to 272 kg more than their "unimproved" counterparts at the same age (Wood, 1980). This of course meant more profitability for ranchers and the possibility of increasing beef production without necessarily increasing land holdings. As Wood (1980) notes, "this blooded stock had begun to work its magic on the range as more and better beef was being produced in a shorter period of time from fewer cattle and on fewer acres of land" (p. 42). Thus, a rather stable and resilient land-use pattern became established in the Flint Hills—summer grazing of transient cattle plus overwintering on grains and hay produced on the cropland. Cattle raising in

Kansas by the end of the 19th century had largely completed its transition from a frontier institution to a ranching industry.

Golden Age of Ranching, 1900–1920

The first two decades of the 20th century are often regarded as the Golden Age of American Agriculture, and this was generally the case for agriculturalists in the Flint Hills as well. Increased demand for feed grains, forages, and beef pushed prices and land values up, increasing the equity and borrowing capacity of farmers and ranchers in the region. The advent of World War I drove demand further, as exports of wheat and beef to Europe increased. This is evidenced in the rapid expansion of acres devoted to wheat production, which expanded more than threefold: from 167,000 acres harvested in 1900 to about 606,000 acres harvested in 1920 (Fig. 6.6). In 1920, nearly a million acres of wheat and corn were harvested. Increased demand from a growing U.S. population and the World War helped to drive this trend.

Higher crop prices also meant that cattle feed for overwintering had become more expensive. Given that the prices of beef were also on the rise, there was upward pressure on land values, and the price of both renting and purchasing pasture increased. The cost of renting grass (pasture) increased from $1/head for 6 months of grazing to as high as $20/head by 1918, reflecting higher beef prices (Wood, 1980). During the same period, the price to purchase pasture went from about $3.25 an acre to roughly $48 an acre (estimated from Wood, 1980). Yet, pasture was in great demand. Even though per-capita beef consumption in the United States was on the decline during this period (Fig. 6.7), a growing U.S. population, plus increased demand from Europe, meant that overall demand for beef was increasing. Malin (1942) estimated the number of transient cattle grazed in the Flint Hills to be between 213,000 and 319,000 annually in the early 1900s, and somewhat higher during the war years.

The acres of land in farms had expanded rapidly from 1860 (about 255,000 acres) to 1900 (about 6.2 million acres). However, the 1900 figure was the peak

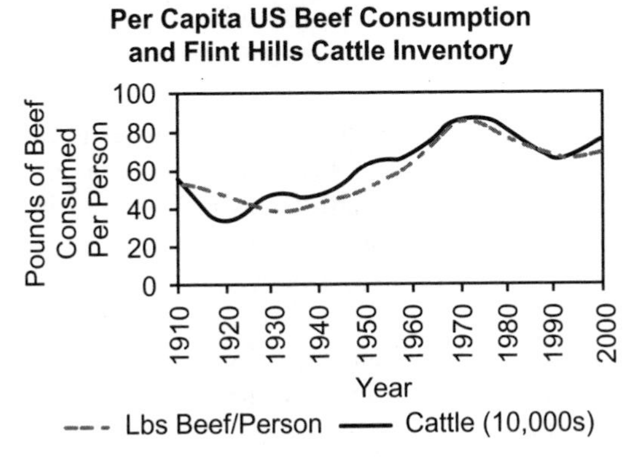

Figure 6.7 Per-capita U.S. beef consumption and Flint Hills cattle inventory. Based on data from U.S. Department of Agriculture and U.S. Department of Agriculture/Economic Research Service, updated 2005.

for land in farms, which then declined by 1920 (to about 5.5 million acres), and remained rather stable through 1950. Average farm size had already begun increasing by 1880, and increased slightly during this period, from about 256 acres in 1900 to 270 acres in 1920.

In sum, the first two decades of the 20th century were relatively prosperous for local agriculture. Less land in farms, increasing prices and costs of production, along with rural population decline all pressed the agricultural industry to push for increases in productivity. Some of this increased production was accomplished through improved feeds (e.g., alfalfa meal), better mix rations, and the continued efforts in breeding to improve beef herds, such that cattle would become more efficient grazers, adding weight more quickly than their predecessors. Land-use patterns remained relatively stable, although some grass and cropland was converted to wheat in response to increasing wheat prices, in part as a result of the demand created by World War I.

Depression and Droughts: The Interwar Years

The story of land use in the Flint Hills during the interwar period is part of the narrative of landscape change in the larger region of the Great Plains. During the 1910s and 1920s in the Great Plains—lands west of the Flint Hills—there occurred a "critical assault on the grasslands" that Worster (2003) and others have called the "Great Plow-up." Combined with prolonged droughts and depression during the 1920s and 1930s, the breaking out of millions of acres of sod helped to produce traumatic ecological and social consequences for the Great Plains.

As mentioned earlier, World War I had created a substantial demand for American wheat as Europe turned to the United States to meet its grain needs. One of the primary means for increasing wheat production was to bring more land under tillage. Between 1914 and 1919, 11,000,000 acres of native grasslands were plowed up in Kansas, Colorado, Nebraska, Oklahoma, and Texas (Worster, 2003). The tilling of this newly broken land continued through the 1920s, as low cattle prices provided further incentive to increase land in wheat and reduce emphasis on cattle, and as farmers became further integrated into the international economy. Thus, when the severe and prolonged droughts of the 1930s hit the Great Plains, it meant that large expanses of open, flat croplands were exposed with little or no vegetative cover, thereby providing the conditions for the dramatic Dust Bowl storms of the 1930s.

The droughts meant that the remaining mixed- and shortgrass prairies to the west could no longer support large numbers of cattle because of their compromised condition. Thus, during the drought years, cattle were pushed into the Flint Hills from western Kansas and other adjacent areas of the Great Plains. Ranchers shipped them into the region to save herds, especially valuable breeding stock (Wood, 1980). Many less valuable drought-stricken cattle were simply sent to market or butchered locally to minimize losses and reduce suffering. Despite the prolonged droughts, there was pasture available in the Flint Hills because of the relatively greater precipitation, and the deep and extensive root system of the bluestem grasses, and its ability to access water well below the surface. The tall-grass prairie ecosystem was thus more resilient than other grasslands and crop-land during periods of severe and prolonged drought. The heavy use of the Flint Hills pastures during the drought years led, in turn, to great concern about over-grazing and the depletion of rangelands in the region.

Professor A. E. Aldous (1935) of Kansas State College (a KSU precursor) observed that pastures in the region had significantly reduced capacity. He noted that some bluestem pastures in the vicinity of Manhattan, Kansas, were reduced by 40% to as much as 75% of their normal stand. Moreover, he documented other serious signs of pasture degradation, including deep ditches cut by increased runoff due to close cropping of vegetation, and his survey found abundant weed invasions in the Manhattan area. He also expressed some optimism because examination of samples of bluestem grass roots showed that some of the root system was alive even though the plant appeared dead.

Anderson (1940), an agronomist for the Kansas Agricultural Experiment Station, also showed that the grazing capacity of bluestem pastures had been depleted by overgrazing and drought. His data showed that in 1900, bluestem pastures could be stocked at a rate of about 2 acres of grass/head (mature cow or steer) per grazing season. By 1933, the same animal needed an average of 5 acres of grass, and 7 acres by 1940, even 10 acres on some severely depleted pastures (Anderson, 1940). He also pointed out that pastures weakened by overgrazing and drought were susceptible to weed invasions. Both Aldous (1933, 1935) and Anderson (1940) argued for deferred grazing strategies as a way to utilize forage without injuring the pasture. The general thrust of the idea was protection of the pasture. Grazing would be deferred for about 6 weeks, allowing enough early spring growth to enable the grass to withstand close grazing for the remainder of the season.

There is little doubt that range depletion and the attendant soil erosion were major concerns at the time. The entire 1946 *Report of the Kansas State Board of Agriculture* was devoted to the topic of soil conservation in Kansas (Kansas State Board of Agriculture, 1946). It contained articles addressing the issue—in cropland as well as range- and pasturelands. In the report, Anderson (1946) reviews the damage of the 1930s, noting that "many [pastures], once thought to be destroyed, now have a cover of grass nearly equal to that before the dry years" (p. 93), both confirming the degradation during the 1930s, but also suggesting that with time, resumed precipitation, and conservation measures (deferred grazing, burning when grass crowns were wet to minimize burn damage, and so forth), pastures could recover from severe depletion. Debate continued about how much of the range depletion could be attributed to overgrazing and how much was the result of drought. Some (e.g., Malin, 1942; Wood, 1980) tended to dismiss the argument that it was the outcome of land-use practices (i.e., overgrazing managed by ranchers and pasturemen), and emphasized that the bluestem pastures would recover with increased precipitation. Others, such as the Kansas State College researchers, were clear about the negative impacts of overgrazing and argued for ameliorative strategies. These debates notwithstanding, the notion of resilience and the restorative ability of the bluestem pastures is an important theme in the environmental history of the Flint Hills.

The agricultural economy was generally depressed during this period, although beef producers generally fared better than those who relied heavily on crops. Average farm size grew gradually during this period—from 270 to 325 acres—then surged after 1940, more than doubling in four decades. Many ranchers and ranch–farm operations took the opportunity to expand their operations as smaller ranchers were forced out of business (Wood, 1980).

World War II and Postwar Growth

Transitions in the Flint Hills during the postwar period were linked to the rapid structural changes in the commercial feedlot and meatpacking industries, which in turn were shaped by larger structural changes in the political economy of the postwar (Fordist) agrifood system. Some of the salient features of this system were (1) increasingly capital-intensive and large-scale production, leading eventually to chronic surpluses; (2) mass consumption of standardized, processed food commodities, increasingly manufactured and packaged for transportability and durability; (3) a shift in focus from nationally produced and distributed commodities to increasingly standardized, globally integrated production and distribution (Friedmann, 1991; Sanderson, 1986); (4) increased cômpetition driving the search for ways to reduce production costs (e.g., new technologies, improved transportation, political pressure for commodity support programs, lower labor costs, new markets); (5) downward pressure on the price of bulk commodities resulting from increased supply; and (6) economic expansion and the increase in wages in the United States after World War II. Related to this last point is the near doubling of per-capita beef consumption in the United States, from about 20 kg in 1940 to more than 38 kg per capita in 1970 (Fig. 6.7).

A key development in terms of land use in the Flint Hills was the growth of cattle feeding during the postwar period. This was followed by the rapid growth and westward movement of the large feedlot system after the 1950s, which was in turn followed by the movement of the large meatpacking plants—historically located near urban rail hubs and stockyards—to the rural areas of beef production in western Kansas and surrounding High Plains areas in eastern Colorado, western Nebraska, the Oklahoma panhandle, and northern Texas. Data on the spatial distribution of the cattle inventory of the state also confirm a dramatic relative shift toward the western Kansas feedlots between 1940 and 1975. These markets increased much more dramatically in the west than they did in the Flint Hills region.

Although there remained some feeding operations in the Flint Hills, their numbers declined after 1960 as it became increasingly difficult to compete with the larger feedlots in western Kansas and the High Plains region. There are a number of reasons why, during the post-1960 period, feedlot development was concentrated in western Kansas. Among these were (1) the strategic location of large feedlots near the source of cattle production reduced shrinkage and bruising incurred during long-distance shipping; (2) much of the area is situated over the Ogallala Aquifer, providing water for cattle and feed crops; and (3) the drier conditions relative to eastern Kansas reduce the difficulties in managing the large amounts of animal waste.

A key factor in the development of the large commercial feedlots was irrigation. The number of acres irrigated in Kansas increased dramatically during the postwar years. In 1945, there were 100,000 acres irrigated, which then expanded ninefold to 900,000 acres irrigated by the mid 1950s. About 90% of this acreage was in the 20 counties in the southwest corner of the state (Wood, 1980). About 2.7 million acres were irrigated by 1997 (U.S. Department of Agriculture, 1997). In the 1970s prefarm crisis period of elevated land values and higher commodity prices, U.S. farmers were encouraged by the Department of Agriculture and lending institutions to plant "fencerow-to-fencerow" to take advantage of the higher prices. Increased demand for both grains and beef meant that prices for these commodities rose to all-time highs.

The meatpacking industry was also undergoing structural change, beginning in the 1960s. They followed the feedlots in their migration to rural areas of the Great Plains, often for similar reasons that had led the feedlots to locate there. In their case, new low-cost competitors began to emerge, building one-level, high-tech plants in rural areas close to beef and pork production. In the 1970s and 1980s, Iowa Beef Packers built a number of plants in the High Plains, and eventually began recruiting from a much broader labor force, mainly from immigrant populations.

Interestingly, the cattle inventory in the Flint Hills—as measured before the spring in-shipments—nearly doubled from 1940 to 1970, from about 460,000 to more than 860,000 cattle. This inventory dipped between 1970 and 1990, but was on the rise again in 2000. Although we are not arguing a direct causal relationship, it does appear that the cattle inventory in the Flint Hills tracks closely with the trends in per-capita beef consumption in the United States (Fig. 6.7). Thus, the Flint Hills, as a grazing and cattle-producing region, still has significance in the beef industry. It seems that this particular landscape has developed a niche role in the industry, primarily by focusing on its uniqueness vis-à-vis the bluestem tallgrasses. The increase in the year-round cattle inventory (i.e., usually local cattle), suggests that the region has developed two primary roles: (1) cow–calf operations that produce young cattle, which are grazed and then sold to feedlots, often outside the region; and (2) breeding herds designed to produce animals for sale to others interested in improving their herd stock. In both cases, it is the weight gains that cattle achieve on the bluestem pastures that are the basic element of the agricultural economy of the region.

Some feedlot operations gained a foothold in the region around 1960, although the number of these dwindled because they were not competitive with areas to the west. Moreover, Wood (1980) points out that the greater precipitation and frequent flooding in the Flint Hills led to substantial manure spills and large fish kills in the Neosho and Cottonwood rivers in the 1960s. This, he suggests, is one of the main reasons for the relatively larger feedlot development in western Kansas and the High Plains.

In terms of cropping, corn production declined throughout the postwar period—no doubt as a result of the difficulty in competing with massive grain operations in the Corn Belt as well as greatly expanded planting of irrigated corn over the High Plains aquifer—although some is still grown for local feeding operations. Wheat is still grown in the region as a cash crop, and in 1940, soybean production began to appear in the data. During the postwar period, soybeans have become a major crop globally for their oil and meal. They are important in the livestock–grain complex and as a cash crop. Corn acres harvested bottomed out in 1980, but have recovered somewhat since then. The increasing presence of corn in recent years is likely attributable to its role in cropping rotations with soybeans.

The most interesting population dynamic in the Flint Hills after 1940 is the contrary trends in population in urban counties and those we identify as deep rural counties. Figure 6.5 illustrates the population dynamics of the 12 counties in the study region. A county-by-county review shows that the increase of population is the result of increases in just three counties. Nine of the 12 counties have experienced some population decline since 1900. This population decline in rural counties is countered by relatively larger population increases in urban counties. The nine counties represented by the rural aggregation are Chase, Chautauqua, Cowley, Elk, Greenwood, Lyon, Morris, Pottawatomie, and Wabaunsee. This line flattens out in 1890, and then the rural areas begin a slow, steady population decline that continues to the present. Just three counties are represented by the urban line: Butler, Geary, and Riley. These three counties are differentiated from the others because of the presence of the cities of Manhattan in Riley County, Junction City in Geary County, and the close proximity of Wichita to Butler County. Junction City and Manhattan are both regional trade and employment centers, including the Fort Riley military base and KSU, respectively. And although the city of Wichita is just outside the edge of the study region, its growth is affecting development patterns in the adjacent Flint Hills. It is also the largest metropolitan area near the study region and is the largest city in Kansas, with a population of more than 340,000 in 2000 (U.S. Census).

Population increases in the urban counties are of special interest to conservationists because of the impact that their social dynamics have on the landscape. In the next section we look at key conservation issues of current importance in the region and examine the impact of social dynamics on the legacy of the tallgrass prairie.

Conservation

Conservation in the Flint Hills has first and foremost taken the shape of land preservation through scientific research on the tallgrass prairie. This effort has been led in the region by researchers at Konza Prairie, a 3,487-ha area of unplowed tallgrass prairie near Manhattan, Kansas, in southern Riley and adjacent Geary counties. Konza was established in 1971 through purchase of the land by TNC, with funding assistance coming from philanthropist Katherine Ordway of New York City. In 1981, Konza became a founding member in the NSF's LTER network. The LTER was formed to "enhance scientific understandings of ecological phenomena and processes operating over broad spatial scales, as well as long-time scales" (Konza LTER, 2004). Buffalo were reintroduced at Konza in 1987, and research is designed to assess the relative roles of climate, fire, and grazing. The experimentation favors examination over long periods, as is evidenced in the design for studying the effects of prescribed burning. Indeed, although some watersheds are burned annually, others are burned at intervals of up to 20 years. Combined with grazing, the landscape at Konza has taken on the appearance of a matrix of experimental plots defined at the watershed level.

Preservation in the form described here was made possible through the private purchase of land for scientific purposes. The public purchase of a substantial area of tallgrass prairie in the Flint Hills has not been as easy, and after nearly 70 years of struggle, only recently (1996) did the Tallgrass Prairie Preserve become part of the national park system. In 1973, a group called Save the Tallgrass Prairie was incorporated, and committed to the task of lobbying for the creation of a national park (Save the Tallgrass Prairie Collection, n.d.). The Tallgrass Prairie Foundation was formed in 1976, and took on the larger goals of conservation beyond legislation. Together, these two groups helped push for protection of the tallgrass prairie by mobilizing citizens and articulating to the public at large the need for preserving the prairie.

Meanwhile, opposition was organizing in the form of the Kansas Grassroots Association (KGA). The KGA opposed government intervention in the protection of the tallgrass, citing two general arguments: (1) ranchers have taken care of the land through management, and (2) the National Park Service would not be able to do as good a job as ranchers. Opponents also argued that a national park would attract tourism and elicit an unnecessary wave of prairie degradation in the form of hotels, vehicular traffic, and other tourism-related activities. Ranchers in the KGA believed they were doing everything they could do to preserve the land, which included management strategies of grazing and prescribed burning.

Proponents for the national park attested that the ranchers were not doing everything in their control to preserve the land, arguing that year-long stocking, which was increasing at the time, was resulting in overgrazing in the area. A compromise between the two groups was achieved in 1994 when the National Park Trust, a private organization, purchased the Z-Bar/Spring Hill Ranch of northern Chase County, in the heart of the Flint Hills. Two years later, the land was passed to the National Park Service to form the Tallgrass Prairie National Preserve. The compromise was accepted by farmers and ranchers in Kansas who, according to a Wichita newspaper, "would rather invite the ghost of Karl Marx to dinner than see even a single acre of private land go to the federal government" (quoted in Miner, 2002, p. 374). Although the land was now part of the park service and available to the public, more than 98% of it would remain in private hands. Of the 10,894 acres, only 180 were allowed under the legislation to be transferred as property of the United States (National Park Service, 2006). The rest would remain under the watch of the National Park Trust. The Tallgrass Prairie National Preserve—a public/private partnership—sees its mission as being "dedicated to preserving and enhancing a nationally significant remnant of the tallgrass prairie ecosystem and the processes that sustain it" (National Park Service, 2006). The compromise in the form of the national park as a government-operated but privately owned entity has succeeded in offering to the public an opportunity to experience and understand the legacy of the tallgrass prairie.

Scientists are also concerned about the long-term effects of biodiversity loss through landscape transformation by competitor species. The tallgrass legacy is currently threatened by the advance of two species: *Sericea lespedeza,* a legume that has invasive characteristics (Fechter, 2000); and red cedar, a woody species that expands in grassland prairie after suppression of fire (Hoch, 2000). The former is an example of invasion by a nonnative species, whereas the latter species is native to the region. Combined, the two species currently represent serious threats to the tallgrass region.

Red Cedar Encroachment

Woody species, such as the red cedar (*Juniperus virginiana*), have been substantially increasing in population in the Flint Hills since around 1970 (Owensby et al., 1973). The majority of this increase has occurred in the northern counties of Riley, Geary, and Pottawatomie. Researchers have shown that the increase in woody species is a result of disruption in the application of fire, which is in turn related to growth in population, residential housing development, and land-use choices (Hoch, 2000). We showed earlier that most of the population growth in the Flint Hills has occurred in the more urbanized counties, a phenomenon that Hoch (2000) identified as a predictor for woody species growth. His analysis found a strong relationship (r^2=.81) between population change after 1970 and woody species growth during the period. Hoch (2000) put forth several explanations for the increase in woody species, and found that although the suppression of fire was the primary determinant, a gradual increase in housing around Manhattan, Kansas, was resulting in fragmentation. Fragmentation around the urban fringe appeared in the form of a matrix of rangeland and housing development, which made prescribed burning more difficult and risky, and, likewise, provided a seed bank of competitor species closer to native prairie.

Early research in woody species invasions in the tallgrass prairie recognized the benefits of regular fire and grazing on the reduction of competitor species and the resulting maintenance of biodiversity (Bragg and Hulbert, 1976; Owensby et al., 1973). In a more recent analysis, Hoch (2000) found that overgrazed areas were susceptible to woody invasion resulting from a reduction in fuel. Overgrazed areas lack adequate dry biomass that can sustain a large enough fire to suspend woody growth. So even if fire remains a part of rangeland management, it may not be sufficient to protect the tallgrass. If an area has been overgrazed as a result of an excessive stocking rate, burning that pasture in the subsequent season may not result in the suppression of woody species. Thus, to prevent woody invasion, responsible grazing through rotation must allow for an accumulation of biomass for the eventual application of fire.

As indicated earlier, woody species threaten the Flint Hills region by transforming the prairie into forest. Estimates are that with 30 years of fire suppression, the prairie can develop a closed canopy (Hoch, 2000). Species richness declines as woody species increase, and the productivity of herbaceous species is reduced.

The fibrous root system of the red cedar species, compared with the deep central taproot of other species, gives the species a competitive advantage on the uplands of the hills that are characterized by shallow, rocky soils. Hoch (2000) concludes by explaining that the transformation to a red cedar forest is not a transitory stage, but represents a climax community.

Sericea lespedeza

We turn now to the history of another invasive species in the region: *S. lespedeza* (*Lespedeza cuneata*, a.k.a. Chinese bush-clover). *Sericea lespedeza* is a perennial legume, native to Asia, that was first planted in the United States by the North Carolina Agriculture Experiment Station in 1896 (Ohlenbusch and Bidwell, 2001). Twenty years passed, however, before the species was recognized as beneficial for agricultural production. After it was planted in 1924 at the Arlington Experiment facility in Virginia, *Sericea* became recognized as having benefits for erosion control, hay production, wildlife cover, and wildlife food. A characteristic that set *Sericea* apart from similar species was that it was able to thrive in damaged areas. Beginning in the 1930s, *Sericea* was planted in areas of southeast Kansas, primarily on strip-mined land. The characteristics of mined soil make it highly susceptible to soil erosion and pose difficulties in regenerating cover.

For this reason, *Sericea* has been used extensively, not just in Kansas, but throughout the southeastern United States. It has been used as land cover on strip-mined land and highway right-of-ways, and on land in numerous state and federal reservoirs. In the 1950s, *Sericea* was spread to prevent soil erosion, and was unintentionally spread through the Conservation Reserve Program after the 1985 Farm Bill. A 1980 USDA farmer's bulletin on the benefits of planting *S. lespedeza* not only recognized its potential as "good ground cover," but also found that it makes "very good hay and pasture" (Guernsey, 1977, p. 1). The planting of *S. lespedeza* continues in several states, but in Kansas it is now viewed as invasive.

Sericea lespedeza threatens native species in Kansas, primarily in the Flint Hills, an area of ecological importance to conservationists because of the pronounced presence of native tallgrass prairie pasture (Eddy et al., 2003). Over time, *Sericea* has spread unfettered into sections of native pastureland and currently threatens not just the ecological integrity of the land, but the economic livelihood of those living there. Because the Flint Hills is home to hundreds of thousands of head of cattle each season, the profitability of operators in the area very much depends on the quality of grass.

Sericea directly threatens native species in several ways (Eddy et al., 2003; Ohlenbusch and Bidwell, 2001). First, established stands of *Sericea* grow over competing plants and pose a photosynthetic drain. Second, *Sericea* plants can develop root structures that extend well over 1.2 m in length, which makes it a competitor for water resources. Third, each *Sericea* plant can produce more than a thousand seeds in a given year, and those seeds can survive up to 30 years before germinating. Fourth, *Sericea* is generally avoided by grazing cattle because of the high presence of tannins in its leaves, making it bitter to the cattle's palate. And finally, *Sericea* is a nonnative species to the North American continent and therefore lacks natural enemies in the tallgrass prairie. Rangeland managers are just now beginning to recognize the serious impacts that *Sericea* can have on species diversity throughout the Flint Hills, and are looking to researchers for methods of abating its spread.

Sericea lespedeza became listed as a county option noxious weed by the Kansas Legislature in 1988, and as a statewide noxious weed in 2000 (Ohlenbusch and Bidwell, 2001). Kansas State University, through its cooperative extension agencies and with the help of other universities throughout the region, has been providing information to grassland managers about the technical elements of *Sericea*, as well as how to control the spread of the species. Control methods developed include biological treatment using webworms (Eddy et al., 2003), and an integrated management strategy (Dudley, 1998). *Sericea* will likely be managed, not eradicated. Biological alternatives to herbicides are judged as safer; yet caution has been voiced regarding overreliance on biological controls, given the possibility of introducing more invasive species (Edwards, 1998). The introduction of goats is one alternative that has shown success in the reduction of *S. lespedeza*, although its widespread adoption in the Flint Hills is not expected because of the reliance of the region on cattle grazing (Dvorak, 1998).

The threat of *S. lespedeza* has propelled those interested in conservation into action to protect the diversity of native species in the region. Terms for controlling *S. lespedeza* have, during the past few years, made their way into lease agreements between landowners and pasturemen. The presence of *Sericea* is a factor in the decisions of those taking care of the land, affecting both the grazing and fire elements of the tallgrass ecosystem. Grazing is obviously affected because of the diminished diversity of native grasses resulting from competition from the foreign *S. lespedeza* species. Indeed, the domination of *S. lespedeza* has the potential for transforming the prairie ecosystem. Fire regimes are also affected by *S. lespedeza*, as scientists have found some evidence to suggest that burning in the late spring may diminish the ability of *Sericea* to produce seeds. On the other hand, the same evidence suggests that an early to midspring burn, a time when nearly all Flint Hills pastures are currently burned, may actually enable the species to advance (Eddy et al., 2003). The survival of the tallgrass, in face of the challenges of *S. lespedeza*, depends upon the work of range managers, legislators, researchers, and owners to negotiate a plan for controlling and preventing further outbreaks of invasive species.

Conclusion

During the past 200 years, the human–ecosystem dialogue in the tallgrass prairie of the Flint Hills has been a continuous, iterative process of the mutual shaping of both the societies that have inhabited the region, and the ecosystem itself. Native Americans managed both the grasses and the bison through their burning and hunting practices, which contributed to the perpetuation of the tallgrasses by preventing succession to woody species and encouraging bison grazing. In turn, the tallgrasses and bison were embedded in the culture and practices of the Kansa Indians. With European American arrival and expansion, the Kansa Indians were more deeply incorporated into extraregional trade networks, leading to the eventual overharvest of buffalo. Additionally, buffalo herds declined as a result of expansionist U.S. policy, which led to the exposure of buffalo to short-term, extractive profit taking in the harvest and trade of hides. Thus, as the ecosystem was altered with the loss of buffalo, so too was society altered. The Kansa lost tribal lands as a result of the social organization, institutions, and policies of the U.S. government, and the elimination of buffalo meant that key components of their cultural system (symbolic practices, protein source, and so on) were also lost. Continuation of Kansa Indian society in a form like that of pre-European contact was impossible after the 1860s.

The European American settlement period was one of dramatic transition for the social ecosystem of the region. Dramatic population increase meant many more farms, more cattle, and increased agricultural production. Although the Kansa Indians had cropped in the bottomlands, their numbers paled in comparison with settler numbers. Thus, much more bottomland was plowed for cropping. Farmers attempted cropping in the uplands, but it was exceedingly difficult to subsist as an upland farmer. In a sense, the upland tallgrass ecosystem resisted human attempts to shape it in the image of eastern croplands. This resistance essentially forced a reorganization of the spatial patterns of social development inherent in the Homestead Act. That reorganization toward larger farm and ranch sizes included, most saliently, the development of a ranching system in which cattle—as many as 700,000 in 1900—were fattened on the upland tallgrasses, putting on 91 to 136 kg in a season.

Trends in the patterns of human use and organization of land holdings were enabled and advanced by the social organization, institutions, and policies of the U.S. government. The Homestead, Pacific Railroad, and other acts were essentially a blueprint to guide westward expansion, and were of course informed by the worldview, experiences, values, and economic culture of the dominant group in U.S. society at the time—white Anglo-Saxon Protestants. It was a blueprint that, presumably, could be superimposed on any lands encountered in the newly acquired West. It was further assumed that where biophysical systems did not cooperate, any obstacles could be overcome through new technologies, tools, and techniques, depending on the requirements of resource extraction. Biophysical limits or setbacks in terms of ecological disturbances would be temporary, because the application of reason, science, and market forces would resolve problems of production and accumulation.

Yet, because ecosystems are also part of the dialogue, they resist their own transformation, to a point. In some cases, for example in grasslands that have been plowed and continuously cropped, the transformation is dramatic. The tallgrasses of the Flint Hills appeared to exhibit remarkable resilience during periods of overgrazing and drought. Especially during the droughts of the 1930s, when cattle were moved into the region from drier grasslands, and when they were heavily overgrazed as a result, the grasses were able to recuperate, even though some thought they had been irreversibly depleted. This resilience also enabled production, employment, and markets to continue during periods when it was not possible in other areas. The notion of resilience and the restorative ability of the bluestem pastures is thus an important theme in the narrative of this environment.

Changes in the extraregional context can also drive change in the "local" social ecosystem. World Wars I and II both drove demand and production in the Flint Hills. Moreover, the emergence of the grain–livestock complex during the post–World War II period, in which feedlots and meatpackers moved to areas over the Ogallala Aquifer, also increased the competitive pressure on the tallgrass prairie. In an era of irrigated, overproduced, and thus relatively low-cost feed grains, and expensive pasture costs, grass feeding appears comparatively expensive. Beef cattle are now much younger at slaughter, resulting from efforts to fatten them more quickly. From the dominant economic viewpoint, cattle are seen as a commodity in which one invests capital, and does not realize a return until the commodity is sold. If this can be accomplished in 18 months as opposed to 2 to 3 years, the return can be realized more quickly. Following this rationale, the sooner cattle can be grain fed in a feedlot, the more beneficial for the investor. Thus, although grass feeding still plays a role in the beef industry, it is now a more specialized role, in contrast to the traditional grass fattening over one or more grazing seasons.

Finally, although it seems clear that the tallgrass prairie exhibits remarkable resilience and recuperative abilities in conditions of drought and overgrazing, it is not yet clear whether those same characteristics will provide defenses to invasive species. In the case of red cedar, fire suppression near urban edges is a social choice related to urban development patterns, profit motives, human understanding of ecosystem processes, and aesthetic sensibilities. Whether policymakers will understand and be able to address this issue effectively through urban planning boards and public policy remains to be seen. In the case of *S. lespedeza*, researchers, ranchers, landowners, and others have now recognized it as a major threat to the tallgrass prairie, and are mobilizing institutional resources to address it. That the approach to *Sericea* is now viewed as one of permanent management—as opposed to eradication—is a vivid example of social–ecosystem dialogue.

References

Aldous, A. E. 1933. "Bluestem pastures," pp. 184–191. In: *The twenty-eighth biennial report of the board of the Kansas State Board of Agriculture to the legislature of the state for the years 1931 and 1932*. Topeka, Kans.: Kansas State Board of Agriculture.

Aldous, A. E. 1934. *Effect of burning on bluestem pastures*. Technical bulletin no. 88. Manhattan, Kans.: Kansas State University.

Aldous, A. E. 1935. "The Kansas pasture situation." *The Kansas Stockman* 19(16): 3, 14.

Anderson, K. 1940. *Deferred grazing of bluestem pastures*. Bulletin no. 291 (October). Manhattan, Kans.: Agricultural Experiment Station, Kansas State College of Agriculture and Applied Science.

Anderson, K. 1946. "Range and pasture." *Report of the Kansas State Board of Agriculture, Soil Conservation in Kansas* 65(271): 92–117.

Bragg, T., and L. C. Hulbert. 1976. "Woody plant invasion of unburned Kansas bluestem prairie." *Journal of Range Management* 29(1): 19–24.

Briggs, J. M., G. A. Hoch, and L. C. Johnson. 2002. "Assessing the rate, mechanisms, and consequences of the conversion of tallgrass prairie to *Juniperus virginiana* forest." *Ecosystems* 5: 578–586.

Bryant, K. 1974. *History of the Atchison, Topeka and Santa Fe.* New York: Macmillan.

Burns, L. 1989. *A history of the Osage people.* Fallbrook, Calif.: Ciga Press.

Cronon, W. 1992. *Nature's metropolis: Chicago and the great west.* New York: W. W. Norton.

Dudley, D. 1998. *Integrated control of* Sericea lespedeza *in Kansas.* Master's thesis, Kansas State University, Manhattan, Kans.

Dvorak, J. 1998. "In fields overrun by weeds, goats gobble up the problem." *Kansas City Star,* September 8: A1.

Eddy, T., J. Davidson, and B. Obermeyer. 2003. "Invasion dynamics and biological control prospects for *Sericea lespedeza* in Kansas." *Great Plains Research* 13(Fall): 217–230.

Edwards, K. 1998. "A critique of the general approach to invasive plant species," pp. 85–94. In: U. Starfinger, K. Edwards, I. Kowarik, and M. Williamson (eds.), *Plant invasions: Ecological mechanisms and human responses.* Leiden, The Netherlands: Blackhuys.

Farney, D. 1980. "The tallgrass prairie: Can it be saved?" *National Geographic* 157: 37–61.

Fechter, R. 2000. *The economic impacts of control of* Sericea lespedeza *in the Kansas Flint Hills.* Master's thesis, Kansas State University, Manhattan, Kans.

Friedmann, H. 1991. "Changes in the international division of labor: Agri-food complexes and export agriculture," pp. 65–93. In: W. H. Friedland, L. Busch, F. H. Buttel, and A. P. Rudy (eds.), *Towards a new political economy of agriculture.* Boulder, Colo.: Westview Press.

Gard, W. 1954. *The Chisholm Trail.* Norman, Okla.: University of Oklahoma Press.

Guernsey, W. 1977. Sericea lespedeza: *Its use and management.* USDA Farmer's Bulletin no. 2245. Washington, D.C.: USDA.

Hayden, B. P. 1998. "Regional climate and the distribution of tallgrass prairie," pp. 19–34. In: A. K. Knapp, J. M. Briggs, D. C. Hartnett, and S. L. Collins (eds.), *Grassland dynamics: Long-term ecological research in tallgrass prairie.* New York: Oxford University Press.

Hoch, G. 2000. *Patterns and mechanisms of eastern red cedar expansion into tallgrass prairie in the Flint Hills, KS.* Master's thesis, Kansas State University, Manhattan, Kans.

Isern, T. 1985. "Farmers, ranchers, and stockmen of the Flint Hills." *Western Historical Quarterly* 16(3): 253–264.

Kansas Applied Remote Sensing (KARS) Program. 1993. "Land cover." Online. Available at http://www.kansasgis.org and http://clone.kgs.ku.edu/land_surface_geology_soils/landcover/. Accessed 2/27/2008.

Kansas State Board of Agriculture. 1946. *Soil conservation in Kansas.* Report. February. Vol. LXV, no. 271.

Kerr, N. A. 1987. *The legacy: A centennial history of the state agricultural experiment stations, 1887–1987.* Columbia, Mo.: Missouri Agricultural Experiment Station, University of Missouri-Columbia.

Knapp, A. K., J. M. Blair, J. M. Briggs, S. L. Collins, D. C. Hartnett, L. C. Johnson, and E. G. Towne. 1999. "The keystone role of bison in North American tallgrass prairie." *Bioscience* 49(1): 39–50.

Knapp, A. K., and T. R. Seastedt. 1998. "Introduction: Grasslands, Konza Prairie, and long-term ecological research," pp. 3–15. In: A. K. Knapp, J. M. Briggs, D. C. Hartnett, and S. L. Collins (eds.), *Grassland dynamics: Long-term ecological research in tallgrass prairie.* New York: Oxford University Press.

Kollmorgen, W. M. 1969. "The woodsman's assaults on the domain of the cattleman." *Annals of the Association of American Geographers* 59(2): 215–239.

Konza LTER. 2004. Rev. October 7, 2005. Online. Available at www.konza.ksu.edu/general/knzlter.html.

Küchler, A. W. 1974. "A new vegetation map of Kansas." *Ecology* 55(3): 586–604.

Malin, J. C. 1942. "An introduction to the history of the bluestem-pasture region of Kansas." *Kansas Historical Quarterly* 11(1): 3–28.

Miner, C. 2002. *Kansas: The history of the sunflower state 1854–2000.* Lawrence, Kans.: University Press of Kansas.

National Park Service. 2006. *Tallgrass prairie.* Rev. June 28, 2006. Online. Available at www.nps.gov/tapr/.

Norgaard, R. B. 1994. *Development betrayed: The end of progress and a coevolutionary revisioning of the future.* London: Routledge.

Ohlenbusch, P. D., and T. Bidwell. 2001. Sericea lespedeza: *History, characteristics, and identification.* Manhattan, Kans.: KSU Agriculture Experiment Station.

Ohlenbusch, P. D., and D. C. Hartnett. 2001. "Prescribed burning as a management practice." Kansas State University Agriculture Experiment Station and Cooperative Extension, publication L-815. Manhattan, Kans: KSRE.

Oviatt, C. G. 1998. "Geomorphology of Konza Prairie," pp. 35–47. In: A. K. Knapp, J. M. Briggs, D. C. Hartnett, and S. L. Collins (eds.), *Grassland dynamics: Long-term ecological research in tallgrass prairie.* New York: Oxford University Press.

Owensby, C. E., K. R. Blan, B. J. Eaton, and O. G. Russ. 1973. "Evaluation of eastern red cedar infestations in the northern Kansas Flint Hills." *Journal of Range Management* 26(4): 256–260.

Parrish, F. L. 1956. "Kansas agriculture before 1900," pp. 401–427. In: J. D. Bright (ed.), *Kansas: The first century.* New York: Lewis Hist Publishers.

Petrowsky, C. L. 1968. *Kansas agriculture before 1900.* PhD diss., University of Oklahoma, Norman, Okla.

Pyne, S. J. 1984. *Introduction to wildland fire: Fire management in the United States.* New York: Wiley-Interscience.

Reichman, O. J. 1987. *Konza Prairie: A tallgrass natural history.* Lawrence, Kans.: University Press of Kansas.

Sanderson, S. E. 1986. "The emergence of the 'World Steer': Internationalization and foreign domination in Latin American Cattle Production," pp. 123–148. In: F. L. Tullis and W. L. Hollist (eds.), *Food, the state, and international political economy.* Lincoln, Nebr.: University of Nebraska Press.

Sauer, C. O. 1975. "Man's dominance by use of fire." *Geoscience and Man* 10: 1–13.

Save the Tallgrass Prairie Collection. n.d. Special Collections, Hale Library. Manhattan, Kans.: Kansas State University.

Sherow, J. E. 1992. "Workings of geodialectics: High Plains Indians and their horses in the region of the Arkansas Valley, 1800–1870." *Environmental History Review* 16: 61–84.

Snipp, M. 1989. *American Indians: The first of this land.* New York: Russell Sage.

Unrau, W. 1971. *The Kansa Indians: A history of the wind people, 1673–1873.* Norman, Okla.: University of Oklahoma Press.

Unrau, W. 1991. *Indians of Kansas: The Euro-American invasion and conquest of Indian Kansas.* Topeka, Kans.: Kansas State Historical Society.

U.S. Census. 1860–2000. American Fact Finder. Online. Available at http://factfinder.census.gov/home/saff/main.html.

U.S. Congress. 1862. *An act donating public lands to the several states and territories which may provide colleges for the benefit of agriculture and the mechanic arts.* U.S. Statutes 12: 503.

U.S. Department of Agriculture. 1997. *1997 Census of Agriculture Volume 1.* Online. Available at http://www.nass.usda.gov/census/census97/volume1/ks-16/ks1_08.pdf.

U.S. Department of Agriculture/Economic Research Service. n.d. "Beef: Supply and disappearance." Online. Available at http://www.ers.usda.gov/Data/FoodConsumption/spreadsheets/mtredsu.xls. Accessed 1/26/2005.

Vinton, M. A., D. C. Hartnett, E. J. Finck, and J. M. Briggs. 1993. "Interactive effects of fire, bison grazing and plant community composition in tallgrass prairie." *American Midland Naturalist* 129: 10–18.

Wibking, R. K. 1963. *Geography of the cattle industry in the Flint Hills of Kansas.* PhD diss., University of Nebraska, Lincoln, Nebr.

Wood, C. L. 1980. *The Kansas beef industry.* Lawrence, Kans.: Regents Press of Kansas.

Worster, D. 2003. "The dirty thirties: A study in agricultural capitalism," pp. 318–333. In: R. Napier (ed.), *Kansas and the West: New perspectives.* Lawrence, Kans.: University Press of Kansas.

7

Water Can Flow Uphill

A Narrative of Central Arizona

Charles L. Redman
Ann P. Kinzig

This narrative of an Arizona case study explores the relationship between ecological and human systems in the Salt River Valley of central Arizona during the past millennium, with a focus on the 140-year period of European American settlement. Our subject matter is the changing mosaic of Sonoran Desert landscape, irrigated farmland, and urban settlement. We examine how evolving patterns of perception, valuation, and use of water and land have been influenced by, and in turn have influenced, climate, environment, settlement, and agrarian landscapes. The logic for studying this region is strengthened by its presence within the CAP LTER project, one of only two urban sites in a national network of 26 sites. Our research perspective derives from two underlying concepts: (1) human management of surface and groundwater resources in this arid region has fundamentally altered both natural and human-modified landscapes, and (2) the technological and social aspects of this relationship have advanced over a relatively short time and are proceeding rapidly. Hence, this locale constitutes an excellent laboratory for examining the interplay between ecological and human systems.

The context for this case study is the rapidly urbanizing region of Phoenix, set in the Sonoran Desert of the United States, within a broad, alluvial basin where two major tributaries of the Colorado River—the Salt and Gila rivers—converge (Fig. 7.1). The Phoenix Basin has known continuous, and often extensive, agricultural activity for more than 1,500 years, when it was first settled by an irrigation-based society known to us as the Hohokam. European American occupation and agricultural activities began in 1867, with excavation and use of former Hohokam canals. By the late 19th century, the pattern seen across the core of the Salt River

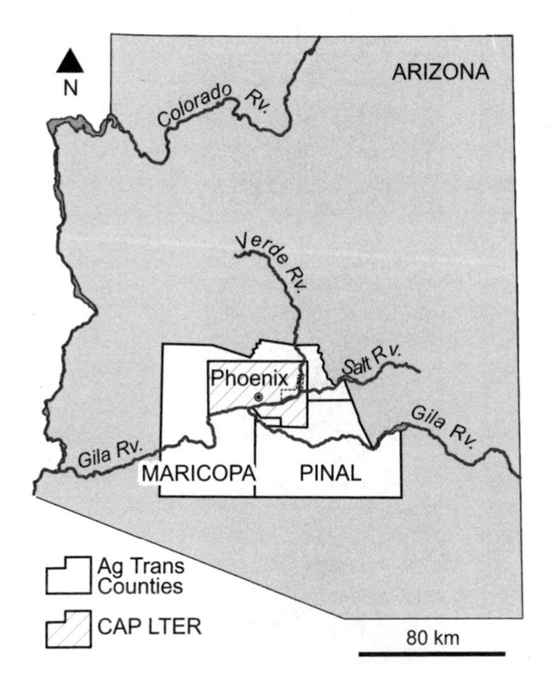

Figure 7.1 Boundaries of the CAP LTER study site. The CAP ecosystem encompasses an area of 6,400 km², whereas the AgTrans study area encompasses 37,993 km².

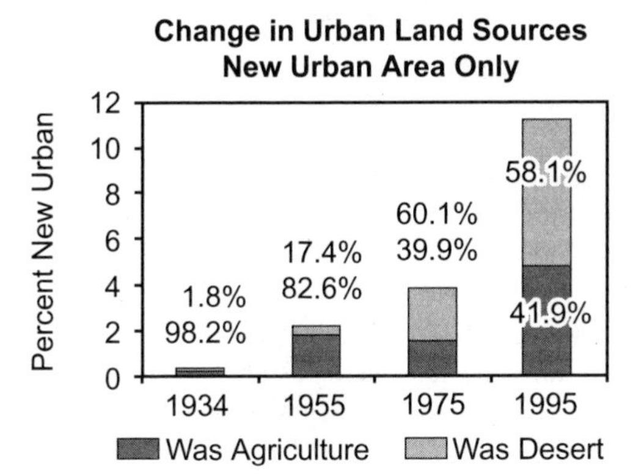

Figure 7.2 The growth of new urban settlement in the Phoenix region, 1934–1995.

Valley was that of a broad extent of farms with access to irrigation water, served by seven small communities centrally located among the farms. During ensuing decades, those towns expanded to form a contiguous metropolitan area, with a concomitant spread of farmland into adjacent flatland farther from the river, yet still within range of gravitation-fed irrigation. The latter half of the 20th century witnessed rapid growth in the region—nearly three times the national average—at the cost of surrounding farms and desert landscapes. During the past 25 years, more than half the region's highly productive farmland has been lost to urbanization (Fig. 7.2), with much of the remaining farmland within commuting distance of the urban fringe expected to be lost during the next 25 years. At the same

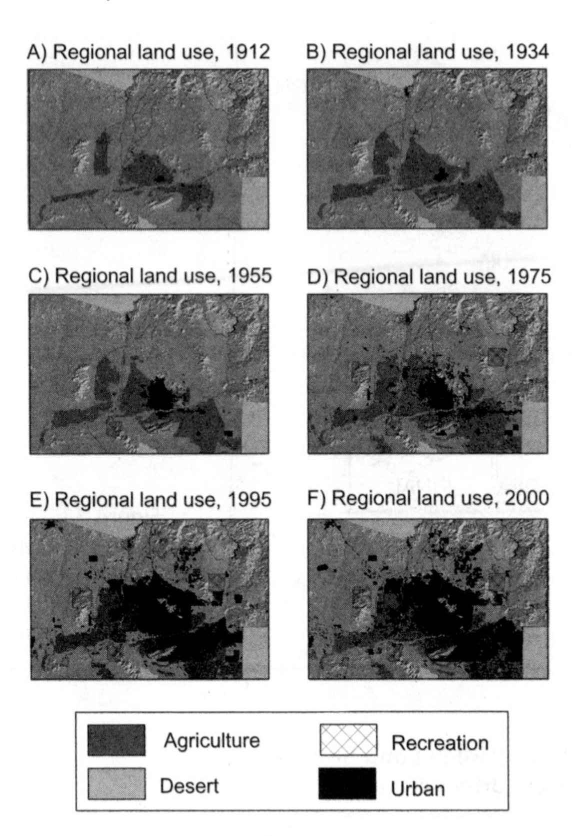

Figure 7.3 Land-use maps of the CAP study area, 1912–2000. (A) Regional land use, 1912. (B) Regional land use, 1934. (C) Regional land use, 1955. (D) Regional land use, 1975. (E) Regional land use, 1995. (F) Regional land use, 2000.

time, farmland conversion has not met the demands of a burgeoning population, with most new housing during the past three decades being established on former desert lands. These historic trajectories have led to significant spatial variation in vegetation and structure of residential and remaining desert landscapes (Fig. 7.3).

The challenge in this narrative, and for CAP LTER in general, is to study rapid urbanization systematically in central Arizona, a region that includes Phoenix, four of the state's five next-largest cities, and more than 500,000 acres of highly productive, irrigated farmland. The conversion of farmland to urban and suburban development is a nationwide issue, with more than 50 acres/hour being lost to this process (USDA, 2001). Therefore, our overarching goal is to build a qualitative and quantitative understanding of the patterns and processes that underlie long-term alterations to the ecosystem and the function of changing landscapes. For an urban ecosystem and its surrounding region, achieving this goal hinges on unraveling the complexities of intense human participation in the system—with attendant economic and social drivers, radically altered land cover, accelerated cycling of materials, and heretofore underresearched ecological impacts of a built environment (Grimm and Redman, 2004; Redman, 1999). The study of agrarian landscapes in transition is an integral component of the urban ecology of the region.

An improved understanding positions us to better grasp the opportunities and challenges facing those crafting conservation strategies for the region. The presence of intensive agriculture for more than a century and rapid urbanization have led some environmentalists to give up on this region as having been subjected to excessive human manipulation. However, the presence of so many people provides a chance to acquaint residents with the value of conserving or restoring desert and riparian habitats, in part because of the services that flow from those habitats to serve regional residents. It also allows a conversation about the role of human occupation and activity in conserved landscapes. An ample prehistoric and historic record of land use and settlement enriches the conservation lessons that can be communicated to the general population. Two special challenges face the conservationist in an urban region, however. First, even if habitats are set aside for conservation, the conservation plan will need to accommodate heavy visitation and recreational opportunities. Second, because humans have modified virtually all landscapes in the region, planners must decide what characteristics to favor when restoring the landscape. These types of decisions are faced in the design of almost all "nature" preserves but, when a preserve is situated on a highly transformed agrarian landscape or close to an urban center, the gravity of these decisions becomes more transparent.

The Ecological and Conservation Context

The Phoenix Basin is situated in the Sonoran Desert, which extends through southern Arizona into southeastern California and covers most of the Baja peninsula and the state of Sonora in Mexico. The Sonoran Desert is lush compared with other deserts, offering the greatest diversity of plant life of any desert in the world. The region houses 2,000 species of plants (about half of which are annuals), 550 vertebrate animal species, and unknown numbers of invertebrates (Arizona-Sonoran Desert Museum, 1999). Despite such diversity, two distinctive plant types are most often associated with this desert—leguminous trees and large columnar cacti.

Phoenix is situated in a subdivision of the Sonoran Desert known as the Arizona Uplands. Characterized by many mountain ranges and interspersed valleys, it is the highest and coldest Sonoran subdivision and is the only one that experiences hard frosts. The dominant upland vegetation types are saguaro–palo verde "forests" on outwash slopes and pediments, with creosote bush dominating lower, flatter areas, and mesquite-dominated riparian communities occurring in the meandering riparian zones of the Salt and Gila rivers and their tributaries (Fig. 7.4) (Turner, 1974).

The region experiences two, roughly equal, rainy periods each year. Winter (December–March) frontal storms originate in the North Pacific and bring relatively gentle rains. The summer (July–mid September) monsoons originate in the

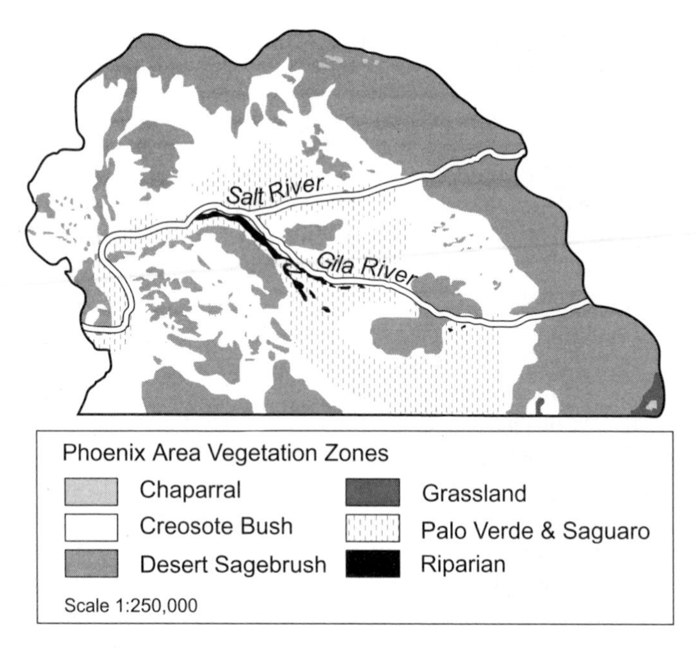

Figure 7.4 Map showing vegetation in the Phoenix area, Arizona. Based on data from Turner (1974).

tropics, with rain often delivered in violent thunderstorms. Rainfall is thus both spatially and temporally variable across the desert. All in all, the region averages less than 200 mm rainfall/year, with more than two thirds delivered across the two rainy seasons.

Limited rainfall in the Phoenix Basin necessitates irrigation agriculture. Modern agriculture concentrates along the Salt River, which is formed in eastern Arizona by a confluence of the White and Black rivers. The Salt River drains an area of about 5,130 sq. mi., with the northern boundary of the catchment defined by the Mogollon Rim, a 2,000-ft. escarpment that forms the southern boundary of the Colorado Plateau. The Salt River is joined by the Verde River, which drains an additional 6,188 sq. mi. Below its confluence with the Verde, the Salt River, on average, delivers 1.23 million acre-ft. (MAF) water/year, with relatively high year-to-year variability. Stream flow reconstructions of the Salt River flow from AD 572 to 1380 and show a slightly lower average stream flow of about 1.15 MAF/year, with a standard deviation of 0.64 MAF/year (Graybill et al., 2006). Below-average stream flow is more common than above-average stream flow (56% vs. 44%), and long periods of drought are not uncommon (Graybill et al., 2006).

The Gila River originates in the highlands of New Mexico and traverses south of Arizona's White Mountains before descending into the floodplain where it is used for irrigation. Although the Gila's watershed is almost as large as the Salt River's, differences in topography and precipitation result in a mean annual discharge less than one third (0.354 MAF) of the Salt's (Graybill et al., 2006). The rivers' annual distribution of discharge also differ markedly. Winter precipitation in the upper elevations (flow increases steadily beginning in November, usually peaking in March or April) determines the annual discharge of the Salt. It has a secondary, but less significant, peak in late summer as a result of seasonal rainfall. In contrast, the Gila River discharge reflects summer rainfall much more and exhibits a markedly bimodal annual distribution (Graybill et al., 2006).

Modern (after 1867) human activities in the Phoenix Basin, including agriculture, have adversely affected many species. In Maricopa and neighboring Pinal and Pima counties, 20 species are classified as endangered (six plant, four fish, four bird, six mammal), and five as threatened (two fish, one amphibian, two bird) (U.S. Fish and Wildlife Service Threatened and Endangered Species, 2008). Agricultural activities that have contributed to the demise of these species include land conversion, removal and redistribution of water from stream courses, alterations of riparian habitats, grazing, pesticide use, and species introduction.

Central Arizona has benefited from a long history of public land conservation that predates statehood. In 1905, the Tonto National Forest (northeast border of the CAP LTER study area) was created to protect the Salt and Verde watershed. Its three million protected acres represent the fifth largest forest in the United States. The Tonto National Forest has preserved, at a low population density, a large proportion of the upland watershed of the Salt River Valley that is the subject of this narrative. With statehood granted in 1912, much of the remaining landscape of central Arizona that was not settled or part of an Indian reservation was allocated to the state, to be held for the benefit of public education, or to the federal Bureau of Land Management (BLM), to be managed for multiple uses. This state land has been leased for commercial uses and recently became the center of controversy regarding how much land the state should sell for residential development or hold for conservation value. The BLM lands have largely remained in public hands, with some lands sold to the city and county park systems, and some transferred into National Monument status. In 1924, prominent local citizens, with the help of Senator Carl Hayden, bought 16,000 acres of BLM lands to establish South Mountain Park in Phoenix, renowned as the largest municipal park in the country. In 1935, the National Park Service developed a plan for the park, with riding and hiking trails, picnic areas, and overlooks of the city. The Civilian Conservation Corps built many of the plan's facilities, and the park now has an annual visitation of more than three million people. The Maricopa County Park System began in 1954 to preserve the mountain areas surrounding the rapidly growing region for future generations. In 1970, the Federal Recreation and Public Purposes Act allowed the county to purchase thousands of acres of parkland from the BLM at $2.50 an acre so that today, through a combination of leases and purchases, the system has 10 regional parks totaling more than 120,000 acres. East of the CAP LTER site are the majestic Superstition Mountains, the subject of many legends of riches and disaster. The wilderness value of the Superstitions has long been recognized, and 160,000 acres were established as a Primitive Area in 1939 and designated a "Wilderness" in 1964, protecting them from urban encroachment. The most recent federal lands to receive additional protection were a series of national monuments created largely on BLM lands in the waning days of the Clinton presidency. Of these, the Agua Fria National Monument, north of the CAP LTER site, is particularly important. Established to preserve a rich archaeological record of prehistoric occupation from urban sprawl, this 71,000-acre monument also serves to protect the rugged canyon and riparian area of the Agua Fria River, which supplies agricultural and residential water to the west side of our study area.

Despite this conservation history, there are forces that limit the capacity to promote significant additional conservation in or near the metropolitan area. Patricia Gober, Arizona State University professor and codirector of the Decision Center for a Desert City, wrote, "The people of Phoenix and Arizona more generally are uncomfortable with the idea of a government-mandated, one-size-fits-all approach to planning and managing future growth. They are unwilling to sacrifice their personal piece of the American dream for the larger regional good, a fact made abundantly clear in the public's reaction to a series of initiatives and proposed legislation between 1996 and 2000" (Gober, 2005, p. 124).

The key ballot initiative was in 2000—the Urban Growth Initiative. The public was asked to vote on urban growth boundaries. Environmentalists and planners proposed the initiative, but the public voted against it by a huge margin. Opposition was led by the development community, joined by farmers who did not want to be constrained in their ability to sell their land, by affordable housing advocates who worried about the price of housing, and by labor unions who were concerned about the loss of construction jobs. The public reaction to this initiative speaks volumes about public sentiments that favor rapid growth and new land development. It is this mentality that leads to the loss in agricultural land, development at the fringe instead of infill, and failure to amend the laws that dictate the sale of state trust land (Gober, 2005).

The Nature–Human Conceptual Conundrum

When people settled permanently in central Arizona and redirected the hydrological cycle to enable irrigation agriculture, separation of the natural environment from its human-engineered counterpart grew complex. During the course of altering land and water to suit economic needs, people often destroy much of what one would consider "natural." Although irrigated landscapes are in some sense artificial, they are constructed with topography, climate, and soil characteristics as guideposts. Hence, these early agrarian landscapes are sometimes thought of as *modifications* of the natural order, whereas more intensive construction is considered to create an anthropogenic landscape. We believe that there is no clear dividing line between the natural and anthropogenic, only a continuum of complex sets of relationships. To simplify the semantics of this narrative, we use the term *nature* to distinguish processes and conditions that have minimal human participation from those where the human impact is substantial. This division is not to suggest that humans are not part of nature, but to permit a contrast that will have intuitive meaning to a large audience. However, we do not intend to suggest that nature is a passive canvas upon which people build their idealized world. Quite the contrary, we assert that nature is a vigorously reactive and creative force. As the construction of an irrigated landscape degrades habitats, it creates others, such as reservoirs, canals, and fields, that provide homes for new species that establish new hydrological relationships. Nature does not disappear in the face of anthropogenic activities; it is transformed into new and, perhaps, novel ecosystems, drawing on age-old relationships and strategies in new contexts.

During the early years of the modern Salt River Valley settlement (beginning in 1867), irrigation agriculture was communicated to prospective immigrants as a great human accomplishment. With irrigation, settlers could transcend nature's limits and farmers could be liberated from their dependence on precipitation. Advocates portrayed early irrigators as modifying the landscape so that the land could reach its potential, referring to farmers in a religious sense as "understudies of the Creator" (Fiege, 1999, p. 23). The farmer need not fear drought; the Master Irrigator could "make water flow uphill" (Fiege, 1999, pp. 11–22). Despite idealized expectations, settlers found a far different reality. Irrigation agriculture involved constant interaction among people (with their aspirations, technologies, and labor) and the uncertainties of nature.

Triggering Events and Human Response

We divide the socioecological history of the region into four eras, representing major transformations in the relationship between residents and their environment. We categorize these eras as the Native American Years, Emergent Years, Boom Years, and Southwest Metropolis Years (see Luckingham, 1989) (Table 7.1). Multiple factors drive agrarian landscape transformations, including climatic uncertainty, soils, demographic trends, economics, politics, technology, values, and evolving perceptions. Rather than discuss these factors in isolation, we treat them as interrelated factors that became *key triggering events* (both crises and opportunities) in each era. The perception, impact, and response to these events acted on local, regional, and national scales; occurred at differing tempos, ranging from rapid to sustained; and led to impacts on and elicited responses from the local coupled human–natural system. In Table 7.1 we present the eras and the triggering events that characterize them. We use the term *events* broadly to encompass actual events, more sustained forces, policy changes, and opportunities upon which settlers could act.

People within a society often (but not always) hold shared ideas about how their socioecological system works and how to respond to a variety of inputs. We refer to these shared ideas as the *mental model* held by that society at that time. These mental models may derive from a variety of sources—textbooks, government agencies, community norms, or personal experience, to name just a few. We concentrate here on the prevailing mental models for each era, although these prevailing models were almost certainly not shared by all members of a society at any given time. Some mental models would be pervasive (e.g., best to plant crops in the spring); others might be held mostly by the powerful (e.g., best to lobby elected officials to pass favorable legislation). These prevailing mental models, though, would have defined the set of appropriate responses to a perceived challenge or crisis. The ecological or social changes derived from these

Table 7.1 Triggering "Events" in the Interaction of Population, Agriculture, and Environment in Central Arizona

Native American Years (500–1866)	
NA1:	Advent of adequate irrigation technology and infrastructure colonized the relatively empty, but potentially productive, Salt and Gila river valleys.
NA2:	A shift in climate and associated surface-water flow prompted the Preclassic to Classic transformation in social, political, and economic organization.
NA3:	Diminished resilience of sociopolitical organization made the system vulnerable to minor perturbations.
Emergent Years (1867–1940)	
E1:	Opportunity presented by an unoccupied, potentially productive agroecological niche (the ability of the lower Salt River Valley to be irrigated) for producing crops to supply local soldiers and miners led to reexcavation of prehistoric canals and success in establishing agriculture.
E2:	Catastrophic floods in 1890 to 1891, followed by severe drought from 1898 to 1904, drew cooperative action and use of federal funds to build the first major reservoir.
E3:	World War I spiked demand for locally mined copper and locally grown cotton, leading to a boom in population, agriculture, and commercial activity.
Boom Years (1941–1970)	
B1:	World War II generated military industry, air bases, and new infrastructure, familiarizing people with the region.
B2:	In 1948, Motorola built an electronics facility in the valley, the first step in a process that would reshape the region's economic base, attract an influx of people looking for good jobs, and transform its rural character.
B3:	The popularity of large housing tracts that provided air-conditioned and economical dwellings created a suburban ambiance that now characterizes the entire urban area.
Southwest Metropolis Years (1971–Present)	
SM1:	National migration to the Sun Belt encouraged population, industrial, and commercial growth.
SM2:	The Groundwater Management Act, enacted in anticipation of a cut in federal funding, reversed the expansion of farming, thus decreasing agricultural acreage and diverting water from agriculture to the urban sector.
SM3:	Increased migration from Mexico and Central America changed the demographic character of the city.
SM4:	The drought of 1999 to the present, combined with concern over quality-of-life issues, may inaugurate a shift in the belief that population growth is a necessary engine driving the city's economic viability.

responses may, in turn, have altered the prevailing mental models (van der Leeuw and Redman, 2002).

The Native American Years (and Spanish Colonization to the South)

The *advent of adequate irrigation technology* (Table 7.1, NA1) around AD 500 prompted the first substantial occupation of the lower Salt and Gila rivers' alluvial plains, signaling the era's first triggering event. Before this time, sparse rainfall and high summer temperatures discouraged settlement in the lowland river valley. By AD 1000, an irrigated subsistence regime had transformed a virtually unoccupied landscape into one of North America's major prehistoric population and cultural centers (Bayman, 2001; Crown and Judge, 1991). Hohokam villages were established across a broad area of the Sonoran Desert, with more dense settlements in the lower Salt River Valley, in the vicinity of modern-day Phoenix. The Hohokam used an elaborate suite of agricultural strategies focusing on water control and irrigation technology. They built hundreds of kilometers of canals, including one canal exceeding 24 km. In addition, we now believe that the Hohokam constructed reservoirs (some based on former canals, others more pondlike) to store water during low-water months. Distinctive styles of craft production, architecture, and mortuary practices marked the crystallization of the Hohokam tradition. During this period, communities of dispersed clusters of pit houses were situated along the canals, distinguished by what archaeologists have interpreted as *ball courts*. Population estimates vary widely, but as many as 10,000 people likely occupied the Salt River Valley and farmed much of the same land that European American settlers would come to use during the late 19th century.

The basic agricultural product in the Salt River Valley was maize, with the region offering higher yields and the possibility of two growing seasons per year. Maize was supplemented with cotton, beans, and squash—crops that show up less in the archaeological record, but were almost certainly key to the agricultural regime. Agave cultivation for both food and fiber was common and required less water than maize (Bayman, 2001). Agave was likely grown both in the upland areas, where irrigation canals could not reach, and possibly as field boundaries in the low-lying maize areas. The Hohokam foraged to supplement their diet with seed-bearing trees (mesquite), fruit-bearing cacti (prickly pear and saguaro), and hunting (rabbit, deer, and bighorn sheep). In addition, fish and rodents were locally available and provided a significant proportion of the protein in the Hohokam diet.

The second triggering event of the Native American Years coincides with the transformation of socioecological landscape during what archaeologists characterize as the Classic period (AD 1150–1450). A *shift in climate and associated stream flow* (Table 7.1, NA2) exacerbated a series of slow changes in the sociopolitical context, thus prompting the landscape transformation. The resulting social and political changes and reorganization involved the rapid abandonment of many ball-court villages and the construction of new monuments—platform mounds and Great Houses (Abbott, 2003). Populations concentrated along perennial water sources in aboveground, enclosed compounds and, with some exceptions, nonriverine desert settlements were largely abandoned. Although the irrigation canal system appears similar during these two periods, it is likely that the sociopolitical organization became more hierarchical and the settlements more tightly aggregated.

Over time, the focus of Hohokam settlement and farming shifted to canals that ran along slightly higher terraces somewhat farther from the river course. With this shift came an increasing reliance for resources on upland and nonriverine areas, including the introduction of irrigation agriculture in the upstream Tonto

Basin. In the Salt River Valley, maize agriculture continued to dominate the diet, but a significant shift in faunal resources may reflect an overexploitation of local resources (James, 2004). There appeared to be a decreasing dependence on large game (deer and sheep) and a decrease in the size of the fish (chubs and suckers) consumed. We are only now detailing the process of Hohokam decline, but there is human skeletal evidence suggesting that their nutrition was worsening and overall health was decreasing (Abbott, 2003). It should be noted that the significance of these factors and the relationship of declines to overexploitation and decreasing agricultural productivity has yet to be securely demonstrated.

It is clear that, by AD 1450, the major river valleys, especially the Salt River Valley, were largely abandoned and the region's overall population had diminished significantly. We consider this response—the *diminished socioecological resilience* of the Hohokam—the third triggering event of the Native American Years (Table 7.1, NA3). The slow-moving aspects of this event likely included straining their resource systems, accumulating excessive wastes (e.g., salinization, soil compaction), and developing vulnerabilities in their sociopolitical responses (diminished flexibility, constrained movement, expanded hierarchy). These factors may have reached a point where an external trigger such as flood or drought, which might have been absorbed under previous regimes, led to socioecological collapse. For the next 400 years, the population of the Salt River Valley was depleted and much of the natural vegetation and hydrological regimes reasserted themselves. Ongoing archaeological investigations focus on the ecological legacies of irrigation agriculture, which the Hohokam practiced for centuries (Schaafsma, 2003; Schaafsma and Briggs, 2007).

The Spanish explored what is now Arizona during the 16th century and established early settlements in the far south of the state late during the 17th century. Settlement focused along the Santa Cruz River, with *presidios* and small settlements established up to the Gila River. With Mexican independence during the early 19th century, interest retreated to the *presidios* and missions in the south. By 1846, there were only 1,000 Europeans living in these settlements (Schmieding, 2003). Pima Indians continued to farm along the Gila River, with early European Americans encroaching upon the surrounding area to trap and mine. Towns were eventually established at the upstream edge of the Gila's alluvial plain at Andersonville and Florence, whose inhabitants diverted much of the seasonal flow to their own fields and thereby denied water in dry years to downstream settlements of Pima Indians.

The image of the region conveyed to prospective settlers was the desert myth of the American West: harsh environment, hostile Indians (Apache and Yavapai), and farming that challenged the imagination. Since the almost complete abandonment by the Hohokam during the 14th or 15th century, central Arizona remained almost unoccupied even as Pima, Spanish, and European American activity continued to the south. The descendants of the Hohokam (admittedly small in number), however, lived on for generations along the Gila River in the so-called Pima villages. The recent Gila River Indian Water Rights Settlement is based, in part, on the long history of Pima settlement and their historic rights to the water.

The mining boom of 1861 to 1862 brought prospectors farther north and, although they did not settle in the lower Salt River Valley, they did locate in the surrounding highlands, laying the foundation for expanded settlement.

The Emergent Years

Our narrative begins in earnest in the Salt River Valley in the 1860s, occupied by a few Pima Indians who farmed along the river and by more mobile Native American groups (Papago and Apaches), and visited by a few European American trappers and miners. The Salt River flowed in episodic fashion, with heavy floods contrasting with low flow periods, via a meandering channel down the center of a large, extremely flat, alluvial valley broken only by isolated outcrops of buried mountain chains. In 1864, the General Land Office commissioned a regional survey that detailed the vegetative communities of that time. There were cottonwood/willow in the riparian areas, mesquite bosques along the margin of the first terrace, following some former canal alignments and up intermittent tributaries, creosote–salt bush in the flats (perhaps less vulnerable to flooding), and palo verde–saguaro on the *bajada* slopes and mountain remnants.

The first triggering event we highlight is the *opportunity presented by a potentially productive, yet unoccupied agroecological niche* (Table 7.1, E1) for producing crops to supply soldiers and miners. The U.S. Cavalry had established the Fort McDowell outpost at the upstream margin of the floodplain to limit Apache movement. Because there was virtually no farming in the valley at that time, the Fort had to import its food and fodder at a substantial expense. In 1867, Jack Swilling, an adventurer who served in the Mexican, Civil, and Indian wars and then tried to promote mining and farming north of the Salt River, recognized traces of prehistoric irrigation ditches along the Salt River. Swilling was quick to realize that these ditches afforded a prospect to grow food and fodder locally (Gammage, 1999; Luckingham, 1989). He and his colleagues dug out one of these canals, grew crops, and initiated the farming settlement that eventually grew into Phoenix (a new settlement arising from the ashes of an old), now the fifth largest city in the United States. During the next 20 years, Swilling and other European American settlers reexcavated ancient canals, dug new canals, and rapidly established farmlands, demonstrating the impressive agricultural potential of the Salt River Valley (Gammage, 1999; Kupel, 2003; Luckingham, 1989; Smith, 1986). The farming community grew quickly, and Maricopa County was established in 1871 with Phoenix as its county seat. Another change affecting the growing perception of the desirability of farming was the Desert Land Act of 1877. This legislation increased the land area of an initial homestead from 160 to 640 acres, recognizing that, to make a go of it, a larger track of land was needed in these arid lands than in the Great Plains states where homesteading had begun.

This first event built upon local ecological conditions and available agricultural know-how, precipitated by an abrupt change in local settlement patterns. This change created an opportunity for economic gain through investment that would establish irrigation agriculture in the region. A shift in perception had occurred: from linking the economic potential of the Salt River Valley to mining, to recognizing that irrigation farming offered vast economic opportunity in a seemingly inhospitable desert. This shift in regional perception may be considered part of a much larger phenomenon that characterized the entrepreneurial spirit and western movement after the Civil War. The local management response, carried out in part by Civil War veterans, was to seek their riches by reestablishing irrigated farmlands in the Salt River Valley. More than planting new crops occurred. The response was to transform the local hydrology, flora, fauna, and human settlement, and thereby redefine the potential opportunities and eventual vulnerabilities of the landscape.

Although early farmers proclaimed their success, their task was difficult, and their transformation of the landscape was not as complete as they wished. Just as farmers took advantage of natural features such as topography and soil, nature took advantage of the changed hydrology engendered by the irrigation canals. In some ways, canals functioned as ecosystems analogous to rivers, carrying sediment and dissolved substances, and supporting fauna and flora of their own. Riparian areas developed along their banks that would house their own characteristic assemblages of fauna and flora. Some of these changes must have provided the landscape with a pastoral charm, especially for new settlers who had arrived from more temperate locations. Algae, cattails, other water plants, and fish would thrive in the canals whereas grasses, shrubs, and willows would have grown along the canal banks, attracting waterfowl, ground squirrels, and other small mammals. Despite the positive aesthetic aspects of this "created nature," these colonizing organisms diminished the efficiency of the canal for agriculture: by absorbing water from the canal, firming the canal banks in some places while weakening them in others, and slowing the flow of water. In addition, evaporation from canal surfaces (the annual pan evaporation rate in Phoenix is now 200 cm/year) and ground seepage (estimates range from 30% to 60% for unlined canals) would reduce the water available for crops and thus alter the local environmental characteristics. Removing water from the main channel of the Salt River also changed the characteristics of the river itself. With diminished volume, the Salt may have meandered more broadly, with heavier sediments, more dissolved nutrients, and a straighter channel coming only during floods. However, the people who built head gates to capture water for their canals and who lived or farmed near the rivers would probably have worked to keep the river within as narrow and fixed a channel as possible by encouraging the formation of levees and other means (Kupel, 2003).

These canals were part of the cultural history of the valley. Most people were farmers and lived in close proximity to the water. Canals were part of public places where people swam, slept, and picnicked. They were shady places that allowed people to escape the desert heat. More than 50,000 trees were planted in the city, and Phoenix boosters began a campaign in the 1920s to "Do away with the Desert." The popular ethic was to remove the desert and create an oasis landscape based on irrigation agriculture (Gober, 2005; Schmieding, 2003).

The redesign of the landscape by irrigators created unintended responses from natural forces and changes in their own social relations. The second half of the 19th century in the United States was a time of unbridled capitalism. Many of those headed west were not only seeking their own piece of land to farm, but were hoping to benefit as entrepreneurial land speculators. Although private property and individual water rights allocated according to the principle of "prior appropriation" were at the very foundation of the early settlement of the region, these principles were not always consistent with the conditions the natural regime established. Prior appropriation was the basis of the legal principle that the first settlers along the river had first call on withdrawing a reasonable quantity of water based on their continued use of this water. Although farmland may have been divisible into finite units for individual ownership, water that gave value to the land did not divide as easily. For example, if a downstream farmer had water rights before the upstream farmer, during a dry year without enough water for both farmers, only constant vigilance (and perhaps armed conflict) would keep the upstream farmer from taking enough water to avoid calamity for his fields and family, thereby depriving water to the downstream farmer with the "superior" water rights. Scenarios such as these led to informal agreements and early legal policy on water distribution, such as the Howell Code and the Kibbey Decision, which provided a baseline for water rights adjudication (Kupel, 2003; Schmieding, 2003).

Other examples of ecological processes that encouraged collective action were the spread of pests overland, algae growth in the canals, or salinization of fields as a result of rising groundwater. In each situation, individual farmers would need the cooperation of all farmers in an area, or perhaps even the entire valley, to redress the problem adequately. Although there was clear competition between major players, cooperative action was an early hallmark of irrigation farming in the Salt River Valley, leading to the creation of social, economic, and political institutions that reflected collaboration as the dominant local solution to these pressures (Kupel, 2003; Smith, 1986). Why did cooperation dominate over conflict? Three possible explanations could be advanced at this point. First, there was more potential farmland and available water than the first farmers could consume. Second, the land that was easily irrigated was distributed in a broad pattern rather than in a narrow line along the river, meaning that considerable farmland could only be reached by canals that passed through farmland owned by others. Third, early leaders may have been sufficiently influential to rally people into collaborative associations.

By 1885, agricultural holdings had spread so that the extent of canals on the north side of the river was insufficient for the demand for land. There remained an enormous amount of flat land available that could easily be farmed if water were brought to it. This demand led to the construction of the Arizona Canal, the most ambitious project to date (62 km long). It was the first major canal that went beyond the geographical extent of prehistoric Hohokam canals, thereby bringing more upstream land and land farther from the river under cultivation (Gammage, 1999; Kupel, 2003; Smith, 1986). The geographical separation of the Arizona Canal from the Salt River channel encouraged the formation of the new town centers of Scottsdale, Glendale, and Peoria north of the river; and the excavation of the Highland and Consolidated canals south of the river sparked the eventual establishment of the towns of Chandler, Gilbert, and Queen Creek. After completion of the Arizona Canal, the first railroad reached Phoenix, facilitating the shipment of local products to more distant customers. The extent of farmlands continued to grow, and by 1890 there were 100,000 acres under cultivation and almost 11,000 people living in Maricopa County, a marked increase from the 300 people recorded in the 1870 census (Sargent, 1988). That the Salt River carried more water than the earlier settled Gila, Santa Cruz, and San Pedro rivers to the south, and that it coursed through a valley with a larger alluvial plain that could be easily irrigated, were important ecological features and permitted Salt River farming quickly to outstrip its neighbors on those other rivers to the south.

This generally held self-image of a relatively sparsely settled region, able to accommodate rapid growth that knew few limits, was altered by the second major triggering event of the Emergent Years: *catastrophic floods in 1890 and 1891, followed by severe drought from 1898 to 1904* (Table 7.1, E2) (Kupel, 2003; Smith, 1986). These serious, but not extraordinary, environmental events threatened the existence of this young farming center, leading the local citizenry to respond on two levels. Locally, many bound themselves together into the Salt River Water Users Association (SRWUA) in 1903, later to be renamed the Salt River Project (SRP), which still manages most of the water distributed in the valley, whereas some of the more influential citizens took it upon themselves to lobby in Washington for a National Reclamation Act that would transform many regions in the West (Kupel, 2003; Smith, 1986). These two responses were efforts to overcome the same set of threats, and worked together at different scales to establish lasting changes in the surface-water regime of the Salt River Valley and its managing political–economic organization. The first response—collective local action—changed the scale of water management in the valley, from individual farmers and small canal companies to basically one supplier of water administered by all users. This semipublic entity, in which 1 acre equaled one vote, reflected a relatively egalitarian attitude. Initially, homestead size was capped, but, over time, this rule concentrated power in the hands of a small number of large landowners. The second response was a classic cross-scale interaction of pressure produced by local lobbyists leading to national-level changes that, in turn, affected a much larger region and population than first promulgated the pressure (Mawn, 1979; Smith, 1986; Zarbin, 1997).

Soon after its formation, the SRWUA convinced federal authorities to build Roosevelt Dam, 60 mi. upstream from Phoenix at the confluence of Tonto Creek and the Salt River. In 1904, a contract was negotiated that promised to repay the federal government the cost of building the dam, using members' land as collateral. Completed in 1911, Roosevelt Dam was then the tallest masonry dam in the world and created what was then the largest artificial lake in the world. The SRP (the descendent of SRWUA) was the first multipurpose reclamation project in the country that provided hydroelectric power, water delivery, and protection from floods for a growing desert metropolis (Gammage, 1999; Smith, 1986).

From the beginning, regional and national perception of the valley was important to the first settlers, who acted as much like entrepreneurs as farmers. They initiated a long tradition of boosterism of the Salt River Valley by those who lived there and potentially would profit by more people joining them. At first, Swilling and other early settlers, who controlled land near the river and the canals that would feed new farms, promoted the region as an untapped agricultural bonanza, but one that took courage to face both the environmental and social uncertainty of the "Wild West" and the extreme climatic conditions. However, with the completion of Roosevelt Dam to control floods and enhance availability of irrigation water during dry years, as well as the admission of Arizona to the Union with Phoenix as its capital in 1912, the Salt River Valley was transformed into a much more developed and predictable place. The mental model of the next generation of boosters was to promote Phoenix as the "Garden City of the West" (Gammage, 1999; Gober, 2005). Although its citizens and the nation perceived Phoenix as an agricultural city with a future intimately tied to agricultural productivity, more and more of the region's livelihoods were beginning to be derived from government, finance, and commerce.

By 1912, much of the landscape of the Salt River Valley had been transformed, with 220,000 acres of farming, a population of 35,000, and seven towns scattered across the plain (Fig. 7.3). Although the Salt River was seemingly controlled, in fact the impact of its variability had been merely altered, not eliminated. The Roosevelt Dam controlled annual variations in flow, but multiyear accumulations leading to large flood events, as well as prolonged periods of below-average stream flow, still threatened the valley. Moreover, one of the large tributaries of the Salt River—the Verde River—enters below Roosevelt Dam, and its flow would remain uncontrolled for 20 more years. The temporal pattern of water availability and threat to valley inhabitants had not been eliminated, only modified.

The first two triggering events in the Emergent Years were local, or at most regional, in scope, and primarily related economics to environmental resources and processes. The third event, *World War I* (Table 7.1, E3), was international in scope and largely political in origin. With the start of hostilities in 1914, copper prices soared and mining towns northwest and southeast of Phoenix boomed. Their success drove demand for agricultural products and brought vast new income to the state. In addition, the war disrupted access to Egyptian long-staple cotton, essential to the manufacture of airplane fabric, balloons, and cord tires. Local farmers shifted their planting patterns so that Pima cotton became the premier crop, replacing the less lucrative, but soil-enriching alfalfa (Sargent, 1988, p. 56). Happily for Salt River farmers, the completion of Roosevelt Dam had changed the agricultural potential of the valley in advance of this event; a more secure source of abundant water enabled the expansion of agrarian production. By 1920, 75% of irrigated land in the Salt River Valley was devoted to cotton production. Despite the momentary collapse in prices of cotton (and copper) at the end of the war, the emerging national economic prosperity of the 1920s restored the importance of cotton and copper to the Arizona economy. Cotton continues to be the most important crop in central Arizona today.

The 1920s also saw the national perception of Arizona evolve to reflect the state's value beyond agriculture and mining. Phoenix promoted itself as a city that could be a year-round vacation place and a Mecca for health seekers, particularly those with respiratory illnesses. Just as affluent invalids had visited the region after completion of the railroad connection in the 1880s, the booming popularity of the automobile encouraged new generations of health and sun seekers to vacation in and move to Phoenix. Local entrepreneurs in the 1920s responded in turn, developing major resorts and hotels, and opening commercial airline connections. Tourism would increase and eventually outpace cotton and copper in its contribution to the state economy.

The global economic depression of the 1930s strongly affected Arizona, with an estimated 50,000 citizens leaving the state to search for employment opportunities elsewhere (Sargent, 1988, p. 58). All was not grim, however. More than most states, Arizona—and Phoenix in particular—benefited from the New Deal's federal investments. These policies funded new parks, schools, and street construction, and were responsible for constructing six additional dams on the Salt, Verde, and Agua Fria rivers. All this building activity transformed Phoenix into a regional retail center and, more important, seeded the infrastructure that the enormous demand, soon to emerge in the coming decades, would require. Alongside these accomplishments were New Deal innovations in banking, especially in construction financing. Although rapid growth was not immediate, these innovations created the financial structure and initial building experiences for entrepreneurs like Walter Bimson and Del E. Webb, who would introduce a new approach to housing construction and marketing as demand increased during the region's next growth phase (Gammage, 1999).

The Boom Years

By the end of the Emergent Years, much of the foundation for Phoenix's subsequent growth was in place, but the dreams of local boosters were yet unrealized. The farming system evidenced the hydrological infrastructure and institutional basis that would serve the valley for the next 50 years. With new dams, reservoirs, and canals, agricultural lands had expanded to almost their maximum geographical extent. During the following decades, they achieved their greatest coverage, with only modest extensions of the canals and filling in of previously unused lands. The region's abundant flat land and easily extended grid of wide streets kept the cost of new residential construction low, yet in 1940, only 65,000 people lived in the city. The triggering event that pushed Phoenix over the threshold and onto a trajectory of rapid, nonagrarian population growth and national prominence occurred far away and was not engineered by the local boosterism.

World War II (Table 7.1, B1) transformed Phoenix, as it did many other parts of the world, and can be seen as the triggering event that initiated the Boom Years. Part of this change was the result of technology and advances in air conditioning and air travel, and part of it was personal experience—exposure to the area by many pilots and other personnel assigned to bases in the valley (Gammage, 1999). Arizona's year-round clear skies and safe distance from the Pacific coast made it a favored location for air bases and manufacturing facilities, and wartime and postwar corporate America responded. Luke Air Force Base opened at the onset of the war, served as temporary home for thousands of young men, and was soon followed by Williams Air Force Base. A growing familiarity with the region increased national awareness of the attractions of Phoenix and led many soldiers to return after the war. The emphasis on military air travel also had implications for the burgeoning commercial air industry. Air travel made Phoenix "closer" to other cities and obviated the need to spend days traveling across the vast, arid western landscape.

Advances in air-conditioning technology were also essential to a new conception of life in Phoenix, making it an attractive place in which to live as well as visit. Air conditioning had existed for some time, but it was only after World War II that residential-size units were produced in quantity. Soon, Phoenix was installing more central air-conditioning systems in houses than anywhere else in the nation. Providing an escape from the intense summer heat, air conditioning transformed a winter haven into an attractive place to live year-round.

Another key triggering event of the early postwar years occurred in 1948, when *Motorola decided to open a manufacturing facility* (Table 7.1, B2) in the region for defense electronics (Sargent, 1988). Motorola saw the advantages of a dry climate where workers were happy to relocate and land remained cheap. They soon were among the valley's most active boosters, and the company expanded until it was the largest employer in the state. Other companies soon established their own high-tech facilities, until the region justifiably became known as *the Silicon Desert*. By 1955, manufacturing had overtaken farming as the main source of income in the valley, with tourism and home construction growing rapidly.

Being a manufacturing center with abundant jobs represented the second aspect of the change in the perception of Phoenix, both locally and nationally. Phoenix was no longer just a successful farming center; it had become a city that was a great place to live and home to fast-growing industries. Phoenix benefited from the national postwar movement of people to the Sun Belt and, in many ways, served as one of the icons of this era. In the 1950s, the city grew by more than 300%—the fastest of any large city in America (Gammage, 1999). Some of that growth came from only annexing neighboring land, but much of it was from newly arrived residents. With the movement of people into the region came a need to build attractive, affordable housing in large numbers. Phoenix was so successful at this that *housing, and the suburban ambiance it spawned, characterized the region and attracted new immigrants* (Table 7.1, B3), and we consider it the third triggering event of the Boom Years. Low-cost land, abundant energy, efficiency in construction, and simple design led to the dominance of the ranch house in Phoenix and elsewhere across America. Its single floor, connecting patios, and outdoor space were well suited to the abundant flat land that had been farmed, but was now increasingly being converted into residential developments. Date farms and citrus orchards came with a water allotment and were considered by subdivision developers to be very attractive sites onto which to build the ranch-style homes of the 1950s and '60s (Gammage, 1999). Phoenix builders were very good at mass production of this type of home and were pioneers in assembling large tracts of land into integrated housing developments. John F. Long introduced this concept to Phoenix with his Maryvale development and was recognized by a national magazine in 1958 as the number-one low-cost homebuilder in America (Gammage, 1999). Along with new residential developments of ranch houses emerged the concept of the shopping center. These centers

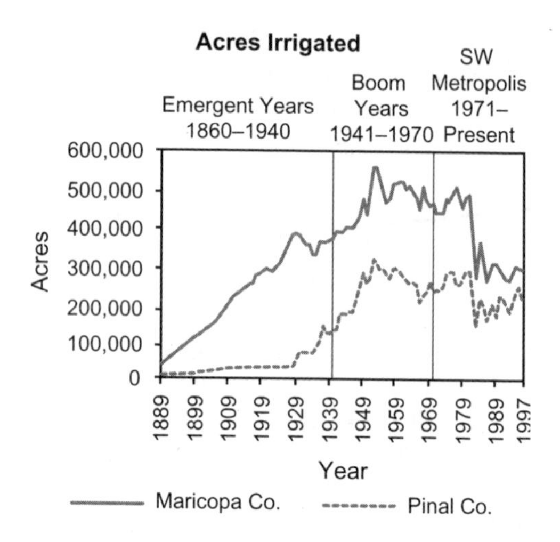

Figure 7.5 Irrigated acres in Maricopa and Pinal counties.

were often built at the corner of section-line roads, surrounded by housing and anchored by a grocery store with abundant surface parking.

These new patterns of growth led to a major transformation in the urban environment. Farmland that had dominated the landscape continued to thrive, but the valley was being converted into new housing developments. The seven dispersed towns separated by farmland and open space had now merged into a relatively continuous spread of streets, businesses, and housing. Empty land was still found within this urban core and a broad fringe area with intermingled farms and housing developments, but the settlement pattern had changed to that of a vast urban sprawl (even by the mid 1950s) surrounded by a broad band of farmlands, all set within the *bajadas* and mountains of the upper Sonoran Desert (Fig. 7.3). Farmland in Maricopa County reached its peak of 550,000 irrigated acres by 1953, an extent that held relatively constant for the next 25 years (Fig. 7.5), despite a more than doubling of the urban population. Cotton remained the primary crop, whereas wheat and orchards increased in importance.

Interesting changes were evident in the local farmlands. Because farms near the center of the valley were being sold and converted into urban areas, new lands were being irrigated, often at the fringe of former farmland or in areas that had been undeveloped in the past. These changes were enabled partly by the completion of Waddell Dam on the Agua Fria in 1927 and Horseshoe Dam on the Verde River in 1948, and partly by the great increase in groundwater pumping that allowed for irrigation in areas away from the canals. The growth in irrigated acres during the 1940s and the maintenance of farmland acreage in the face of expanding cities during the 1950s, '60s, and '70s meant a filling-in of farming areas with more continuous irrigated lands.

The 1950s also saw the greatest number of farms in Maricopa and Pinal counties, peaking at more than 7,500 farms compared with just fewer than 3,000 in the region both at the completion of Roosevelt Dam in 1911 and in a recent agricultural survey in 2002 (USDA, 2002b). Much of the variability in farm numbers

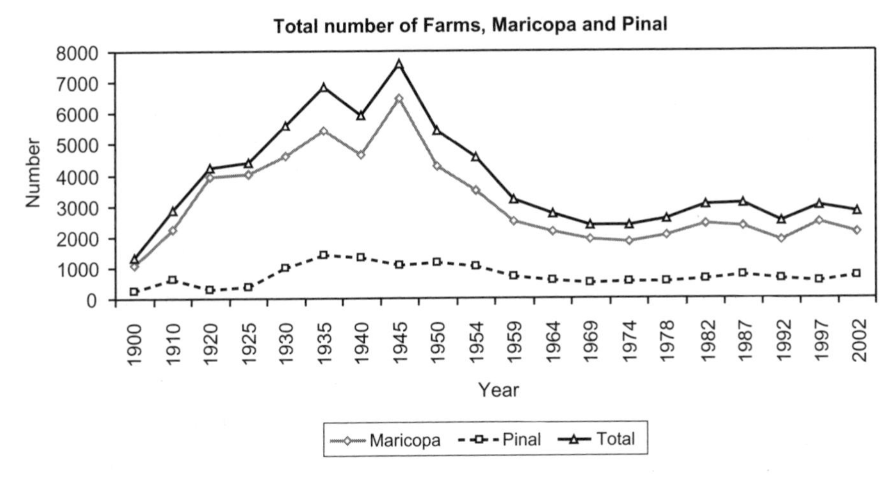

Figure 7.6 Total number of farms in Maricopa and Pinal counties, 1900–2002.

comes from Maricopa County, where agricultural growth exploded during the early years (from 1,000 in 1900 to 6,500 in 1950), decreased dramatically as the city consumed the nearby farms, and leveled off at about 2,000 farms from 1970 until the present (Fig. 7.6). In the case of Pinal County, the number of farms did not vary as much as their size. By 1940, Pinal contained its maximum number (about 1,300), a figure that slowly and steadily has decreased until the present-day 800 farms. Interestingly, the average farm size in Maricopa has changed gradually—from only a few hundred acres from 1900 to 1950, until the 1950s and 1960s when the average grew to just more than 1,000 acres, and now back down to about 300 acres (Fig. 7.7). This number contrasts sharply with Pinal County, where there were always fewer farms, but where they quickly grew as the Depression ended and the boom began—up to an average size of almost 5,000 acres throughout the 1960s and 1970s. During the past 20 years, Pinal County farms have become smaller, but they still average more than double the size of those in their neighboring county. Clearly, the urban pressure has never been substantial in Pinal County and, in fact, some Maricopa County farmers who sold their holdings shifted their efforts to farms in Pinal. The relatively large number of small farms that have characterized Maricopa County reflects the originally small homestead size and the fact that many were family farms and often run by people in the city involved with other pursuits. The maintenance of numerous small farms into the latter 20th century also relates to the importance of land speculation; those who owned the farmlands hoped to sell their land at a high price.

During this era, especially the 1940s and 1950s, the climate experienced an extended dry cycle with below-average stream flow in 17 of the 20 years. This long, dry cycle could be considered another sustained triggering event, pressuring the system. Although drought concerned some, the common perception was that technology could meet the challenge. Without derailing the meteoric population

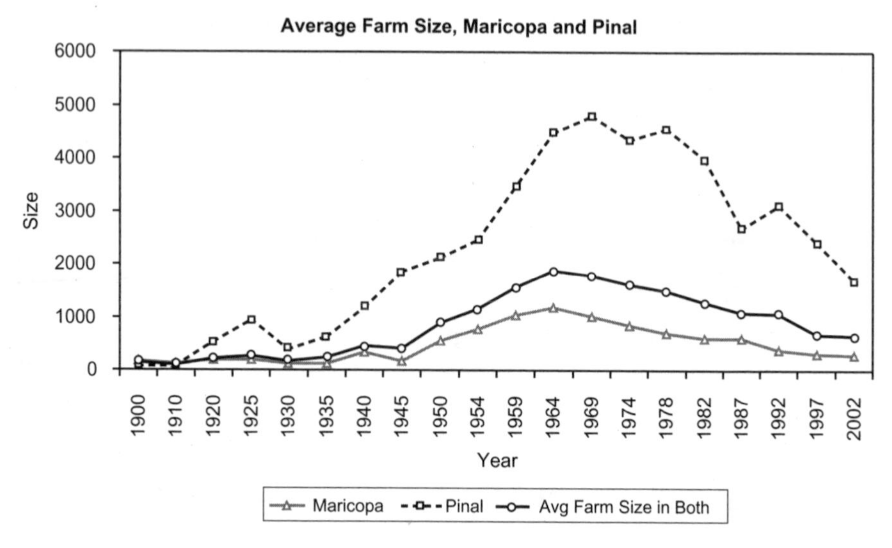

Figure 7.7 Average farm size for Maricopa and Pinal counties, 1900–2002.

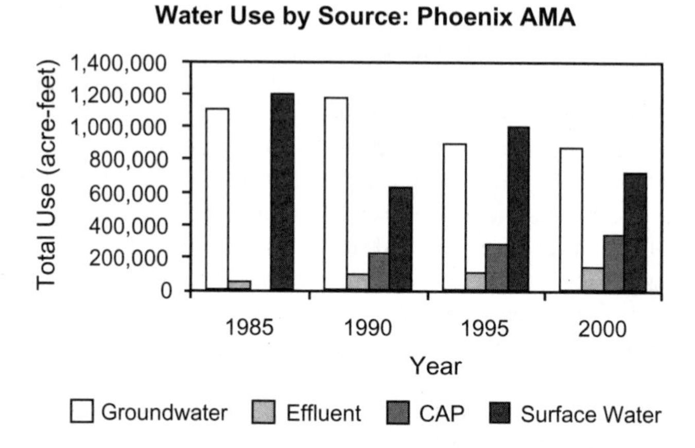

Figure 7.8 Water sources for the Phoenix Active Management Area.

growth of the era, water needs were met by dramatically increasing the amount of groundwater pumped (Fig. 7.8) and by making the water distribution system more efficient. Major canals were lined to limit water lost to percolation, and secondary canals, especially those in urban areas, were covered or put into pipes partly to save water from evaporation and seepage and partly to enhance the urban nature of the landscape. Diminishing the amount of water "lost" from the canals seriously affected the ecology of the canals and the man-made riparian areas that had dominated the canal-laced landscapes during the Emergent Years. Canal banks no longer supported trees and shrubs, because all the water was reserved for human use. Within the growing cities, much of that water use was for mesic landscaping, where bermudagrass lawns, fast-growing trees, and exotic species such as the African sumac tree, citrus trees, and the Mexican fan palm dominated the new master-planned subdivisions (Gammage, 1999; Weworski, 1999).

The ecological imprint of urban growth is revealed in myriad ways. First, irrigated farmland fills in and is pushed out from the center to the practical limits of gravitation-fed irrigation and soil. Second, the patchwork of towns interspersed within a largely agrarian landscape is transformed into a large urban core with mesic landscaping, with a fringe where farms and housing tracts intermingle, all surrounded by a periphery of farmland where the topography allows, and desert beyond that. Third, efforts to make water use more efficient would progressively transform canal ecosystems from a more streamlike system with a riparian zone to one more like an open pipe where water is confined for human consumption. This sequence has occurred because the shared perception of most people is that in the competition for water use, human-directed uses have a higher value than nonhuman-directed uses. Fourth, and finally, demand for water is so great that all available water is withdrawn from the Salt River before it passes through Phoenix. What was once a river with regular flow and associated wetlands was transformed into a dry gravel strip that only held water during flood episodes. During the span of this narrative, priorities among human-directed uses changed order, with mining, agriculture, and now municipal uses being the most highly valued

How did changes in the landscape and activities of inhabitants alter their perception of the urban ecosystem in which they participated? First, the hydrology of the irrigated agriculture was becoming more engineered and less visible to most citizens. Canals were being lined and losing their riparian appearance, other canals were placed underground, and groundwater was extracted to supplement and, in some places, replace surface water. As the city limits spread, an increasing number of citizens did not have contact with farmland or even see it as part of their region. In tandem with this visual separation, farming lost its paramount position in the local economy and politics. Manufacturing was number one, and other industries such as tourism, construction, retail, and government employment were soon to surpass agriculture as well. It was not that farming was failing—quite the opposite; during the Boom Years, farming was booming as well. With the introduction of more chemical fertilizers, farm machinery, and supplemental groundwater, the productivity of most crops grown in the region more than doubled in relation to national averages. The key was that farming interests were still seen as important, but only as one segment of the overall interests of Phoenix. For most people, especially the boosters, Phoenix now represented industry, affordable housing, a higher quality of life, and the excitement of (and profit to be derived from) growth. Phoenix was now the largest city in the Southwest and the fastest growing in the nation. Its boosters promoted a "can-do" attitude and were able to push through Congress the ambitious Central Arizona Project to build a 300-mi. canal to bring Colorado River water to Phoenix and Tucson. Residents' self-image was also on the upswing. For example, Phoenicians insisted on passing in 1958—despite the reticence of the Board of Regents—a referendum to change the name of the local college from Arizona State Teachers College to Arizona State University (Gammage, 1999).

Southwest Metropolis

In the past 30 years Phoenix has grown to become the fifth largest city in the nation and the fastest growing among the largest cities (U.S. Census Bureau, 2002). It is now the undisputed central city of the Southwest with the sixth busiest airport in the nation, more than double the number of people of any other city in the region, and a growth trajectory that has not faltered even during tough economic times. What perceptions have driven that growth and what triggering events have occurred to modify these perceptions and trajectory?

Unlike the Boom Years, the first triggering event of this era was not a specific event of short duration, but more like a continuous series of pressures and opportunities. First, the region continued to benefit from *the national migration of interest, companies, and people to the Sun Belt* (Table 7.1, SM1). Jobs, affordable housing, an outdoor lifestyle, all brought people to Arizona. Second, Phoenix has the highest rate of home ownership among the large cities of America (Brookings Institution, 2003). Finally, there was also a shared vision of Phoenix as a place not only to get a job, but to get rich quick, as reflected in wide participation and frequent success of real estate entrepreneurs and private citizens turned speculators. This continued immigration, especially of people who hoped to own their own homes, meant that new houses would need to be built in unprecedented numbers (currently more than 50,000 per year). The spread of new housing into former farmlands continued apace, but was soon equaled by the conversion of former desert terrain into residential developments—one of the results of this triggering event that transformed the regional landscape in new ways. Real estate entrepreneurs encouraged this pattern. They sought to purchase large tracts of inexpensive land, often looking well beyond the current fringe of the city, developing them into integrated, master-planned communities. This leapfrog pattern of development beyond current city limits had the effect of increasing the rate of spread and only encouraging infill just behind the leapfrogged development, rather that staying within current urban limits and increasing core densities (Fig. 7.3). This pattern of ever-spreading urban settlement consuming the remaining farmlands may be challenged by the countervailing force of increasing commuting distance and the heat relieving property of surrounding farmlands.

The shift to building in the desert did not yield surplus water that could be dedicated to the new housing developments, as had the conversion of previous farmland. The tendency when old farms were sold for housing had been for a few farmers to retire, but more often to establish new farms on unused Maricopa County lands or to move south to Pinal County, hence maintaining total farming acreage in the region. This continuing need for agricultural water, added to the demand of the growing urban population, led to developing tensions about water supply in the 1970s, despite relatively favorable precipitation patterns. The perception, especially to outsiders, was of Arizonans disregarding appropriate limitations on water use in their desert city. The focus became falling groundwater levels, as a result of 30 years of a higher rate of withdrawal than recharge. Possible subsidence, extra expense of drilling wells and pumping water from greater depths, environmentalists' concerns of diminished water for native vegetation, and the ultimate fear of just plain running out of water were considerations for differing segments of the local population, part of their shifting mental models of life in the desert.

These fears led to the second triggering event of the era: Secretary of Interior Cecil Andrus threatened to stop funding the Central Arizona Project unless the people of Arizona took seriously the maintenance of groundwater levels (Kupel, 2003). The Central Arizona Project was an expensive, federally funded project to bring Colorado River water to central Arizona, regarded locally as the "final" insurance against running out of water. Under enormous political pressure from then-governor Bruce Babbitt and others, rival interest groups cooperated in creating the *Groundwater Management Act of 1980* (Table 7.1, SM2), which transformed expectations and actions in terms of water use. The goals were to balance groundwater extraction with the rate of natural recharge by encouraging surface-water use, by prohibiting new farms or increased consumption by existing farms in the Phoenix and Tucson regions, and by setting conservation standards for municipal users (Kupel, 2003).

The Groundwater Management Act provoked diverse reactions. First, prohibiting irrigation of new lands around Phoenix meant that the maximum extent of farmlands was fixed at boundaries from the 1980s; hence, total irrigated acreage in the region would diminish as urbanism swallowed farms. This trend has continued during the past 20 years (Figs. 7.8 and 7.9), and most farmers on the original SRP lands are expected to have sold their farms for residential construction by 2010. Some of these farmers have moved their operations to Pinal County, where the Act's restrictions are different, but many have ceased to farm, diminishing the state's overall agricultural acreage. For the remaining farmers, limitations on water use—allocated by acre, not by crop—has encouraged them to select crops with high economic value per quantity of water used, such as seed crops and vegetables. Alfalfa, alternatively, requires more irrigation but less labor and capital, and is preferred by many farmers with sufficient water allocation who are waiting to be bought out.

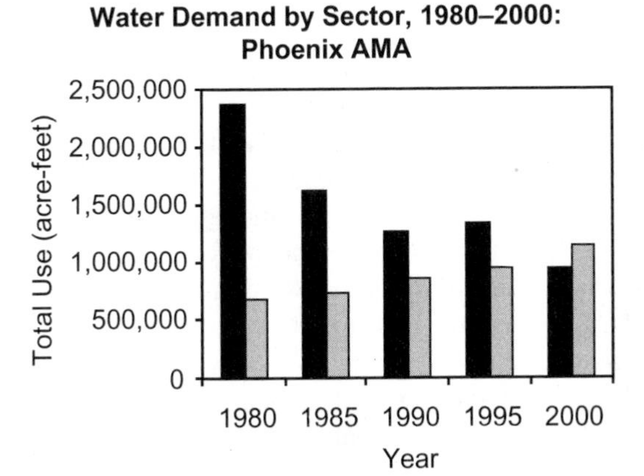

Figure 7.9 Water users for the Phoenix Active Management Area, 1980–2000. AG, agriculture; M & I, mining and industry.

Another reaction came from lawyers and managers aiming to rationalize water consumption without jeopardizing rapid growth in Arizona's cities. The regulations they devised for the Groundwater Management Act allow developers or municipalities not adjacent to canal water sources to join an association (replenishment district). This arrangement permits them to withdraw groundwater in their locality in return for payment to the Central Arizona Project, whose water is purchased and pumped back into the ground elsewhere, in an attempt to replenish the aquifer. Although the principle here seems sound, there are many questions concerning the "banking" scheme, particularly related to spatial variations in drawdown and replenishment, and to water-quality issues. The goal of the Groundwater Management Act was to reach "safe yield"; meaning, the amount of water entering the aquifer roughly equaled the amount withdrawn. The assumption behind these regulations was that water allowed to seep into the ground in one part of the basin would be stored there for later extraction anywhere in the basin. Not only is this a legal regulation, but it is a widely held perception of water managers, providing the rationale for new residential communities as long as the developers and residents pay for the surface water available in the Central Arizona Project. These legal institutions provide a mechanism for continued urban development into areas far from the current surface-water infrastructure, in some ways contradicting the law's intent. These arrangements usually affect desert locations and allow leapfrog developments to remain practical even under Groundwater Management Act guidelines (Fig. 7.3).

The third triggering event of the Southwest Metropolis Years involves *the changing demographic character of the city.* In a study conducted by the Brookings Institution (2003), it was found that Phoenix has a relatively low number of ethnic minority residents (44% vs. 62% for all cities in the study), yet this number is growing rapidly. Between 1990 and 2000, Phoenicians of Hispanic origin increased from one fifth to one third. We identify this as a gradual but significant change to the urban fabric, and a triggering event of as-yet-to-be-determined impact (Table 7.1, SM3). How the changing ethnic mix of residents will affect decisions about water use, farming, and related economic and quality-of-life issues is difficult to determine; however, it will tinker with the mental model of the region.

The very changes in flows of basic resources that have made Phoenix hospitable and economically successful have radically altered species composition in the city from what occurs in the surrounding Sonoran Desert. Supplemented resources, particularly water, increase and stabilize productivity, dampening seasonal and interannual fluctuations in species diversity, elevating abundances, and altering feeding behaviors (Faeth et al., 2005). Relative abundances have greatly increased select native and nonnative generalist bird species (sparrows and doves), ground arthropod generalists (ants, springtails, and mites), plant-feeding arthropods (aphids, white flies), generalist pollinating arthropods (honey bees), jumping spiders (Lycosidae), and fence lizards at the expense of more specialized species in these respective groups (Hostetler and Knowles-Yánez, 2002; McIntyre et al., 2001; Shochat et al., 2004). Most of these changes in species abundances have not been caused by purposeful human extirpation, but are the result of conversions of habitats from desert to agriculture and urban residential with the accompanying changes in water and other flows.

Several patterns that characterize the Southwest Metropolis Years affect the environment of central Arizona and the agrarian and urban landscapes of the Phoenix region. First, a number of actions have allowed and perhaps encouraged the continued geographical spread of urban development. The Groundwater Management Act and subsequent legal institutions of replenishment districts and water-banking authority have enabled the construction of new housing developments in areas that were not traditionally served by surface-water sources. During this era, voters underwrote the vast expansion of the region's freeway system. The migration of primarily young people from other states and from south of the border continued at a rapid pace, driving the demand for new housing. Second, the Groundwater Management Act, which prohibited new agricultural uses of water and promulgated a pricing strategy for new residential water that subsidized the remaining farmers' use of surface water, has stabilized and, in some cases, gradually replenished local aquifers. As a result, during 1980 to 2000, irrigated acreage in Maricopa County has diminished by more than 50% whereas the overall water use has remained relatively constant despite a doubling of population. Further easing the potential water problems was the opening of the Central Arizona Project canal in 1988, which brings more than 1 million additional acre-ft. water/year to the Phoenix–Tucson region (Fig. 7.8). Although the availability of supplemental water has allowed the continuation of mesic landscapes, the combination of lawns and broadleaf trees is largely confined to older neighborhoods at the center of the metropolitan area and scattered developments in newer sections. Desert plant communities still prevail in their native locations outside the city, and interestingly have been maintained in various urban locations with elevations that are above the floodplain. In an effort to re-create desert vegetation in newer urban developments, landscapes have been designed that provide aesthetic aspects of a Sonoran Desert habitat, but with a mixture of native and exotic plants, gravel surfaces, and hoped-for low water requirements. Homeowners with these re-created desert landscapes sometimes, but not always, use less water than their mesic neighbors. Surveys have also shown that the preferred landscape form for most Phoenicians is a pattern that combines mesic and xeric areas in what we refer to as an *oasis design* and it does not use significantly more water than a completely xeric yard.

These developments have led to a strange paradox in public perception of water in central Arizona. On the one hand, the lack of rainfall (an average of <200 mm/year), the dry riverbeds that course through the city, and the surrounding desert landscapes encourage residents to worry about water availability. This fear has caused restrictions on use of potable water for golf courses and decorative lakes, hence encouraging the growth of water reuse in selected situations in the valley. Xeriscaping in public areas has become the norm, and several water treatment plants are using a wetlands approach that promotes groundwater recharge. Nevertheless, concern about the adequacy of the water supply has not invoked conservation restrictions on individuals or serious pricing disincentives on residential or industrial water use. Moreover, despite the limitation on new decorative lakes, the largest of them all—Tempe Town Lake (90 ha)—was constructed during this era. Contrary to public perception, water managers and those who are most involved with water decisions do not believe there is a local water problem, nor is one likely to develop in the near future. Hence, they are not inclined to push for serious legal requirements for residential conservation or limits on new residential construction given the potentially dire effects a "water scare" would have on public perception and possibly on the economy. But is there enough water to support continued population growth and relatively high per-capita consumption? And will the available water be of sufficiently high quality for residential use? We do not know, and new research and data compilation are necessary to answer these questions.

Into what appeared to be a relatively sanguine situation, a fourth triggering event occurred, this time prompted once again by an extreme, but not unforeseeable, climatic pattern—*drought* (Table 7.1, SM 4). Since 1999, there has been a period of very low rainfall and associated surface flows in the region. This period came after a 5-year span of relatively low rainfall, meaning that the reservoirs were already low when the intense drought began. The reservoir behind Roosevelt Dam, the main repository for storing valley water, decreased to 12% of its 1,653,043 acre-ft. capacity, and the two largest reservoirs on the climatically "independent" lower Colorado River system were down about 50%. Although this was not an extraordinary situation historically, it has attracted media attention, resulting in front-page coverage on a number of occasions (Arizona Republic, 2005). Ramifying this perception was the fact that, for the first time, SRP (which relies on local reservoirs) cut its water allocation to customers by 33%. Municipalities, having adequate financing, were able to adjust their water sources and did not pass on this water restriction to customers. However, cuts to farmers, operating at the margin of economic profitability, led them to respond by not planting some of their lands.

The recent drought and reduced SRP deliveries, occurring alongside considerable media interest in the implications of unchecked urban growth, have sparked discussion about the potentially dire long-term consequences of rising and inflexible (i.e., urban) water demand in an arid environment. Responding to heightened public concern, Arizona governor Janet Napolitano created a Drought Task Force that received the unprecedented assignment of developing plans for dealing with potential water crises in the state. Would the change in public perception during this drought foster substantial changes in public policy on growth and water use when it had failed to before? Those anxious about growth and quality-of-life issues may combine with advocates of water conservation to prompt serious policy changes, as well as modify the individual's regard for water resources. In addition, Arizona State University's Decision Center for a Desert City, funded by NSF, is analyzing the decision processes used to plan and manage water resources and urban growth, and is assisting water resource managers as they plan for increased urbanization in the region (Decision Center for a Desert City, 2007).

Ironically, serious water management may not lead to positive conservation outcomes. Water-use restrictions would likely affect the remaining natural riparian vegetation first, followed by public landscaping vegetation, and eventually residential vegetation. In Phoenix's environment, rainfall supports only a small proportion of the current urban vegetation, most of it being exotic and requiring purposeful supplements. With virtually all surface-water flow taken out of its natural courses and managed toward other objectives, an extreme scenario of water conservation would be for the urbanized region to be changed from an oasis to a landscape largely devoid of plants unless people converted their landscaping to location-appropriate native vegetation at appropriate densities. Will this happen? We doubt that the citizenry will either convert to a totally native regime or give up their urban oasis, but what environmental, economic, perceptual, and legal changes will have to happen to avert the desiccation of valley landscaping? The rate of agricultural decline in Central Arizona will largely be determined by competition for water and demand for land by residential construction. Balancing those forces will be the desire of local residents to maintain a remnant of former farming lifeways, the ameliorating climatic effect of nearby farmlands, and emerging trends toward favoring local food production.

Prospects for Conservation

How does one join the growth patterns of a burgeoning southwest metropolis with the need to conserve key habitats, preserve ecosystem function, respect biodiversity, and educate a largely urban population regarding the rewards of being environmentalists? Several pathways are possible for funding or mandating conservation activities in central Arizona—public, private, or some combination. Arizonans are pursuing all three approaches, but combination approaches are clearly gaining the most momentum. Before examining these funding streams, it is useful to address the more fundamental issue in central Arizona: What do we want to conserve?

As suggested earlier, traditional environmentalists are loath to apply their energies to conserving the urban–agrarian landscapes of Phoenix and its environs. Part of the problem is that farming transformed so much of the land. Land either remains in that use or has been further transformed by urbanization. This trajectory accounts for all the flat, low-lying terrain within reach of canal irrigation. Flat areas beyond farming have what aesthetically appears to be uninteresting vegetation and may have been denuded by cattle grazing. Beyond those areas are mesas, ridges, and small mountain chains with potentially more aesthetic appeal that have been the focus of conservation efforts in other parts of the state. Much of that land is already under public ownership and, to some degree, is being conserved as county parks, state lands, or National Forest. This gradient logically leads local conservationists to focus on the small tracts of uplands that are in private ownership, beyond the current reach of urbanization. This approach has largely relied upon the creation of local land trusts or the acquisition of land by local chapters of national organizations such as TNC. Land trusts often emerge from the efforts of a few dedicated individuals who seek to preserve a particular landscape from development. One of the most successful in the region is the McDowell Sonoran Land Trust, a grassroots organization that serves to preserve, protect, and celebrate the McDowell Sonoran Preserve. Founded in 1990, this organization has raised private funds and eventually placed a bond issue on the municipal ballot in Scottsdale that passed by a 73% margin. They contribute funds to the city's efforts to expand the preserve, which currently encompasses more than 17,000 acres, with plans to acquire more than 19,000 additional acres (once these purchases are made, the Preserve will total 57 sq. mi.). Similar efforts to conserve land, but through a private land trust, are occurring around the periphery of the Superstition Wilderness Area. Another example is the Desert Foothills Land Trust, which seeks to preserve tracts of land at the northern periphery of the metro area. Each endeavor has been successful at preserving land at a small scale and then broadening their reach by seeking partnerships with public entities or access to public funds through ballot initiatives.

Recent conservation achievements have received direct support from municipal or state leaders. The Phoenix Mountain Preserve grew out of a grassroots movement during the early 1960s in which citizens were deeply involved. As the scale of the opportunities increased, it became apparent to city leaders that a more formal approach would be beneficial. A Phoenix Mountain Preservation Council was formed and sparked a momentum for city leadership (former Mayor Skip Rimza, in particular) to spearhead a ballot initiative to earmark a small portion of the city sales tax for the purchase of lands that would expand and, in some cases, connect the various mountain preserves. Another initiative was enacted by Arizona's relatively conservative state legislature in 1996. The Arizona Preserve Initiative was designed to encourage the preservation of select parcels of state trust land in and around urban areas for open space, scenic beauty, protected plants, wildlife, and archaeology to benefit future generations. This bill was a good start, but it did not come with its own funding, leaving a tremendous financial burden on any organization or municipality that wanted to purchase parcels for preservation. A ballot initiative was passed 2 years later that allocated significant state funds for this purpose, but only if matching funds could be raised. Because of this burden, the bill has not led to large-scale land preservation and there is still a strong sentiment that more must be done to preserve key parcels of state land near urban areas.

Key business and environmental leaders began to meet unofficially in an effort to forge a compromise strategy for attaining the sometimes conflicting goals of various stakeholder groups. Among the stakeholders were real estate developers, who wanted to ensure a steady stream of new land available in urban areas for future construction; educators, who stood to reap maximum financial benefit from the sale of state lands (whose proceeds go to education); the business community, who wanted to see new housing continued to be built, but also wanted to enhance the quality of life by preserving open space to attract new industry; environmentalists, who were primarily concerned with rural land preservation; and ranchers, who wanted to maintain their inexpensive leases of rural state lands. Interestingly, the most difficult conflict was between the environmentalists (most of whom dropped out) and the ranchers, who eventually scuttled the tentative deal after 2 years of negotiations. The other unlikely opponent of inexpensive land for conservation included the education lobbyists, who saw this as diminishing their long-term financial benefit. A smaller group that focuses solely on urban lands is still meeting, and there is hope for a compromise that will lead to the immediate preservation of substantial tracts and an easier method for securing additional land for preservation over time.

An even bolder approach is being implemented just south of our study area, in Pima County around the city of Tucson. In May 2004, voters approved the Sonoran Desert Conservation Plan, which integrates all natural resource protections and land-use planning activities into one plan, rather than keeping these activities separate and often conflicting. The Pima County Board of Supervisors initiated the discussions 6 years earlier and insisted that the science-based planning should be the top priority rather than political considerations. There has been widespread participation, with a citizens' steering committee of more than 80 members, more than 400 public meetings, and more than 150 scientists contributing their expertise. In addition to the comprehensive plan, the ballot issue funded a bond authority of $174 million to support conservation land purchases. Although supporters might point to the broad participation as key to the success of this initiative, it needs to be recognized that land developers are not against conservation as much as they are anxious to have a steady stream of reasonably priced land available in desirable locations. Under the threat of possible limitations resulting from an endangered species (the pigmy owl), developers saw it was in their best interest to find a way in which both conservation and development could succeed.

The success of the Sonoran Desert Conservation Plan and its accompanying bond issue has emboldened some Maricopa County residents to consider a similar initiative. The Nature Conservancy, which has been looking for a way to capture the interest and financial support of area residents, is in the early planning stages of developing such a plan. It is already being discussed whether Arizona State University scientists involved in the CAP LTER and the AgTrans Project could aid in developing science-based plans for conservation within the county. Can a comprehensive plan in a rapidly urbanizing region like central Arizona aid in mitigating habitat fragmentation, ensuring watershed integrity, and encouraging remaining biodiversity? These important ecological questions require a rational, comprehensive approach to conservation and development needs—one that understands how urbanization and human activities can be guided to maintain maximum environmental values. At the same time, our own work on this case study suggests additional lines of inquiry that could lead to innovative approaches to conservation.

First, high-priority endangered habitats or threatened species—the target of traditional conservation efforts—should be identified, and strategies for their protection devised. Given the long-term human use of landscapes near the metropolitan area, it is unlikely that there will be significant quantities of these types of targets.

Second would be landscapes that could be protected or restored so that they could serve as substitute habitats for threatened or valued species, once again provide substantial ecosystem services, and offer aesthetic rewards to the visitor. This research area has great promise, especially considering the federal funds available for local efforts to provide replacement habitats for the southwestern willow flycatcher and other endangered species. Of special interest here would be the recognition that the restoration was not re-creating a truly "natural" habitat, but a designed habitat where the designers decided what species, food webs, and material flows to favor in the restored habitat. We believe this is an essential exercise for modern conservationists and could lay the foundation for a rational approach to land transformation in the coming generation.

The third line of inquiry is a variation of the second, but one in which education and recreation are the primary objectives. As the globe becomes more urbanized and as many of those urban residents have medium or low incomes, it is likely that most citizens will not experience "nature" that is distant from their homes, such as national parks or wilderness areas. Yet, for a wide variety of reasons it is essential that meaningful environmental exposure is available to them. We need to seek readily accessible landscapes that can be protected or restored to a point where meaningful environmental experiences can be had. Because access is a priority, compromises will be necessary and supplemental inputs will have to be made continually to maintain valued aspects of the ecosystem.

The fourth type of conservation target derives directly from AgTrans research and attempts to integrate the human and ecological aspects of landscape transformation and function. At one end of the spectrum, it would mean preserving historically important human–nature relationships, such as early farms, ranches, and rural towns. This effort is already evident in culture parks and historic monuments. The landscapes that we do not think are sufficiently targeted are those that reflect the "struggle" between people and natural forces in the environment. An example from our narrative is traditional irrigation canals. First is the element of having to follow the natural topography and soils in the construction of the canals that are little more than supplemental streams. Next is the way natural forces take hold of this human transformation and infest it with plants and fish, creating an anthropogenic riparian habitat unlike anything in a prehuman world or human designed. Yet, in microcosm, this example represents much of the world we are creating and offers a classroom for human–environmental interactions.

Conclusion

The context of this case study is the rapidly urbanizing region of Phoenix, which is set in the Sonoran Desert of the United States. Here, conservation of desert land, animals, habitat, riparian areas, farmland, and water has become vital. The decisions made by planners for conservation will have implications for the appearance and survival of life in central Arizona. Conservation was accomplished through collaboration and cooperation instead of conflict during early-20th-century central Arizona. Farmers and local citizens in 1903 came together to form SRWUA (later renamed SRP). The droughts that prevailed during the 1940s and the 1950s necessitated water conservation. Major canals were lined to prevent percolation, and secondary canals were covered or put into pipes to prevent evaporation. The landscaping that had grown up naturally along the soil banks of the canals was no longer supported, because water was conserved for human use.

With the reassertion of irrigated agriculture onto the natural landscape of the Sonoran Desert, people modified nature and became part of the definition of what was natural for the area. Desert gave way to irrigated farmland. Some animals thrived within the new environment created by agriculture, and some did not. World War II brought changes and people and urban pressures to central Arizona. Farmland became expensive and much was sold for housing development. However, agriculture is still vital in central Arizona. By 2002, 50,354 acres of baled cotton was still being harvested in Maricopa County (USDA, 2002a). Nowadays, agriculture holds its position of importance along with all aspects of economy, although this importance is diminishing as farmland is sold. The future of water availability and conservation looms as a most important challenge for the future of central Arizona.

To survive, farmers on the urban fringe have adapted to new forms of urban farming. They offer hayrides and pumpkin festivals, and promote peach picking and other activities that encourage people to experience an agrarian lifestyle. The reality is that this marketed experience may be the only kind of agriculture that will survive in rapidly urbanizing regions such as Phoenix (Gober, 2005). Families that have owned farms for generations have found that financial security lies more in selling their farms to subdivision developers and leaving their old way of life behind. However, some farmers have discovered ways to have the best of both worlds. The head farmer of Gilbert's Agritopia subdivision, John Milton (personal communication, 2007), surveys the 7-acre farm that is surrounded by homesteads and reflects that residents are tasting both urban and rural life as they enjoy local organic produce picked almost in their own backyards. Others, like artist/farmer Matthew Moore, whose photograph of a housing pattern mown into his alfalfa field is pictured on the cover of this book, ask us to step back and reflect on a vanishing way of life.

References

Abbott, D. R. (ed.). 2003. *Centuries of decline during the Hohokam Classic period at Pueblo Grande*. Tucson, Ariz.: University of Arizona Press.

Arizona Republic. 2005. "A long dry streak." *Special Report*. Rev. May 22, 2005. Online. Available at www.azcentral.com/specials/drought.

Arizona-Sonoran Desert Museum. 1999. *A natural history of the Sonoran Desert*. S. J. Phillips and P. W. Comus (eds.). Tucson, Ariz.: University of California Press and the Arizona-Sonoran Desert Museum.

Bayman, J. M. 2001. "The Hohokam of southwest North America." *Journal of World Prehistory* 13(3): 257–311.

Brookings Institution. 2003. "Phoenix in focus: A profile from Census 2000." *Living Cities: The National Community Development Initiative*. Rev. July 7, 2007. Online. Available at http://www.brookings.edu/reports/2003/11_livingcities_phoenix.aspx.

Crown, P. L, and W. J. Judge (eds.). 1991. *Chaco and Hohokam: Prehistoric regional systems in the American Southwest*. School of American Research Advanced Seminar Series. Santa Fe, N.M.: School of American Research Press.

Decision Center for a Desert City. 2007. Science and Policy of Climate Uncertainty. Rev. July 4, 2007. Online. Available at http://dcdc.asu.edu.

Faeth, S. H., P. S. Warren, E. Shochat, and W. Marussich. 2005. "Trophic dynamics in urban communities." *BioScience* 55(5): 399–407.

Fiege, M. 1999. *Irrigated Eden: The making of an agricultural landscape in the American West.* Seattle, Wash.: University of Washington Press.

Gammage, G., Jr. 1999. *Phoenix in perspective. Reflections on developing the desert.* Tempe, Ariz.: The Herberger Center for Design Excellence, College of Architecture and Environmental Design, Arizona State University.

Gober, Patricia. 2005. *Metropolitan Phoenix: Place making and community building in the Desert.* Philadelphia, Pa.: University of Pennsylvania Press.

Graybill, D. A., D. A. Gregory, G. S. Funkhouser, and F. L. Nials. 2006. "Long-term streamflow reconstructions, river channel morphology, and aboriginal irrigation systems along the Salt and Gila rivers," pp. 69–123. In: J. S. Dean and D. E. Doyle (eds.), *Environmental change and human adaptation in the ancient American Southwest.* Salt Lake City, Utah: University of Utah Press.

Grimm, N. B., and C. L. Redman. 2004. "Approaches to the study of urban ecosystems: The case of central Arizona–Phoenix." *Urban Ecosystems* 7: 199–213.

Hostetler, M., and K. Knowles-Yánez. 2002. "Land use, scale, and bird distributions in the Phoenix metropolitan area." *Landscape and Urban Planning* 62(2003): 55–68.

James, S. R. 2004. "Hunting, fishing, and resource depression in prehistoric southwest North America," pp. 28–62. In: C. L. Redman, S. R. James, P. R. Fish, and J. D. Rogers (eds.), *The archaeology of global change.* Washington, D.C.: Smithsonian Books.

Kupel, D. E. 2003. *Fuel for growth: Water and Arizona's urban environment.* Tucson, Ariz.: University of Arizona Press.

Luckingham, B. 1989. *Phoenix: The history of a southwestern metropolis.* Tucson, Ariz.: University of Arizona Press.

Mawn, G. P. 1979. *Phoenix, Arizona: Central city of the Southwest, 1870–1920. Vol. I.* PhD diss., Arizona State University, Tempe, Ariz.

McIntyre, N. E., J. Rango, W. F. Fagan, and S. H. Faeth. 2001. "Ground arthropod community structure in a heterogeneous urban environment." *Landscape and Urban Planning* 52: 257–274.

Redman, C. L. 1999. *Human impact on ancient environments.* Tucson, Ariz.: University of Arizona Press.

Sargent, C. (ed.). 1988. *Metro Arizona.* Scottsdale, Ariz.: Biffington Books.

Schaafsma, H. 2003. *Modern Sonoran plant communities: Reflections of prehistoric agriculture.* Master's thesis, Arizona State University, Tempe, Ariz.

Schaafsma, H., and J. M. Briggs. 2007. "Hohokam field building: Silt-fields in the northern Phoenix Basin." *Kiva* 72(4): 431–457.

Schmieding, S. 2003. *Five historical periods from the study of water use and agriculture in pre-statehood Arizona.* Manuscript on file at Global Institute of Sustainability, Tempe, Ariz.

Shochat, E., W. L. Stefanov, M. E. A. Whitehorse, and S. Faeth. 2004. "Urbanization and spider diversity: Influences of human modification of habitat structure and productivity." *Ecological Applications* 14(4): 268–280.

Smith, K. L. 1986. *The magnificent experiment: Building the Salt River Reclamation Project, 1890–1917.* Tucson, Ariz.: University of Arizona Press.

Turner, R. M. 1974. *Map showing vegetation in the Phoenix area, Arizona.* Denver, Colo.: Department of the Interior, United States Geological Survey. [Folio of the Phoenix area, Arizona Map I-845-I; Base from U.S. Geological Survey, Phoenix and Mesa 1954–69, Ajo 1953–69, Tucson 1956–62.]

U.S. Census Bureau. 2002. *PHC-T-5 ranking tables for incorporated cities.* Online. Available at www.census.gov/population.cen2002/phc-t5/tab04.xls.

USDA. 2001. Policy Advisory Committee on Farm and Forest Land Protection and Land Use. *Maintaining farm and forest lands in rapidly growing areas.* Report to the Secretary of Agriculture. Washington, D.C.: U.S. Department of Agriculture.

USDA. 2002a. *Census of agriculture.* Online. Available at http://www.agcensus.usda.gov/Publications/2002/index.asp.

USDA. 2002b. *National agricultural statistics service.* Online. Available at www.nass.usda.gov/census.

U.S. Fish and Wildlife Service Threatened and Endangered Species. 2008. Online. Available at http://www.fws.gov/southwest/es/arizona/Threatened.htm#CountyList.

van der Leeuw, S., and C. L. Redman. 2002. "Placing archaeology at the center of socio-natural studies." *American Antiquity* 67(4): 597–605.

Weworski, R. 1999. *Residential landscape development in Phoenix, Arizona: Past, present and future.* Unpublished master's thesis, Arizona State University, Tempe, Ariz.

Zarbin, E. 1997. *Two sides of the river: Salt River Valley canals, 1867–1902.* Phoenix, Ariz.: Salt River Project.

Conclusion

Ted L. Gragson

Despite differences in the onset, intensity, and scale at which changes across time occurred in each of the six sites, process similarity far outweighs any differences in underlying mechanism in comparing the narratives. For example, many of the regions had periods of increased soil erosion as a result of farming practices even if topsoil was lost by wind action in the Shortgrass Steppe and Konza regions during the drought of the 1930s (Bennett, 1947) or through excess runoff in the Coweeta region in consequence of forest clearing during the early 20th century (Price and Leigh, 2006). The increased availability of fossil-fuel energy after World War II both locally and nationally led the inhabitants of all six regions to subdue the land in novel if varied ways. In the Harvard Forest and Coweeta regions, increased availability led to transportation improvements; the CAP region benefited from the introduction of chemical fertilizers and supplemental groundwater; whereas in the Kellogg region it resulted in mechanical, chemical, and biological intensification of agriculture. The transformational path is not always direct or intentional, as in the case of the introduction of invasive species such as the hemlock woolly adelgid in Harvard Forest, and *S. lespedeza* and red cedar (Fechter, 2000) in the Konza region.

There is no longer debate that humans leave their footprint on the land irrespective of whether their impact is large or small. In studying land use, it has become obvious that humans have been altering the land for thousands of years, even if transformation of the land through the agricultural practices of the 19th and 20th centuries is the dominant legacy of the modern era. What is clear is that land once altered can never be restored to its exact previous condition. For example, the Shortgrass Steppe region appears to have recovered from overgrazing and fire suppression, and the Harvard Forest region from timbering and agriculture. However, the species that now dominate are not the same known to have previously dominated these regions, and presumably there are functional consequences of subtle and not so subtle structural differences, even though these are less well understood.

Recognizing that the land can never be restored completely to its previous state does not mean there is no debate between conservationists and environmentalists over what the land *should* be restored to. For example, we could protect habitats for the sake of endangered species such as the 20 mentioned in the CAP narrative. Alternatively, flora and fauna could be protected in the measure they are beneficial—financially, culturally, aesthetically—to humans, such as the case of certain tree species noted in the Harvard Forest narrative. The sheer diversity of focus of the organizations active in southern Appalachia as described in the Coweeta narrative points to the spectrum of conservation practice accepted as legitimate.

From an agricultural standpoint, the challenge is to determine how to save the land and its biotic assemblages while simultaneously providing the food and materials humans require or demand to sustain their current standard of living. It is a challenge because it entails making choices even before the complete answer is known. For example, many farmers in the Kellogg region realized their use of toxic pesticides and herbicides was leading to a loss of biotic diversity. This led them to gravitate away from the use of toxic chemicals even though these chemicals allowed them to grow in abundance the crops they desired. The challenge increasingly faced in the United States is how to sustain the land as the extent of agricultural area shrinks and the urban surface expands. Whatever else is true, it is clear that public initiatives and public organizations help a great deal. For example, although the CAP region experienced first a farming boom followed by an urban boom, large areas such as the Phoenix Mountain Preserve and the Superstition Mountains Wilderness were set aside as areas intangible to both farming and urbanization.

Each site offers unique insights on the agricultural transformation of the land as well as how the legacies of this transformation manifest themselves currently. We examined transformations and legacies across these six sites using a multiscalar approach to consider how natural and human site attributes affected the land across time. For example: How did short versus long droughts affect soil erosion? How did new irrigation practices affect the use of space? How did the removal of a native species such as the wolf affect predator–prey relations? As mentioned earlier, although many of these attributes are the same or similar at each LTER site, the timing of their interaction and the relative strength of their additive and synergistic effects often result in variable outcomes. Such variability introduces the risk of simply telling interesting stories with no attached lesson, so we next identify the five themes that cross-cut our scalar approach and serve as the basis for drawing conclusions from site comparisons.

Interpretive Themes

We judge these interpretative themes as characterizing the process responsible for the historical trajectory of transformation and perhaps underpinning the future sustainability of the process. Each of these themes resonates with material discovered in several of the cases, if not in all of them. The six case studies provide ample empirical evidence for these five broad themes, which allows the reader to assess to what extent the U.S. agrarian experience is uniform or idiosyncratic across space and time. They also provide ample lines of inquiry to be pursued in the future.

Theme 1

Through its establishment, readjustments, and abandonment, agriculture can be better understood when evaluated as an adaptive, problem-solving process. It is obvious that agriculture responds to the available conditions and variability in the biophysical context. Crops need adequate soil fertility, precipitation, sunlight, and frost-free days. Within any particular region, certain parts of the landscape will be more amenable to cropping as a result of favorable values of the aforementioned characteristics or can be made more so by human alteration. With barns and fodder, animals can be more flexible in the landscapes they inhabit; if they graze for a major part of the year, location can be very important. Even in similar environments and time periods in the same location, very different agricultural regimes may emerge as a result of sociocultural values of the participants and other factors in the broader market economy. Many agrarian decisions have been culturally driven—that is, people bring with them particular knowledge, use a selection of current technologies, and have traditional preferences for elements of alternative agricultural regimes.

In the Shortgrass Steppe region, the indigenous Pawnee adapted corn, squash, and beans to the bottomlands of the Republican River basin (near the border between Kansas, Nebraska, and Colorado) before the severe drought of the 13th century drove them away from the High Plains. Centuries later, the Apache also cultivated corn, pumpkins, and plums, although the arrival of horses, guns, and smallpox encouraged the Plains people to abandon horticulture, and increase their hunting range, thus migrating away from European disease vectors. European Americans later brought their complex of small grains and midwestern corn varieties, but soon shifted to winter wheat in most areas. Corn was grown in areas with stream-fed irrigation along the South Platte River. Prior to the Green Revolution, dryland crop regimes were integrated with livestock raising, but since the 1960s continuous cropping regimes divorced from animal husbandry have prevailed.

Although we use the term *nature* to differentiate processes and locations where human involvement has been minimal from those such as agriculture, where human management dominates the ecosystem, we do not intend to imply that nature is passive or that humans are not part of the natural system. We have observed nature to be a resilient and vigorously creative force. As the construction of an irrigated landscape degrades habitats in some ways, it also creates new habitats, such as reservoirs, canals, and fields, that provide homes for new species that establish new hydrological relationships. Nature does not disappear in the face of anthropogenic activities, but is part of a transformation, creating new and novel ecosystems. In this way we consider that nature has agency and the form this agency takes must be examined in the different regions and under the varying human activities. The results of this type of analysis can provide fundamental insights for what is possible under alternative conservation initiatives.

Adaptation does not only concern landscapes and crops, but also involves the human actors in the agrarian regime. Historically, households have been the fundamental unit of farming activity. The size and demographic structure of households have taken different forms in response to a variety of social and economic factors. Moreover, farmers have aggregated into larger cooperative or corporate groups to perform their tasks more effectively. The dominance of one form or another seems to have varied over time and may be related to regions as well.

Theme 2

Resource regimes and their heterogeneities within a region and across time mirror the theme of the overall book. Because each of the six regions examined exists in a significantly different biome and was settled at a different time, we refer to the agricultural adaptation at any particular place and time as the *resource regime.* This regime comprises the soil, water, climate, animals, and crops exploited as well as the strategies used by humans. The regimes used at different time periods and in different parts of the country do vary, but there is a question regarding whether contemporary and historical perceptions about agricultural regimes within a region actually portray them as more homogeneous than they really are. Does the mix of activities within regimes change in a regular pattern over time or in response to external or internal events? Do these patterns provide insights that might help guide socioecological decision making to ensure food security in the face of growing global population and finite fresh water and arable land resources? For example, the Shortgrass Steppe effectively comprises two regimes divided between the South Platte and Republican River watersheds as a result of the variation in the availability of stream-fed runoff close to the Rocky Mountains. The division holds true for both the pre- and post–European American settlement periods, until irrigation by groundwater pumping made possible more intensive farming of corn and sorghum in the high tablelands of the Republican River basin. The heterogeneities evident in resource regimes are cultural as well as physical. The Pawnee cultivated corn, squash, and beans whereas the Apache grew corn, plums, and pumpkins in the Shortgrass Steppe region. The early American settlers who followed in time hand-irrigated gardens with melons, peas, beans, potatoes, lettuce, wheat, rye, and oats.

Theme 3

The asynchronous succession in social, economic, and agrarian landscapes evident across these sites reflect differences in population and technological capacity at the time of initial settlement. In New England, more than 150 years separated the maximum extent of farmland from the time of initial settlement. In southern Appalachia, farmland peaked 80 years after European Americans first occupied the Cherokee homeland. In southwestern Michigan, it took half a century for settlers to identify the maximum extent of agricultural land. In eastern Kansas it took two generations, but increasing aridity moving west slowed the pace to 70 years in northeastern Colorado and 80 years in central Arizona.

Overall, there is a compression of the transformational stages moving east (passed through slowly) to west (passed through quickly). This is not particularly surprising in that farming by settlers of European descent was initiated at very different time periods in the six case studies (beginning in 1620 to as recent as 1867). Over time, learning and the increasing speed of communication accelerated the process. However, there is also something internal to the process that increases the pace of change moving across the continent. Some of the early farming locations (e.g., Harvard Forest and Coweeta) eventually became uncompetitive vis-à-vis the more recently occupied western regions as a result of soil exhaustion and the opening of more productive lands. It is important not to assume the operation of deterministic Turneresque forces such as Manifest Destiny.

Theme 4

Beyond the differences themselves, there is a need to identify the triggers that move these systems from one regime to the next by altering local path dependencies. A trigger could be an event or force that is strictly local or that reflects a regional, even global, action scope. World War I stimulated the growth of commercial agriculture in the western United States and accelerated the decline of cropland agriculture in the eastern United States. The Green Revolution was not confined to developing world areas because it allowed land use to intensify everywhere. In the Shortgrass Steppe region, it led to a simplification of the crop system as crop farms transitioned to continuous wheat growing and either abandoned or specialized in livestock raising.

A trigger sufficient to promote a regime change could be a specific, relatively rapid (pulse) event, such as a drop in market prices, legal or regulatory change, or the beginning of a war. Alternatively, it could be the culmination of a gradual process crossing a threshold, such as salinization or loss of soil fertility, increasing land costs, or growing population density. Triggers are often finite, easily measurable phenomena (e.g., a changing market price of a commodity), but in other cases it could be a change in perception that led to change in the regime. An example of the latter would be the change in perception or mental model that resulted in Phoenix transitioning from being a farming center to a quality place to live because of tourism, health, desert recreation, and high-tech employment.

Whether any particular event develops into a trigger depends on a variety of factors and the system itself. A system may develop in such a way that it is more susceptible to change, which will be viewed as vulnerability if the change is perceived as negative or opportunity if the change is seen as positive. In many ways, a better understanding of what makes a system vulnerable to change (less resilient) and what characteristics of a potential trigger make it effective in prompting change provides helpful explanations to scholars and is useful for decision makers. A key characteristic of the system in each of the case studies is the region's connection to the broader national and increasingly global marketplace. This often involves a significant shift in prices of goods sold, nature of customers, available investment capital, interest from potential immigrants, and changing self-image.

Theme 5

Among the key insights we sought from the six case studies are the kinds and effectiveness of institutions that emerge to organize activities and solve problems. These institutional units guide human-to-human and human-to-ecosystem interactions and themselves become legacies, sometimes adapting their function to new activities, sometimes losing contact with current issues and becoming a force against change. Of particular interest are the institutions that emerge to promote or organize agricultural activities and those that focus on land conservation efforts.

Land policies in the United States are responsible for creating widespread individual ownership of land, and established the institutional framework for good stewardship and/or conservation. Policies enacted during the New Deal emphasized land stewardship through retiring or setting aside from production marginal land. In the Kellogg region, for example, farmers internalized the merits of conservation practice and generally welcomed the stability brought to agriculture by compliance with a policy limiting the physical extent of cropland agriculture.

In the Harvard Forest region, organizations such as TNC, the Massachusetts Audubon Society, and the Natural Heritage Programs try to save many of the riparian forest bird species (Scheller, 1994; Sharp, 1994); whereas in the CAP region, public and private groups are attempting to save the wilderness through protected parks and forests, and by creating habitats in urban areas for wildlife. Many organizations, such as The Sierra Club and The National Forest Protection Alliance, are active in southern Appalachia, working on forestry, air quality, and education issues.

In conclusion, this volume revolves around a novel domain of inquiry about how environmental sustainability can be achieved on agrarian landscapes in transition. European Americans have been on the North American continent for a relatively short amount of time, and only within the past few decades has it become possible to consider how a near-pristine landscape stood up to intense modification. In this volume we examined landscapes that are very different biotically and culturally from each other, yet are united by a common transformational pattern: They were all first used productively (i.e., crops, cattle, lumber, mining), then effectively abandoned. The common truth that emerges from these narratives is that regardless of what humans do to the land, it will never revert nor be restored to its original state—our footprint on it is ever present. Changing one aspect of nature will always have a ripple effect that, regardless of whether the imprint is dramatic (as in the case of mountaintop removal) or subtle (as in the practices of Native Americans), the changes wrought will forever endure in some fashion and affect how humans move forward in using the land. How we choose our future rests ultimately on avoiding the hubris of domination that so frequently characterized our relations to the land in the past.

References

Bennett, H. H. 1947. *Elements of soil conservation*. New York: McGraw-Hill.
Fechter, R. 2000. "The economic impacts of control of *Sericea lespedeza* in the Kansas Flint Hills." Master's thesis, Kansas State University, Manhattan, Kans.
Price, K., and D. S. Leigh. 2006. "Morphological and sedimentological responses of streams to human impact in the southern Blue Ridge Mountains, USA." *Geomorphology* 78: 142–160.
Scheller, W. G. 1994. "The politics of protection." *Sanctuary Magazine* November/December: 17–19.
Sharp, B. 1994. "New England grasslands." *Sanctuary Magazine* November/December: 12–16.

Index

Accelerated Mass Spectrometry, 53
agrichemicals, 184, 187, 189. *See also* toxins
agricultural abandonment, 66, 75, 90
Agricultural Depression, 167, 169–170, 176–177, 184
agritainment, 21, 174, 181, 191
aquatic ecosystems, agricultural effects on, 8, 75, 103, 105–106, 109–111, 198
Armstrong, W., 144

biological change, 21, 52, 156, 158–159, 182
bison, 122, 127–131, 145–146, 206, 208–209, 211, 213–216, 219–220, 232–233
Blue grama, recovery of, 122, 127, 130
Blue Ridge Province, 94
Boone, D. M., 215
buffalo. *See* bison
Bureau of Agricultural Economics, 135
Bureau of Land Management, 243
Butz, E. L., 143

Canary Coalition, 112
cattle, 57, 62, 98, 100, 103, 130–132, 192–197, 208–209, 213, 218–227, 233–234
  Texas longhorn, 131, 219–220, 222
  transitory, 220, 222–223
Central Mississippi Valley, 97
Central Plains Experimental Range, 123
Civil War, 100, 101, 103, 131–132, 175, 216, 219, 250
Civilian Conservation Corps, 243
Clean Air Act, 112
climate change, 51, 65, 72, 74, 127, 155
  deglaciation, 51, 155
  Ice Age, 75, 127, 163
Coastal Lowlands, 47–48, 63
Connecticut Valley, 47–48, 50, 56–57, 73
Conservation Trust for North Carolina, 113

contour cultivation, 133, 135, 143
corn. *See* maize
Corn Belt, 165, 171, 211, 228
Coronado. *See* Vasquez de Coronado, F.
cotton, 100, 102, 247, 249, 253–254, 256
Cotton Belt, 165
crop rotation, 176, 198

Desert Land Act of 1877, 139, 249
diseases
  cholera, 215
  smallpox, 55, 129, 215
Dogwood Alliance, 212
drought
  causes and effects of, 122–123, 126–128, 130–132, 137, 140, 143, 145, 219, 224–225, 233, 242, 245, 247–248, 251–253, 257, 263–264
  Dust Bowl, 122–123, 144, 224
  horizontal centrifugal pump technology, 141
dryland farming, 132–133

eco-villages
  Earthaven Eco-Village, 112
  Narrow Ridge Earth Literacy Center, 112
edge effect, 67
endangered species, 65, 112–113, 243, 267–268
Enlarged Homestead Act, 133
exotic pests. *See* invasive species
exurban rural development, 153

family labor, 59
farm/field abandonment, 104–105, 109, 132
federal payments for farmers, 172
Federal Recreation and Public Purposes Act, 243
"fencerow-to-fencerow" planting strategies, 166, 172, 227
fertilizer, 103–104, 108, 172, 177–178, 181, 198, 219, 259

Food Security Act of 1985, 144
Forest Cutting Practices Act, 71
forest vegetation, 47, 49, 50, 65, 67–69, 73–74, 76, 94, 126, 156,
Fruit Belt, 182–190

gentrification, 105
Gila River Indian Water Rights Settlement, 248
glocalization, 174
Golden Age of American agriculture, 167–169, 223–224
Granville, L., 99
grassland corridors, preservation of, 122, 144, 146
grassland management practices, 123
Grassland Reserve Program, 144
Great Depression, 135–136, 140, 167, 169–170, 176–177, 179, 184, 224–226, 254
Great Migration, 55
Great Out Migration, 104
Great Plains Conservation Program, 143
Great Smoky Mountains National Park, 112–113
Greeley, H., 139–140
Green Mountain Uplands, 47
Groundwater Management Act, 261–263

Hatch Act of 1887, 103, 217
Hayden, F., 123, 131
herbicides. *See* toxins
Heritage Programs, 45, 80, 277
High Plains aquifer, 130, 228
Homestead Act, 133, 161, 217, 219, 223
Hudson Valley, 53
hunting and gathering, 51, 157, 208

Indian Homestead Act of 1872, 161
Industrial Revolution, 61
Insect and Plant Disease Act of 1875, 183, 185
insecticides. *See* toxins

International Biological Program, 123
invasive species
  balsam woolly adelgid, 91
  brown-headed cowbird, 107
  chestnut blight, 65, 91
  European corn borer, 176, 180
  gypsy moth, 76
  hemlock woolly adelgid, 65, 76
  Peach Yellows virus, 183
  red cedar, 211, 230–231, 234
  Sericea lespedeza, 230–232, 234
irrigation technology, 123, 132–133, 137–143, 182, 188, 227–228, 238–253, 256, 259, 261, 268

Kansas Grassroots Association, 229
Kansas/Nebraska Act of 1854, 216
Katuah Earth First, 112
Konza Prairie, 209, 213, 228–229

Land Ordinance Act, 144

maize, 52–57, 97–98, 103–104, 122, 127–129, 137, 158–160, 175–180, 214, 221–222, 247–248
  hybrid corn, 178
malnutrition, 102
market revolution, 59–61
Massachusetts Audubon Society, 67, 71, 80
Massachusetts Forestry Association, 71
Massachusetts Slash Law, 71
Mead, E., 142
mechanization, 63, 104, 168–170, 178, 186–187
Michigan fruit industry, challenges to, 184–186
Milton, J., 269
monoculture plantings, 185
Moore, M., 269
Morrill Acts of 1862 and 1890, 217

Napolitano, J., 264
National Forest Protection Alliance, 212
National Park Service, 45, 229, 243
National Park Trust, 229
National Parks Conservation Association, 112
National Reclamation Act, 252
Native American activity, 45, 53, 55, 75, 98–99, 101, 106–107, 122, 128–131, 137, 158–161, 206–208, 214–216, 219, 233, 246–249
  Paleo-Indians, 51, 95–96, 127, 156
Native American lands, justifications for the taking of, 55, 99
Nature Conservancy, 45–46, 71, 80, 209, 228, 265, 267
New England Forestry Foundation, 71
New River Community Partners, 111
nomad, 52–53, 95, 128–130, 156–159, 214, 249

Ogallala aquifer, 142, 226, 234
organic farming, 63, 173, 180–181, 269

Pardo, J., 98
Pawnee National Grasslands, 123
pest resistance, 187, 198
pesticides. *See* toxins
Powell, J. W., 131–133, 139
Preemption Law, 139
Public Lands Commission, 119

Quabbin Reservoir, 49, 62, 72
Quejo, de, P., 98

range management practices (burning and grazing), 206
Resettlement Administration, 142
riparian zones, 106, 110, 211, 241–244, 249–250, 259, 264, 268

Salt River Water Users Association, 252
Sampson, A. W., 143
Sargent, A. A., 139
Schneider, M., 98
Sierra Club, 212–213, 277

Smart Growth Partners of Western North Carolina, 113
Smith-Lever Act of 1914, 103, 217
Soil Conservation Service (SCS), 122, 142–143
soil erosion, 57, 61, 76, 94, 103–104, 122, 141, 176, 178–179, 198, 207, 210, 225, 231
soil exhaustion, 176, 276
soil types, 56, 58, 126, 134–135, 148–149, 154–155, 206, 209–212
Sonoran Desert Conservation Plan, 267
Soto, de, H., 98
South Mountain Park, 243
Southern Alliance for Clean Energy, 112
Southern Appalachian Biodiversity Project, 112
Southern Appalachian Forest Coalition, 112–113
State Cooperative Extension Service, 188
surface evaporation, 133

Tallgrass Prairie Foundation, 229
Tallgrass Prairie National Preserve, 229
tallgrass verses shortgrass, change in presettlement range, 136
Tamarack Ranch Natural Area, 144
Timber Culture Act, 139
TNC. *See* Nature Conservancy
Tonto National Forest, 243
toxins (herbicides, insecticides, and pesticides), 104, 134–138, 170, 178–179, 180–181, 184–191, 198, 232, 243, 273
  dichlorodiphenyltrichloroethane (DDT), 178, 186
  Roundup, 180
  transgenically modified crops (GMOs), 180
transportation, 59–60, 63, 101, 104, 132, 161, 168, 182, 213, 217–220, 252–254
Treaties of Medicine Lodge, 131
Treaty of Fort Laramie, 131
Treaty of Washington (1836), 160
tree-ring analysis, 128
tropical cultigens (maize, beans, squash), 52–53, 128, 156, 158–159, 166, 214, 247, 274–275

Ulibarri, de, J., 128
Urban Growth Initiative, 244
urbanization, 59, 128–129, 137, 174, 239–241, 264–267
urban–suburban sprawl, 54, 65, 90, 111, 113, 146, 153, 171, 174, 181, 189, 190, 199

Vasquez de Coronado, F., 128
vegetation type. *See* forest vegetation
Véniard de Bourgmont, E., 128–129
Virginians for Appropriate Roads, 113

Weeks Act, 70
World War I, effects of, 63, 169–170, 171, 223, 224
World War II, effects of, 102, 104, 165, 170–171, 177–179, 186, 188, 198, 226, 234, 254–255, 269

Yucatan peninsula, 45